PHILOSOPHY AND MEDICINE

VOLUME 7

ORGANISM, MEDICINE, AND METAPHYSICS

ORGANISM, MEDICINE, AND METAPHYSICS

ESSAYS IN HONOR OF HANS JONAS
ON HIS 75TH BIRTHDAY
MAY 10, 1978

Edited by

STUART F. SPICKER
University of Connecticut Health Center, Farmington, Conn., U.S.A.

D. REIDEL PUBLISHING COMPANY
DORDRECHT : HOLLAND / BOSTON : U.S.A.

Library of Congress Cataloging in Publication Data

Main entry under title:

Organism, medicine, and metaphysics.

(Philosophy and medicine ; 7)
Includes bibliographies and index.
1. Medicine–Philosophy–Addresses, essays, lectures. 2. Biology–Philosophy–Addresses, essays, lectures. 3. Metaphysics–Addresses, essays, lectures. 4. Jonas, Hans, 1903- –Bibliography. I. Jonas, Hans, 1903- II. Spicker, Stuart F., 1937-
R723.073 610'.1 77–28255
ISBN-13: 978-94-009-9785-1 e-ISBN-13: 978-94-009-9783-7
DOI: 10.1007/978-94-009-9783-7

Published by D. Reidel Publishing Company,
P.O. Box 17, Dordrecht, Holland

Sold and distributed in the U.S.A., Canada, and Mexico
by D. Reidel Publishing Company, Inc.
Lincoln Building, 160 Old Derby Street, Hingham,
Mass. 02043, U.S.A.

Softcover reprint of the hardcover 1st edition 1978

HANS JONAS, Dr. phil.
D.H.L. (h.c.), Hebrew Union College, 1962
D.L.L. (h.c.), The New School for Social Research, 1976
D. theol. (h.c.), Philipps-Universität Marburg, 1976

Alvin Johnson Professor of Philosophy, Emeritus
The Graduate Faculty of Political and Social Science
The New School for Social Research
New York City

EDITORIAL PREFACE

This *Festschrift* is presented to Professor Hans Jonas on the occasion of his seventy-fifth birthday, as affirmation of the contributors' respect and admiration.

As a volume in the series 'Philosophy and Medicine' the contributions not only reflect certain interests and pursuits of the scholar to whom it is dedicated, but also serve to bring to convergence the interests of the contributors in the history of humanity and medicine, the theory of organism, medicine in the service of the patient's autonomy, and the metaphysical, i.e., phenomenological foundations of medicine. Notwithstanding the nature of such personal gifts as the authors' contributions (which, with the exception of the late Hannah Arendt's, appear here for the first time), the essays also transcend the personal and serve to elaborate specific themes and theses disclosed in the numerous writings of Hans Jonas.

The editor owes a personal debt of gratitude to many, including Hannah Arendt, who offered their assistance during the preparation of the volume.

On November 25, 1974, just over a year before her death, Hannah Arendt, lifetime friend of Hans Jonas (whom he met in a seminar on New Testament interpretation offered by his beloved teacher Rudolf Bultmann at Marburg in 1924) wrote to Mrs. Lore Jonas and indicated that although she did not normally accept invitations to contribute to *Festschriften*, she was considering what she might write for inclusion in this one. Regrettably she died before she could act on her decision to submit a contribution. Her concern that the project go well was indicated by her enclosure of two short articles which suggested "how Festschriften are made and how one goes about the whole business." The editor has followed Hannah Arendt's advice, by way of posing the questions raised by Keith Thomas in his "No Monolithic School," which appeared in *The Times Higher Education Supplement* of August 11, 1974, and in "A Passionate Realist," a book review by Martin Gardner which appeared in the October 14, 1974, issue of *The New Leader*. Specifically, one had to raise and resolve the following questions: Who is to contribute? How does one assure a "real book," and not an embarrassing miscellany of articles? How does one avoid a superficial unity compiled with mere verve and *panache*. What are to be the standardized editorial procedures, and how is one to eliminate all misprints? From the beginning all such questions were posed. More importantly, since Hans Jonas had already achieved international recognition for his work on Gnosticism, how could one volume acknowledge this monumental work *and* his accomplishments in the philosophy of organism, biomedical ethics, and post-Kantian metaphysics? The answer was fortuitously given without asking; for a decision had been made in Germany to edit and publish a second *Festschrift* for Hans Jonas, dedicated solely to his pioneering work in Gnosticism. I therefore wrote to the contributors to the American *Festschrift* on April 24, 1975: "The European *Festschrift* will contain papers from eminent scholars, most of whom reside in Europe; it will appear in German and be published in Europe. The *Festschrift* which is being edited here will not, therefore, include papers on Gnosticism. Rather it will be devoted to a theme that encompasses issues in Philosophy and Medicine, the latter terms taken in the widest senses. . . . The main point is that this *Festschrift* is somewhat different from most *Festschriften*, since it cannot be a compilation of essays whose themes and theses are selected only by the authors. . . ."

By the time all the essays were completed, Hannah Arendt's suggestions became the hallmark for all who tender a permanent gift to a lifelong friend and colleague.

Although it was decided to publish only original papers, it seemed improper to publish this book without a contribution from Hannah Arendt. For this reason a slightly modified excerpt from *The Life of the Mind*, Vol. I, *Thinking*, (which has just recently come from the press) has been included. Acknowledgment is gratefully made to Roberta Leighton and Rita Vaughan of Harcourt Brace Jovanovich, Inc. for securing permission to reprint Chapter II, Part 13 of *Thinking*. I also wish to express my appreciation to Mary McCarthy West, Trustee of Hannah Arendt's literary estate for her support, and to Professors Larry May and Jerome Kohn who suggested that 'Metaphor and the Ineffable' be selected for inclusion in this gift.

Another consideration during the preparation of this volume was whether or not to include a statement from Hans Jonas himself in the form of a personal and intellectual autobiography. I have elected not to do so since an excellent personal statement appears as the Introduction to Jonas' collected papers: *Philosophical Essays: From Ancient Creed to Technological Man*[1] and in 'A Retrospective View'[2] listed in the complete 'Bibliography of the Works of Hans Jonas' which appears at the end of the volume. In addition, though still unpublished, Ioan Petru Culianu's doctoral dissertation, *Gnosticismo E Pensiero Contemporaneo: Hans Jonas*,[3] completed at the Università Cattolica del Sacro Cuore in Milan, Italy, contains a seventeen page transcript of a comprehensive interview with Hans Jonas which took place on April 26–27, 1975, in New Rochelle, New York, entitled 'Without Some Religious Survival . . . .' This transcript one can only hope will one day be published for all to read. It is an extremely enlightening account of the man and the influence of other scholars on his work.

When the decision to prepare this *Festschrift* was first made in Summer, 1974, Bethia S. Currie, a student of Hans Jonas, had graciously initiated the undertaking. Subsequently she sent out invitations to those eminent scholars and friends whose contributions now appear in its pages. Due to tragic personal circumstances Professor Currie was compelled to seek additional assistance. She shares equally the role of Co-editor, however, not only because she offered this editor the privilege of completing this volume, but because, in truth, her unfailing spirit motivated the entire enterprise. This editor takes the liberty of relocating the opening page of Bethia Currie's contribution in order to place it here as part of the Editorial Preface

Hans Jonas was my professor at The New School for Social Research during my work for the M.A. and Ph.D. in philosophy, and for three years I was privileged to be his assistant. During the time I was at The New School, Jonas was doing important work, including *The Phenomenon of Life*[4] and many of the later sections of *Philosophical Essays*,[5] and I had the unusual pleasure of attending his course in Phenomenon of Life and Organism the year before the final version of *The Phenomenon of Life* was prepared for publication, and again the year after its publication. The notes from these two series of lectures on the same subject are evidence of one of the facets of Jonas as thinker and teacher that make him so unfailingly inspiring to students, that is, the never-ceasing evolution of his philosophical perception. The theme of the two series is the same, but no two lectures are the same, and even the list of philosophers dealt with in the two series is different.

Jonas never stales, and he never stands still, and it is an honor and a privilege to be invited to contribute to this volume of essays on the occasion of his seventy-fifth birthday.

For the very reason that he does not stand still in his thinking, it is difficult to choose from among the ideas Jonas has contributed to philosophy of organism, ethics, philosophy and medicine, the life sciences and technology. Jonas has ideas the way the rest of us have mice. . . .[6]

The editor is personally indebted to the Connecticut Humanities Council for its material support which sustained the research for his personal contribution, as part of the project, 'Colloquium on the Humanities and Death.' The original, unpublished paper, 'The Rebirth of the Clinic: Palliative Treatment, Euthanasia, and the Moral Imperative of the Hospice Care Context,' was presented before the public in New Haven, Connecticut on March 19, 1977. Although the paper does not cite any of Hans Jonas' publications – as my teacher he always encouraged me to go the way of my own thought – discussions with him during the time of the writing of the paper contributed greatly to its thesis and general line of argument; in this sense he has influenced virtually every page.[7] I will not in all likelihood write on this topic again – at least not for some time – since I shall always recall with affection a conversation with him in which, having just completed his lecture on 'The Right to Die,' he remarked in typical fashion: "Since death takes care of itself, one shouldn't take too much time in life with it."

While this volume is in press Hans Jonas is working on still other papers. He is presently completing two essays entitled 'Philosophical Aspects of Technology' and 'Parallelism and Complementarity,' the latter taking up Spinoza's solution to the psychophysical problem as well as an alternative suggestion by Niels Bohr.[8] Yes, "Jonas has ideas the way the rest of us have mice."

In closing, it is appropriate to acknowledge the fact that Strachan Donnelley's contribution was adapted from his dissertation, *The Living Body: Organism and Values – Whitehead and the Interpretation of Value Experience*, guided by Hans Jonas and submitted to The Graduate Faculty of The New School for Social Research in partial fulfillment of the requirements for the degree of Doctor of Philosophy. The editor's appreciation is also extended to Professor David Parent of Southern Illinois University, who graciously undertook the task of translating Michael Landmann's contribution into idiomatic English. Personal thanks also to Susan Engelhardt and H. Tristram Engelhardt, Jr. for their diligence in checking the proofs with great care and for making additional, helpful editorial suggestions. Closer to home, the editor registers his gratitude to Mrs. June Madden, whose meticulous devotion to and labors with this volume – as with everything else – were indispensable to its completion.

On behalf of all the contributors and those whose efforts enabled the final preparation of this *Festschrift*, we register our appreciation to the officers and staff of D. Reidel Publishing Company, whose unfailing constructive advice and assiduous care are reflected in the expeditious preparation and publication of the manuscript in time for the celebration of Hans Jonas' seventy-fifth birthday. We thank the publishers for preparing one specially bound leather copy which immediately stands as a rare book, and take delight in presenting this copy to Hans Jonas and his family as a handsome addition to their personal library.

We are all deeply grateful for the loving care, tenderness, and attention given unconditionally by Mrs. Lore Jonas during the entire time this gift was in preparation. Without her commitment to its completion and her impeccable good sense, the contributors would not have been able to have had the privilege of honoring their friend, teacher and colleague in this indelible way.

S.F.S.

December 13, 1977
West Hartford, Connecticut

NOTES

[1] See p. 318 this volume, 'Books,' No. 10, pp. xi-xviii.
[2] See p. 324 this volume, 'Articles and Reviews,' No. 61.
[3] [*Gnosticism and Contemporary Thought: Hans Jonas*], Tesi di laurea in Lettere presentata a la Facoltà di Lettre e Filosofia de l'Università Cattolica del Sacro Cuore di Milano, Ottobre–Novembre, 1975.
[4] See p. 317 this volume, 'Books,' No. 7 A and B.
[5] See p. 318 this volume, 'Books,' No. 10.
[6] Bethia Currie's contribution focuses on the proposed redefinition of death as 'irreversible coma,' a "subject of importance to both of us," she writes regarding Jonas and herself, "and one which we have discussed on more than one occasion."
[7] Hans Jonas delivered the Keynote Address, 'The Right to Die,' (unpublished) at a jointly sponsored conference, 'The Right to Die: Expectations and Obligations,' before an audience on the Johns Hopkins University campus, Baltimore, Maryland, on October 13, 1977, at the invitation of Leivy Smolar, President of Baltimore Hebrew College.
[8] See p. 324 this volume, 'Articles and Reviews,' No. 56. Also see Stuart Spicker, 'The Spurious Psyche-Soma Distinction: Comments on "On the Power or Impotence of Subjectivity",' in *Philosophical Dimensions of the Neuro-Medical Sciences*, Volume 2 of the series 'Philosophy and Medicine', S. F. Spicker and H. T. Engelhardt, Jr. (eds.), D. Reidel Publishing Co., Dordrecht, Holland / Boston, U.S.A., pp. 163–177.

TABLE OF CONTENTS

EDITORIAL PREFACE vii

A MESSAGE FROM HIS STUDENTS: *Gewidmet* Hans Jonas xv

INTRODUCTION xvii

ADOLPH LOWE / Prometheus Unbound? A New World in the Making 1

SECTION I / HUMANITY, HISTORY, AND MEDICINE

MICHAEL LANDMANN / The System of Anthropina 13
PAUL OSKAR KRISTELLER / Philosophy and Medicine in Medieval and Renaissance Italy 29
OTTO E. GUTTENTAG / Care of the Healthy and the Sick from the Attending Physician's Perspective: Envisioned and Actual (1977) 41
ERIC J. CASSELL / The Conflict Between the Desire to Know and the Need to Care for the Patient 57
STUART F. SPICKER / The Execution of Euthanasia: The Right of the Dying to a Re-Formed Health Care Context 73

SECTION II / PHILOSOPHY OF ORGANISM

LEON R. KASS / Teleology and Darwin's *The Origin of Species*: Beyond Chance and Necessity? 97
MARJORIE GRENE / Individuals and Their Kinds: Aristotelian Foundations of Biology 121
CHARLES HARTSHORNE / The Organism According to Process Philosophy 137
STRACHAN DONNELLEY / Whitehead and Jonas: On Biological Organisms and Real Individuals 155
BETHIA S. CURRIE / The Redefinition of Death 177

SECTION III / SCIENCE, INFIRMITY, AND METAPHYSICS

RICHARD KENNINGTON / Descartes and Mastery of Nature 201
WILHELM MAGNUS / The Philosopher and the Scientist: Comments on the Perception of the Exact Sciences in the Work of Hans Jonas 225
MURRAY GREENE / Life, Disease, and Death: A Metaphysical Viewpoint 233
RICHARD M. ZANER / Ontology and the Body: A Reflection 265
JITENDRA NATH MOHANTY / Intentionality and the Mind/Body Problem 283

EPILOGUE

HANNAH ARENDT / Metaphor and the Ineffable: Illumination on "The Nobility of Sight" 303

BIBLIOGRAPHY OF THE WORKS OF HANS JONAS 317

NOTES ON CONTRIBUTORS 325

INDEX 327

A MESSAGE FROM HIS STUDENTS: *GEWIDMET* HANS JONAS

It is never simple for a student to summarize what he owes to his teacher. When the subject is philosophy, the task of appreciation is even more difficult, for the labor of thinking, apparently of little use and esteem, makes great and inexorable use of its practitioners and of all those who let themselves be touched by its powerful simplicity and plenitude. Both teacher and student stand before the supreme works of philosophical reflection in the hope that their own thinking, as small and limited as it may be, can be challenged, uplifted and enhanced by this undiluted force.

Nothing can serve this end more directly than the thrill of genuine insight into the text of a great thinker, and nothing is so extraordinary and rare for a student as such an experience. Yet without the encouragement of this kind of understanding, no one would trouble himself to continue along the path of philosophy. This is why the student rejoices (quite literally) when in lectures and seminars he or she is suddenly brought to see, often for the first time, the real meaning of what a great philosopher has written.

We have had only one teacher who habitually brought us to such insight and elation, and we are certainly not the only students who owe this kind of enlightenment to Hans Jonas. If we were asked to sum up the dominant experience of all those who have studied with him – an experience enjoyed again and again over many semesters – we would say very simply: he understands; he understands philosophy, and he understands how to communicate this knowledge so that its true essence is preserved.

Philosophy in its essence, as strange as it may sound at first, is not primarily a matter of ideas, concepts, theories, doctrines or world views. Rather, philosophy finds its meaning in insight, and insight comes only to those who think, and thinking is the activity of those who question and wonder. It is impossible to comprehend a philosophy outside of the reflective movement of thinking in which it was born and nourished. And precisely here lies the gratitude which so many students feel toward Professor Jonas.

It is extraordinary indeed to encounter a teacher who can let the philosophers speak to us, even across centuries, as if their thinking were once more actual and alive. It is rare indeed when a professor shuns the traditional doctrines in handbooks and histories in order to present a thinker's work as

S. F. Spicker (ed.), Organism, Medicine, and Metaphysics, xv–xvi.

part of a ceaseless project of reflection and questioning into which we are invited to enter. It is far from usual when a teacher preserves in philosophy what is most intrinsically its own, and hands it over to his students, not as the story of past concepts, but as an example of what can emerge in thinking. And all of this Professor Jonas accomplished as a matter of course, and he did so with a penetration, an earnestness, a literary elegance and, if we may say so, a humility which could not fail to touch anyone deeply for whom philosophy was in some way the object of serious concern.

We can never remember an imperious tone in his lectures; never a doctrinaire assurance; never a selectivity designed to promote a particular world view; never a grand scheme of interpretation. There were no sweeping dialectical reconstructions leading to his own standpoint; no abrupt dismissal of metaphysical and theological ideas; no intellectual trends or bandwagons which diverted us from asking our own questions, or which served only to fill the emptiness which the simplicity of genuine thinking often brings.

We know that we are not alone in voicing this appreciation of our teacher, Hans Jonas. Our commitment to philosophy in its classical scope and richness goes back in no small measure to his capacity to understand and to help us also to understand. We are truly grateful for having had at least one teacher who rescued philosophical thought from the "history of ideas" and returned it to its own true milieu of wonder and questioning. To praise a man for such an accomplishment might not sound like much of a commendation, but when it comes to philosophy and to the depth of insight possible in such thinking, we do not know what more we can say.

PAUL SCHUCHMAN

On behalf of the graduate students who studied philosophy with Hans Jonas at the Graduate Faculty, the New School for Social Research

INTRODUCTION

'Metaphysics' is no longer a popular word. 'Medicine' is. Yet metaphysical questions survive, though they remain with us somewhat tempered or transmuted by our contemporary positivist and analytic temperament. In fact, the puzzlings of metaphysics thrive in medicine. It is in medicine that one asks questions such as: When do persons begin? (i.e., Are fetuses persons?), When do persons end? (i.e., When may one declare a person dead?), How unified are persons? (i.e., Does forebrain commissurotomy create two persons, when there was previously one?), and How does psychosurgery bear on our view of ourselves as free agents? (i.e., Does it show that mental states are reducible to physical states and/or that we are free precisely because we can shape our own embodiments?). Though such questions concerning the unity of the soul and the place of freedom are now domesticated by analytic trainers, one still senses the urgency of the quest for human self-understanding that lies behind them. After all, medicine tends to all the passages of life, the junctures at which we are struck by wonder, awe and consternation at the meaning of man. T. S. Eliot has Sweeney remind Doris:

> Birth and copulation and death.
> That's all the facts when you come to brass tacks. [6]

As a result medicine is involved in the crucial junctures of life and the ethical quandaries they evoke. Medicine is the all-encompassing, all-pervasive science, art and technology.

This ubiquitous (and ambiguous), central place of medicine in our culture has been clearly recognized by Hans Jonas. In his reflecting on technology and man, medicine and the biomedical sciences have provided a *leitmotif* of interest. It was natural, therefore, to bring these essays on the organism, medicine, and metaphysics together in honor of Professor Hans Jonas. On the one hand, he has contributed richly to the fabric of concerns out of which modern interests in the philosophy of medicine have been drawn. His works have touched the writings of all the contributors to this volume. On the other hand, the essays that comprise this volume speak to concerns that he has embraced and addressed.

These essays also speak to issues in their own right. An understanding of medicine requires that it be viewed in terms of our general concerns with

S. F. Spicker (ed.), Organism, Medicine, and Metaphysics, xvii–xxvii.

biology and technology, with *homo vivens* and *homo faber*, with man as a living entity and man as a maker and fashioner of tools (and finally of technology). Because medicine acts to restore "normal" conditions and cure or ameliorate "abnormal" conditions, the technologies of medicine presuppose an appreciation of the goods of the human organism. Doing work in the philosophy of medicine brings one at once then to background issues in the philosophy of biology. Medicine is a technology that intrudes into all dimensions of our lives; it raises salient, in fact, inescapable questions. Hence the range of essays in this volume and their unity: an examination of how medicine involves issues in the philosophy of biology and in philosophical anthropology.

In raising background problems in the philosophy of medicine these essays provide a scholarly appraisal in the mode of Hans Jonas towards understanding medicine. One must remember that Professor Jonas has recurrently pressed the analysis of philosophical issues in medicine back to their conceptual roots. Moreover, he has never been a prisoner of fashion or of prevailing clichés. On the one hand, he has called the contemporary, brain-oriented definition of death into question, asking whether it does not presuppose a Cartesian view of man ([12], p. 31). On the other hand, in the use of human subjects in experimentation he has raised questions concerning the probity of employing humans in non-therapeutic research. In doing so, he adds a remark, however, which cuts to the presuppositions of all current discussions on prisoner research. "Prison inmates are, with respect to our problem, a special class. If we hold to some idea of guilt, and to the supposition that our judicial system is not entirely at fault, they may be held to stand in a special debt to society, and their offer to serve from whatever motive – may be accepted with a minimum of qualms as a means of reparation" ([12], p. 30). Professor Jonas reminds us that reflections concerning the moral permissibility of using prisoners in human research require a prior analysis of the meaning of punishment, and an assessment of the rectitude of our judicial system. Medical ethical issues cannot, as Professor Jonas thus points out, be understood apart from an analysis of fundamental issues in ethics and ontology. For example, it is Professor Jonas who has raised the issue of whether the very being of certain entities does not impose obligations upon us, ontological oughts [15]. Similarly, the essays in this volume address background, fundamental issues, which frame, undergird, and structure the discussion of philosophical issues in medicine.

The essays by Adolph Lowe and Michael Landmann open the volume with a view of mankind in the latter part of the twentieth century. A contemporary philosophy of man or philosophical anthropology, as they indicate, is one marked by a view of man as scientist and technologist, as a self-fashioning, self-creating organism. For his part, Landmann sketches the basic structures of human existence, the lineaments of creativity, of culturality, of the interplay of creativity and culturality, and of knowledge. Both authors evoke the image of Prometheus and signal a human philosophy of biology marked by questions bearing on our culture's celebration of the progress effected by science and the belief that science can lead us to a golden age – two shaken cornerstones of the modern faith. In this regard, Lowe restates Jonas' ontological imperative, the duty of humanity to continue to exist. He sees it as requisite for regaining faith in civilized life. Moreover, in this view the physician as "biological engineer" receives a special duty to give witness that a communal ethics dedicated to the preservation of life and to candor in reporting scientific data (one should here recall Bronowski and Monod) can sustain faith in the modern age. Medicine can instruct the modern ethos.

To this view of medicine's bearing upon our current view of ourselves, Paul Kristeller adds a view of the history of the relationships between philosophy and medicine into the Middle Ages and the Renaissance, particularly in Italy. In Bologna, especially in the 14th century, there was a marriage of philosophy and medicine. It was not uncommon for university students to acquire first a degree in arts and philosophy, and then one in medicine, followed by a teaching career in logic and natural philosophy, and only later to finish with a career in medicine [16]. As a consequence, many professional philosophers were also trained physicians. This led to contributions by humanists to medicine including new translations of Greek medical writings, particularly in the fifteenth and sixteenth centuries. As a result, medicine entered the modern age, having received many explicitly philosophical inspirations.

Otto Guttentag provides an example of Kristeller's philosophical physician. Guttentag addresses the significance of health, disease, life, living, self, death, and embodiment in the context of medical practice. In doing so, he demonstrates the role of ideas in framing the nature of medical practice and the ways in which medicine is experienced. Moreover, he indicates that the philosophical vistas in medicine extend way beyond the confines of bioethics. Eric Cassell develops these issues further in an examination of the conflict between the desire to know and the need to care for a patient. He does this through a portrayal of the intrinsic conflict within the medical profession, between the role of the physician as scientist and the role of the physician as healer. As Cassell indicates, these two roles come into conflict when there is little likelihood that the acquisition of further knowledge will cure, though there is a substantial likelihood that the process of acquiring knowledge will place the patient at further risk of both morbidity and mortality.

Cassell's analysis focuses as well on the nature of medical knowledge and the benefits and the pitfalls of classifications. Labeling entities appears in its own right to have a therapeutic effect. Patients often need simply to be given a name for their problems. On the other hand, classifications of diseases, and the often rigid notions of disease entities which they introduce, may obscure the fact that disease processes are always idiosyncratically expressed in a particular patient. There is in addition simply a desire to do, to act. As William Cullen, the great eighteenth-century classifier of diseases put it, "It is sufficiently probable that very soon after the first beginnings of human society, some art of physic and some knowledge of remedies arose amongst men; and accordingly, no country has been discovered, among the people of which, however rude and uncultivated in other respects, an art of physic and the knowledge of a great number of remedies has not been found." [4] It is to these needs to classify, to act, and to cure which Eric Cassell speaks. By recognizing the erotic allure of knowing (which usually involves a doing, an intervention, even when a cure is not likely), Cassell reminds us that one should always look critically at purported needs for further knowledge concerning a patient's condition. In doing so, Cassell attempts to focus the drive to acquire knowledge.

As Stuart Spicker indicates, the human drives displayed in medicine are structured within institutions, (a point made as well by Cassell). Spicker focuses upon the hospital and the modern hospice for the dying patient in order to examine how contemporary value suppositions structure health care so as to make euthanasia an issue of considerable contemporary concern. "Physicians and other health professionals are now obliged to examine the moral and value-laden principles underlying humane care and superimpose that reflection on the contemporary American hospital care setting. Once this is done, we will see that we have far too long ignored the social setting, the milieu, the full context in which the physician, other health care professionals and the patient meet." [20] In fact, Spicker concludes that "the typical American hospital ... is in essential conflict with the proper needs of the terminally ill and the chronically ill patient." [20] Euthanasia enters as a practical moral issue because of the tensions (contradictions) of the modern hospital setting. The force of the analysis given by Stuart Spicker is that one can avoid many of the moral quandaries raised by euthanasia by changing the character of the institutional setting for care given to the dying patient. By analyzing background assumptions (i.e., how and for what purposes or structures health care institutions exist), one comes to a clearer understanding of bioethical quandaries than one would have had, had one simply attended to those questions in isolation.

The essays in the first third of this volume portray the nexus of individual, society, and medicine. They show the cement of ideas that holds the interplay together. What follows is much more narrowly focused on the meaning of 'organism'. The second section brings us to the consideration of such cardinal terms as 'normal', 'abnormal', 'disease', and 'illness'. In particular, with the papers of Leon Kass and Marjorie Grene, the essays in this volume turn to even more fundamental issues in the philosophy of medicine: the analysis of the meanings of 'health' and 'normality' and the role and place in biology and in medicine of teleological or purposive explanations. That is, both Kass and Grene provide comparisons of the Aristotelian and the Darwinian views of teleology in nature, and their implications for the assessment of function and dysfunction, health and disease. Kass does this in part by examining the teleological notions remaining in Darwin's works (e.g., the teleological character of 'self-preservation' and the tendency of living things to stay alive and increase their numbers). He identifies normative terms used in Darwin such as "the preservation of the favored species" (where 'favored' could mean nothing more than the 'preserved races') [15] and 'higher' in "higher animals" While wishing to reject the Aristotelian theses of the stability and permanence of the *eide,* Kass wishes to suggest a sense of *eide* and *tele,* even though species, individuals and their organs may be the product of a non-teleological nature. Moreover, against that backdrop, Kass wishes to recapture a legitimate sense of 'higher', as used in the term "higher animal".

Marjorie Grene takes these issues a step further and with a much more explicit focus on medicine. Professor Grene begins with a rejection of an eternal and undeveloping Aristotelian universe. She, however, suggests how a revivified Aristotelianism can aid us in coming to terms with the significance of biological classifications and with our use of terms such as 'normal' and 'abnormal'. Out of a criticism of a Humean account of impressions and a Kantian account of subjects, she forges a suggestion that one must, as did Aristotle, find a way of speaking about nature which recognizes natural kinds. To do this, she envisages a concept of the kind as norm, one that does not entail that all members of a kind have all the characterizing conditions of the taxonomist's key for that kind. Organisms can fall into species by more or less exactly exemplifying such characteristics. Moreover, within such a view those characteristics do not indicate eternal types, nor are there clearly defined essences that mark out a particular kind. Finally, she applies this approach to the analysis of terms such as 'normal' and 'abnormal', 'health' and 'disease'. Members of species may lack some of the taxonomist's characteristics for a species, and be abnormal without being unhealthy. She thus dissects out three often intertwined senses of normal, 'the healthy', 'the most frequent', and 'the standard for a certain kind of process, thing, organ or tissue'. In doing so, Grene enters into the already considerable debate about the nature of disease, illness and health language [3, 8, 10, 14, 18]. Moreover, she indicates that understanding the nature of organisms, and the meaning of function and dysfunction, lie at the core of understanding the meaning of pathology and the character of the biological deviances which are properly treated by medicine.

The second section of the book closes by moving to a consideration of the definition of death, through two general treatments of the ubiquity of consciousness in the universe (and the senses in which animals may be unities, organisms). Charles Hartshorne and Strachan Donnelley contrast Jonas' account of organisms with that of process philosophy in general, and Whitehead's works in particular. Both Hartshorne and Donnelley indicate the virtues of viewing the world in general as being composed of elements or unities each of which are conscious. Donnelley, however, sides with Jonas, and against Whitehead, in favor of the real individuality of animal organisms. Yet, though Jonas rejects Hartshorne's and Whitehead's panpsychism, he is still quite cautious in drawing the line between the self and the world. As Bethia Currie shows, Jonas holds tenaciously to a skepticism regarding our ability to distinguish between the life of persons and mere biological life (i.e., in terms of a brain oriented definition of death). Though Jonas does not embrace panpsychism, he, with Currie, raises a critical question with respect to the localization of the self in the brain (one is here reminded of the Thomistic dictum – *tota in toto* and *tota in qualibet parte*.)

Even where one may disagree with the arguments and conclusions of Professor Jonas and Bethia Currie, one is pressed to reiterate more clearly the theories of embodiment which are presupposed by contemporary brain oriented concepts of death. Moreover, as the Currie article suggests, extreme care is required in applying any test for the presence of death, in that false positive tests are extremely costly and therefore are to be avoided as far as is prudently possible.

Each of the essays in the final section of the volume deals with the impact of René Descartes upon our contemporary views of the relationship of man to nature, and of mind to body. One is provided with a series of explorations of the sense of man's mastery of nature, moving finally to explore the sense of self mastery in relationship to one's body. The reflections carry important implications for understanding health and disease language, especially reports of private states (i.e., "I am depressed"). In this regard, one must appreciate the force of Descartes' notion of mastery over nature. One usually associates the

Cartesian metaphor in medicine with the mechanistic view Descartes presented [5] and engendered in medicine [17]. However, beyond those mechanistic views Descartes inspired individuals who framed our current paradigm of medicine. He gave them a view of medicine as a key to solving general social problems. For example, Rudolf Virchow in an 1849 essay published in the second volume of his famous *Archiv* stated that "Let us recall the words of Descartes, who said that if it were at all possible to ennoble the human race, the means for this could be found in medicine. In reality, if medicine is the science of healthy as well as of the ill human being (which is what it ought to be), what other science is better suited to propose laws as the basis of the social structure in order to make effective those which are inherent in man himself?" [21, 22]. Such views of nature as malleable and restructurable in terms of human goals and intentions lie behind much of the more ambitious interpretations of medicine.

In this final section, Richard Kennington directs special attention to the concept of mastery of nature starting from the assertions in *Discourse VI* that one should replace the speculative philosophy of the schools with a practical philosophy which will make us "like masters and owners of nature". From a comparison of the Cartesian, Băconian, and Christian views of the relationship between man and nature, Kennington suggests that the Cartesian dualism supports the notion that nature stands outside of us to be reshaped and brought to our purposes. In addition to the conceptual problems ingredient in such a dualism, there is a difficulty in providing canons for guidance in the mastery of nature (think of Virchow's all-embracing view of medicine). Towards this end, Kennington suggests that we need an account (which Descartes does not provide) of the relationship between goodness and the expectation of happiness, and of evil and misfortune in nature and in history. One needs, so Kennington holds, a new theodicy.

Such a modern theodicy is offered by Murray Greene in his account of life, disease and death within the Hegelian perspective. One is given an account of the significance of life and disease in terms of conceptual categories with spirit seen as the realization of nature. In this fashion a telos is provided for the development of nature (albeit not a historical development). One should note Hegel offers an account similar to Aristotle's in contrast with mechanistic accounts. Unlike mechanistic views which would see health as optimum performance, and disease as malfunction, Hegel saw health as a form of self-activity and self-relation, and disease as a conflict (a contradiction or negativity in that relationship). Health, thus, has a goal or purpose: it can be understood teleologically as self-relation. "In Hegel's concept of organic life, disease cannot be comprehended as malfunction merely, as in the case of a mechanism. Nor death as destruction beyond possibility of restoration. Disease is for Hegel a disrelation not of parts to one another, but of the animal's subjectivity to itself." [9] Though this reflection on Hegel may seem to be obscure, the intent is quite the opposite. Hegel's goal is to provide a language, a rationale in terms of which death and human finitude can be comprehended. Individuals are always in jeopardy, for only the truly universal endures, "The life of the animal shares in the vicissitudes of the external universal life of nature and, consequently, it merely alternates between health and disease... [The milieu continually subjects] animal sensibility to violence and threat of dangers ..., consequently the general life of the animal appears as a life of illness, and the animal's feelings are those of *insecurity, anxiety*, and *misery*." [11] By contrasting the essential finitude of individual animals with the enduring significance of spirit (e.g., the trans-individual goods of the human community), Hegel offers an understanding of health and disease, not simply a description, as do the mechanists. In this way, Greene presents an important development in our reflections on ways of reconciling mechanist and Aristotelian views of organism and teleology. He also contributes to the nascent literature concerning Hegel's importance for our understanding of health, disease, and death [2, 11].

Richard Zaner and Jitendra Mohanty conclude the third section of this volume with reflections on the problem of embodiment from a phenomenological perspective. They survey the arguments of Roderick Chisholm, Aron Gurwitsch, Edmund Husserl, Gabriel Marcel, Maurice Merleau-Ponty, Jean-Paul Sartre, Wilfred Sellars, and Peter F. Strawson, with respect to the development of an adequate account of mind/body. It is to these discussions that Hans Jonas has contributed importantly, especially through his contention that an understanding of embodiment requires an understanding of metabolic activity. And yet strains of dualism remain in the contemporary language of embodiment for which Mohanty suggests that only a more strictly phenomenological approach could be of use. For example, a phenomenological analysis would signal the various senses in which the body appears (e.g., as physical object, as living organism, as expressive object, and as cultural object). By attending to the different meanings of body, one could extricate oneself from at least some of the specious quandaries of a dualism that portrays the mind-body relationship as a relationship between a thinking entity and an otherwise life-less body. In particular, Mohanty claims that it is the neglect of the body as an expressive object and as a cultural object that leads to many of the misdescriptions of the mind-body relationship. It is here where the body is expressive of symptoms not merely of signs, where it is understood as one's presence in a cultural world, not simply a physical world, that one can provide an account of the rich sense of embodiment encountered in medicine (consider how the body appears to a plastic surgeon, to someone investigating the body images of obese patients, to a sex therapist, etc.). This brace of articles provides a point of departure for serious reflection on the role of notions of embodiment in medicine. As the history of medicine has shown (consider the views of the body of the iatro-mechanists and the iatro-chemists), an understanding of the nature of embodiment is core to the way in which medicine views itself and the ways in which it views patients [19]. In fact, one can at times clearly substantiate the fact that views of the mind-body relationship led to the development of whole areas in medical science and clinical research [7].

The volume ends with an epilogue by Hannah Arendt. In her paper, she sketches the limits of metaphors and how they touch upon the ineffable character of thinking. She compares, for example, the languages of sight and hearing, and notes, "We know from the Hebrew tradition what happens to truth if the guiding metaphor is not in vision but hearing (in many respects, more kin than sight to thinking because of its ability to follow sequences). The Hebrew God can be heard but not seen, and truth therefore becomes invisible... *The invisibility of truth in the Hebrew religion is as axiomatic as its ineffability in Greek philosophy* " [1] Her remarks provide suggestions concerning the differences that the various perspectives of phenomenologies of touch, sight and hearing would make in our understandings of embodiment (and of the role of the body in medicine). Human existence and the status of the organism are variegated

These several essays by compassing basic issues in ontology and metaphysics give a sketch of the meaning of organism and man in medicine. They should remind us of the fundamental issues that lie below the surface of bioethical debates. As has been stressed in the previous volumes of the series, questions in medical ethics or bioethics cannot be understood simply in terms of the moral problems they address. Such problems exist, rather, within a theatre of being, structured by such crucial concepts as mind, body, mind-brain relationship, health, disease, illness, normality, abnormality, deviance, organism, and death, to name only a few. These are in the end the epistemological, in fact, ontological questions that lead us back to the ancient reflections of metaphysics. Medicine touches man to his core. The philosophy of medicine is constrained to come to terms with illusive but essential issues which frame any culture's view of medicine and the human condition. It is to Professor Hans Jonas' tribute that he has addressed these issues and has inspired others to address them as well. This volume is one issue of that inspiration.

Washington, D.C.
January, 1978

H. TRISTRAM ENGELHARDT, JR.

BIBLIOGRAPHY

1. Arendt, H.: 'Metaphor and the Ineffable: Illumination on "The Nobility of Sight",' in this volume, p. 310.
2. Bole, T., III: 1974, 'John Brown, Hegel and Speculative Concepts of Medicine', *Texas Reports on Biology and Medicine* **32**, 287–297.
3. Boorse, C.: 1975, 'On the Distinction Between Disease and Illness', *Philosophy and Public Affairs* **5**, 49–68.
4. Cullen, W.: 1808, *Treatise of the Materia Medica*, Mathew Carey, Philadelphia, p. 9.
5. Descartes, R.: 1972, *Treatise on Man*, transl. by T.S. Hall, Harvard University Press, Cambridge, Massachusetts.
6. Eliot, T. S.: 1952, 'Sweeney Agonistes, Fragment of an Agon', in *The Complete Poems and Plays 1909–1950*, Harcourt, Brace and Co., New York, p. 80.
7. Engelhardt, H. T., Jr.: 1975, 'John Hughlings Jackson and the Mind-Body Relation', *Bulletin of the History of Medicine* **49**, 137–151.
8. Engelhardt, H. T., Jr.: 1976, 'Ideology and Etiology', the *Journal of Medicine and Philosophy* **1**, 256–268.
9. Greene, M.: 'Life, Disease and Death: A Metaphysical Viewpoint', in this volume, p. 251.
10. Grene, M.: 1977, 'Philosophy of Medicine: Prolegomena to a Philosophy of Science', *Proceedings of the 1976 Biennial Meeting of the Philosophy of Science Association*, Frederick Suppe and Peter D. Asquith (eds.), Philosophy of Science Association, East Lansing, Mich., pp. 77–93.
11. Hegel, G. W. F.: 1970, *Philosophy of Nature*, transl. by M.J. Petry, George Allen and Unwin, London, #370, p. 179.
12. Jonas, H.: 1970, 'Philosophical Reflections on Experimentation with Human Subjects', in *Experimentation with Human Subjects*, Paul A. Freund (ed.), G. Brazillier, New York, pp. 1–31.
13. Jonas, H.: 1977, 'The Concept of Responsibility: An Inquiry into the Foundations of an Ethics for Our Age', in *Knowledge, Value and Belief*, H. Tristram Engelhardt, Jr., and Daniel Callahan (eds.), the Hastings Center, Hastings-on-Hudson, N.Y., pp. 169–198.
14. Kass, L. R.. 1975, 'Regarding the End of Medicine and the Pursuit of Health', *The Public Interest* **40**, 11–42.
15. Kass, L. R. 'Teleology and Darwin's *The Origin of Species*: Beyond Chance and Necessity?' in this volume, p. 109.
16. Kristeller, P. O.: 'Philosophy and Medicine in Medieval and Renaissance Italy', in this volume, p. 34.
17. La Mettrie, J. O. de: 1748, *L'Homme Machine*, Zucac, Leyden.

18. Margolis, J.: 1976, 'The Concept of Disease', *The Journal of Medicine and Philosophy* **1**, 238–255.
19. Rather, L. J.: 1965, *Mind and Body in Eighteenth Century Medicine*, University of California Press, Berkeley.
20. Spicker, S.: 'The Execution of Euthanasia: The Right of the Dying to a Re-Formed Health Care Context', in this volume, p. 90.
21. Virchow, R.: 1849, 'Die naturwissenschaftliche Methode und die Standpunkte in der Therapie', in *Archiv für pathologische Anatomie and Physiologie und für klinische Medizin* **2**, p. 36.
22. Virchow, R.: 1971, 'Scientific Method and Therapeutic Standpoints', in *Disease, Life and Man: Selected Essays by Rudolf Virchow*, transl. by Leland Rather, Stanford University Press, Stanford, Calif., p. 66.

ADOLPH LOWE

PROMETHEUS UNBOUND? A NEW WORLD IN THE MAKING

I

The second half of the twentieth century seems to mark another turning point in the history of humanity. Whether the decades ahead of us will raise the world to a higher level of civilization or throw it back into an era of primitive savagery or even destroy humankind physically, we cannot know. But this very ignorance reveals more than the trivial fact that in some manner the future is always unknown. This ignorance itself is a consequence of the transformation which is upon us. Less and less, the future will be dominated by impersonal natural and social processes, the course of which we could anticipate by extrapolating past experience. In quite a new sense it will be the product of human choices – even if we choose to remain passive.

Speaking with such emphasis of an historical turning point is an assertion that calls for factual evidence and explanatory proof. It will be the task of the following comments to offer some of both. But before doing so we must remove a more general doubt. Is our accepted image of history not one of uninterrupted flux? In what sense is it then permissible to speak of turning points and radical transformations?

Now it is true that students of history have habitually used such caesurae, e.g., when separating antiquity from the Middle Ages or demarcating the beginning of the modern era. But when did antiquity really end? With the dethronement of the last West-Roman emperor? With the consolidation of the Byzantine empire a hundred years later? Or only with the crowning of Charlemagne after two and a half centuries? When, on the other hand, did we enter into the modern era? In the thirteenth century when the first national states overthrew the overloadship of the German emperor? Or two hundred years later when Luther nailed his ninety-five theses on a Wittenberg church door, and Copernicus destroyed traditional cosmology? Or perhaps only when, another three hundred years later, Darwin gave the *coupe de grâce* to the belief in humanity's singular creation?

Clearly the answers depend on a criterion, and all criteria derived from the historical phenomena themselves beg the question. Still, there is a formal criterion for such periodizations, which also supports the claim of the unique-

S. F. Spicker (ed.), Organism, Medicine, and Metaphysics, 1–10.

ness of our own era. It can be illustrated by the well-known controversy whether there ever was an "industrial revolution" and, if so, when it started.

Again, so long as we focus on the individual strands that form the complex totality of the event under discussion – population increase, urbanization, rise of a factory system and of a labor market, growing capital formation, and, above all, technological progress in agriculture and industry, to name a few – their respective origins can be traced back to anywhere from the fifteenth to the eighteenth century. And yet it makes good sense to pinpoint the last decades of the eighteenth and the early decades of the nineteenth century as a fundamental break with the past. What happened in that short period was a simultaneous accumulation of changes in every one of these spheres – so dramatic that their very concurrence sets the material civilization of the subsequent era apart from any other preceding it.

The similarity with the second half of the twentieth century is striking. Reaching back to the outbreak of the First World War but gaining its full momentum with the end of the Second, a convulsion has seized practically all spheres of human experience, this time over the entire globe. Its universality has not its equal in any earlier period of known history even if, again as in our earlier example, the roots of most of these developments can be traced much further back.

Let me submit here a representative sample of these radical transformations. First, there is the rise of Communism in its several variants as a rival to the sociopolitical system of the West, the reawakening of the Asiatic civilizations, and the end of Colonialism. Second, there is the acceleration of social equality for the non-caucasian races and for all other strata of the former "underlying population" of the globe, among them, not least, women and children. Third, we witness a worldwide takeoff into economic development and, in the advanced countries, the conquest of mass poverty. Next, we have the release of atomic energy and the push into space and also cybernation and its effect on the organization of production. Last but not least, the main concern of the contributors to this volume, there is the breaking of the genetic code, the expanding control of birth, and even of death, and all the other advances in biomedical science and technology – events that can be fully understood only in the wider context I am placing them here.

At another level lie the subtler transformations in the realm of the mind – new techniques of communication and instruction, greatly extending the range of planned socialization of the individual; the opening up of the unconscious and its symbolic expression in non representational forms of art; a philosophy and even theology centering on human existence and the numinosity of immanent experience. Ultimately, underlying all these changes, there is the most subversive force of all – a progressive spread of information among ever widening strata of advanced and even primitive societies, exposing to rational and critical thought what was formerly accepted as absolute truth or as inscrutable.

What is even more significant than the mere concurrence of these changes is the fact that this concurrence is by no means coincidental. Once we break through their surface appearance, they reveal themselves as engendered by one and the same human aspiration – *emancipation*. More specifically, they should be understood as another stage in the perennial struggle against the fetters which, since time immemorial, have tied humanity to the stern rule of a harsh nature, of even harsher human masters, and of still harsher implacable gods.

I fear this interpretation will meet with strong reservations unless I make it clear from the outset that it is meant as a diagnosis and not as a value judgment. Nor is it to imply that the mere breaking of ancient bonds will by itself lift us to a higher level of civilization. Let us examine both qualifications more closely.

To the extent to which the deeper meaning of these changes has been grasped, they have met with diametrically opposed evaluations. To some, and in particular to many scientists, they open the road to a New Jerusalem, or, to use a Marxian term, they mark the "end of prehistory." To them an age-old dream, narrated in fairy tales and rationally expounded in ancient and modern utopias, approaches realization – man as master of his fate in global association with other free persons.

But there are others to whom the headlong rush into new knowledge and revolutionary action means the dissolution of the ties which have held together society through past millenia. To them the prospect is one of self-destruction in atomic wars, ecological catastrophes, and, the ultimate penalty inflicted on such a Promethean rebellion, the madness of universal anomie.

Neither of these visions should be lightly dismissed. A detached observer might hold both side by side. What their coexistence points to is the open-endedness of the present or, if you like, its "dialectics." It shows, to give a few examples, in the simultaneous prospects offered by the new science and technology of mass welfare and mass incineration, of an expanding span of individual life, and of population explosion, in the rising expectations of the peoples in the underdeveloped regions and the coincident threat of an international class war.

These examples can be multiplied. What they indicate is expressed in the second qualification of our diagnosis. The recent quantum jump of emancipation will prove compatible with the preservation of civilized life only if we can reap its potential benefits without falling prey to the multiple dangers which this transformation of our nature and socio-political conditions poses. This, however, implies that strict limits must be set to the use of our new freedoms, substituting new constraints for the discarded ones. But does this qualification not contradict the very essence of emancipation? Are we to cast off the fetters of the past only to enter into new bondage?

II

In tryint to answer this question we shall come to realize that successful emancipation is more than the negation of past limitations. It is the affirmation of positive imperatives on the part of the emancipated, which at one and the same time prompt and restrain their actions. The nature of these imperatives will reveal itself when we focus on a common characteristic of the constraints which past history imposed on human behavior. Rather than the result of free volition, they were forced on humanity by nature and social organization and, for this reason, were *inescapable*.

This is still manifest in the primitive societies of today where the anonymous pressures of nature and custom predominate. It is no less true of the higher civilizations of the past in which control operated through political and technological organization. Pre-industrial technologies, rather than disturbing the ecological equilibrium of natural forces, adjusted to their rhythm, leaving the physical and biological environment in essence inviolate. Political institutions, on the other hand, were seen as sacrosanct, beyond the reach of intentional transformation. Although rulers and ruling factions frequently dislodged one another, the prevailing social order remained unaffected. All real threats to communal life came from outsiders. But even then the survival of the group was at stake only in the rare cases when the combatants considered each other as beyond the pale of humanity. At any rate what was, to the dawn of the modern age, unthinkable was a threat to the natural or sociopolitical foundations of a society – not to speak of the human race at large – brought on by deliberate action unthinkable because it was *impossible*.

This is the historical background against which the consequences of the present emancipatory revolution loom large. This revolution is about to case off essential objective constraints and subjective taboos which through the millenia ensured the metabolism between Man and Nature and safeguarded sociopolitical stability.

We might watch this obliteration with equanimity if we could be confident that the energies thus released in individuals and groups would spontaneously promote the welfare of all – the anarchist dream of humanity's innate sociality. But even if we distrust trans-historical tenets about the "nature of Man," we should probably all agree that humanity is "socialized" rather than "social." In contrast with that of ants or bees, its sociality is only partly instinctive but is largely the product of a highly complex process of acculturation. Without subscribing to extreme versions of Freud's dictum that repression is the price of civilization, we have every reason to doubt that all unconstrained strivings of individuals and groups are in pre-established harmony with the goals of acculturation.

If the protagonists of emancipation make light of this fact, this is easy to understand. They are more concerned about the opposite danger of "oversocialization," as the modern totalitarian regimes practice it and to which in our own midst rapidly expanding social controls alert us. But in order to face up to the challenge of emancipation, we must distinguish between what Herbert Marcuse calls "surplus repression," namely, "the restrictions necessitated by social domination," and "the modifications of the instincts necessary for the perpetuation of the human race in civilization" ([3], p. 35).

In a similar vein we can contrast redundant with functional constraints – the former stemming from misplaced fear of human initiative and from sociopolitical oppression, the latter serving the consummation and preservation of emancipation itself. However, and this is the most radical transformation of all, less and less will such functional constraints be imposed from without, nor will they be protected as in the past by our impotence to undo them. As the emancipatory process continues they will ever more be placed in the precarious custody of our own power. No longer are they "just there" – they will have to be contrived by deliberate action springing from rational foresight. And even if we "know," the collective will to act accordingly may be absent.

This then is the ultimate reason for the "dialectics" of emancipation, for its open-endedness toward the fulfillment of the highest aspirations as well as toward the decline of civilization, if not the physical annihilation of the race. From now on the preservation of civilization is conditional on insight into the long-term consequences of our actions and omissions and on the adoption of a moral principle which guides our choices toward the constructive.

III

Among these two conditions – rational foresight and moral imperative – the latter holds the strategic position, not as if it were an easy task to anticipate the long-term consequences of any action taken today. Contemporary history is full of examples in which the immediate benefits of undertakings, inspired by the most humane intentions, were thwarted by unpremeditated secondary effects. The biomedical field, in particular, abounds with such experiences of which I have already mentioned the outstanding example – worldwide population explosion as the direct result of the medical triumph in drastically reducing the death rate. Still it may not be a chimera to expect that these very experiences will help in strengthening the intellectual discipline and analytical subtlety required for preparing us for the distant effects of our actions.

The moral problem, as it poses itself in the new situation, is much more complex. What I have in mind is not the eternal conflict between what we think is right and what our selfish interests tempt us to do. A more profound dilemma confronts us – our traditional ethics tells us no longer what is right and what is wrong in strategic situations.

Reinhold Niebuhr's dictum of "moral man and immoral society" expresses our problem in a flash. It states expressly what all former systems of ethics proclaimed implicitly, namely, that moral commands refer primarily to personal relations among contemporaries living in proximate contact. To express it in the words used by Hans Jonas, who was the first among contemporary thinkers to draw attention to this issue:

> The good and evil about which action had to care (to wit: in the past) lay close to the act, whether in praxis itself or in its immediate reach, and were not a matter for remote planning . . . . The ethical universe was composed of contemporaries, and its horizon to the future was confined by the foreseeable span of their lives. Similarly confined was its horizon of place, within which the agent and the other met as neighbor, friend or foe, as superior and subordinate, weaker and stronger, and in all other roles in which humans interact with one another. To this proximate range of action all morality was geared [1].

It is true, political philosophy since Plato and Aristotle stepped outside the purely private scope of ethics generally. But in its quest for "good government" it confined itself to the ordering of in-groups. Even when from the seventeenth century on the idea of a "law of nations" spread, in practice this amounted to little more than the application of the rules of chivalry to international con-conflicts. The decision of whether to enter such a conflict continued to rest on "raisons d'Etat," if not on the maxim that necessity knows no law.

Nations passing through the present emancipatory revolution can no longer afford such ethical neutrality. But where are they to find the basic principle on which their leaders could orient themselves? This is starkly illustrated by the manner in which they try to meet the foremost concerns of this age – the triad of ecological threats (population explosion, gradual exhaustion of essential resources, environmental pollution) and, above all, the danger of atomic war. These threats have introduced quite new space and time dimensions into political decision making, pitting the interest of the present generation against that of its grandchildren or the honor of some nation against the survival of all. What is to be the criterion for these communal choices?

To be specific, it is possible that a world population of six billion, which we must expect by the end of the century, cannot survive with the present distribution of resources. Does such a possibility impose on the rich nations the moral duty of initiating now a gradual redistribution? If it is true, as a UNESCO study predicts, that at the present rate of pollution wide areas of the globe will be uninhabitable within two decades, are those who benefit from the present mode of production and consumption morally bound to change their life style? Or take as the most extreme example a fact of past experience rather than of speculation about the future. Was the manner in which President Kennedy handled the Cuban missile crisis, by going to the brink of atomic war, in accord with the ethical commands of the new age? Or was Bertrand Russell right in proclaiming "better red than dead," meaning that even total surrender of a revered way of life was the lesser evil when compared with the extermination of humankind?

IV

No answers seem to be possible to these, the life and death questions of our time, unless we bring to light the basic principles of what may be called a "communal ethics." From these principles novel but quite specific commandments will address themselves to the guardians of public policy, that is, to political leaders and administrators, but also to scientists and technologists, whose thinking and action control the fate of society at large, including that of its future members.

So far the vision of such a communal ethics is dim, and in the rare cases when the established trustees of moral principles – theologians and philosophers – have spoken at all to it, they have confined themselves to generalities.[1] Indeed elaboration of such a moral code may well be the task of the "day after tomorrow," when institutions and attitudes have adjusted themselves to the requisites of an emancipated order. But where does this leave us during the critical interval of transition when safe landing on the other shore depends on proper steering today and tomorrow?

Paradoxically, the very state of emergency in which we find ourselves provides the answer. There are immense potentialities of amelioration of the material and intellectual state of mankind. But they are today in constant jeopardy by the threat of universal destruction. At the same time, the natural boundaries of our spaceship Earth and the psycho-sociological limitations of its crew make all amelioration relative and subject to slow development. But catastrophe may strike at any moment, and destruction may well be absolute – involving not only humankind but most other organic species.

It has been Hans Jonas's important insight that this threat of total disaster yields a supreme ethical command of stark simplicity and intuitive persuasiveness. Though divided by material inequality and ideological conflict, ordinary people everywhere go about their business peacefully, bestowing even in this age of turmoil an astounding stability on the humdrum processes which sustain their daily existence. Were we to ask them on what code of ethics or system of values they based their behavior, they would be deeply embarrassed. Their response might be no more than the avowal that, in spite of all, life is worth living. Raised to the level of philosophical abstraction, their words, and even more their deeds, express the *self-affirmation of civilized life.*

Like the biblical commandments this principle is ultimately negative. In Hans Jonas's formulation, "There is an unconditional duty of humanity to exist," and therefore, "Never must the existence or the essence of humanity as a whole be made a stake in the hazards of action" [2].

V

There we have the supreme commandment which, from now on, is to guide not only political decision making but also all social and economic practices with long-term consequences. Moreover, and this gains special importance in the context of this volume, it sets definite limits to scientific procedures and experiments, irrespective of whether they are technologically feasible. But to perceive and to affirm this commandment as a philosophical postulate is only the first step. If survival is to be assured, it must become the action directive of researchers no less than of politicians, administrators, and the members of the body politic at large. How can this be achieved?

To give an honest answer we must grasp the nettle firmly. There are, in principle, only two ways through which an action directive can become the actual beacon of orientation for the many – spontaneous social learning, resulting from insight and experience, or autocratic imposition. We need not argue which of these alternatives is preferable. However, the current discussion of the ecological issues, to take a significant example, does not inspire much faith in social learning. It highlights the modern tendency of overestimating the present at the expense of a by-no-means distant future. Moreover, the critical time spans – in the ecological realm hardly more than two generations – may be too short for the slow-working process of rational persuasion. Our best hope may lie in a succession of small calamities which may teach a more convincing lesson than the preaching of Doomsday prophets. But let us not be deceived by reassuring appeals to our individualistic-democratic tradition. Once a true emergency is upon us, no government will hesitate to impose the measures functionally necessary to meet it. And it may then well foist upon us constraints beyond what is functionally necessary.

Against this gloomy background the scientific community, national and even international, offers a somewhat brighter picture. In its own way it established centuries ago a kind of communal ethics in its unconditional commitment to the search for truth and to all the procedural implications of this search. From this commitment a canon of scientific behavior has grown, voluntarily accepted and broken only at the risk of excommunication. Moreover, the practicing physician – the "biological engineer" – binds himself through a solemn oath to a principle of affirming life which, in the personal sphere, is the equivalent to the first commandment of a communal ethics.

Of course, even unconditional commitment to the preservation of humanity does not answer the many specific problems which the new power of knowledge and action is posing to the guardians of the biomedical field. But the seriousness and humane concern with which these questions are being discussed and of which this volume is a living testimony justify the belief that the traditional code is still inviolate. If we dare to be optimistic, we may even hope that traces of its spirit may filter into the wider realms of human activity in which ultimately also the fate of science and technology will be decided.

New School for Social Research,
New York, New York

NOTES

[1] Cf. 1937, *The Churches Survey Their Task*, Report of the Ecumenical Conference at Oxford; and Pope John XXIII: Encyclical on *Pacem in Terris*.

BIBLIOGRAPHY

1. Jonas, H.: 1973, 'Technology and Responsibility: Reflection on the New Tasks of Ethics', *Social Research* **40(1)**, 35–6.
2. Jonas, H.: 1976, 'Responsibility Today: The Ethics of an Endangered Future', *Social Research* **43(1)**, 93.
3. Marcuse, H.: 1955, *Eros and Civilization*, Beacon Press, Boston.

SECTION I

HUMANITY, HISTORY, AND MEDICINE

MICHAEL LANDMANN

THE SYSTEM OF ANTHROPINA[1]

1. Humanity has varied historically; from nation to nation, from culture to culture, from age to age it has assumed an ever different form. Therefore some have said that there are no universal, permanent, "ontological" structures of humanity as such. To want to limit humanity by such fixed structures would be static thinking and a failure to recognize the historical mobility, openness, and abundance of possibilities. But such an argument jumps to conclusions. It ignores that timeless structures and the historically varying form are two coexisting strata of man. The structural stratum constitutes, as it were, the underlying genotype; the various forms, the actualized phenotype. To be an historically variable being is itself a supra-historical fixed structure of human existence. The fact of variability is not subject to history; each particular form is a product of history. That is no contradiction.

As elemental common structures, the *anthropina* have a merely formal character. By *anthropina* we mean the unchangingly fixed, "timeless" basic structures of human existence. Their power of achievement must not be overestimated. They apply to the whole; they cannot explain particulars. They do not, for example, tell us why society is structured now so, now otherwise.

Individual *anthropina* gain in prominence at different times in history. The Greeks stressed theoretics; the Renaissance and the Age of Goethe, creativity. Both intensified and heightened individuality. So there is an historicity of the *anthropina* themselves. While during one phase, certain *anthropina* may dominate, at other phases they may be of less obvious importance. But this historicity is only secondary. As a basic constitution, all the *anthropina* exist from the first and at all times.

2. Although existentialism originated in the 1920's concurrently with the revival of anthropology, the two most important existentialists, Heidegger and Jaspers, both rejected anthropology. This, however, has little significance for the history of ideas. The origin of the two philosophical movements in the same decade can be no accident. Internal differences are always overestimated from too close a distance; the common epochal structures are recognized only from a distance. Each movement focuses on basic determinations of man: the *anthropina* of anthropology correspond to Heidegger's "existentials," with

S. F. Spicker (ed.), Organism, Medicine, and Metaphysics, 13–27.

which he wants to supplement Kant's nature-referrent transcendentals. Anthropology, however, is derived partly from biology, for it compares humanity with the animal, partly from the historical human and social sciences, which regard man as the cultural-historical being. Its theme is not the individual, "the human race." But existentialism has its roots in theology, ethics and psychology. It is concerned with the individual, his/her soul, religio-ethical basic questions of existence, genuineness, and authenticity. It is blind to "objective spirit."

3. According to Kant, Aristotle connected the categories "rhapsodically," but he did not combine them into any coherent system. Therefore, Kant himself seeks to present their "transcendental deduction" after the fact. To derive and interlink the categories is no slighter philosophical task than their first discovery, but it is a completely different kind of task.

The same is true of the categories of human existence, the *anthropina*. Philosophical efforts to define the nature of humanity since Protagoras, Diogenes of Apollonia, and Anaxagoras, and with a new impetus since Pico della Mirandola and Herder, and in our time Plessner, Portmann, Rothacker and Gehlen, have unearthed a great number of anthropina. The following generation must not simply enumerate the *anthropina* disconnectedly. As always, it faces the problem of coordinating and reconciling traditional principles that first stood isolated from one another. The task at hand is to arrange the *anthropina* into groups, to determine the relation between these groups, and, within the groups, to distinguish between fundamental *anthropina*, the most underlying stratum, which provides the conditions of possibility, and the secondary *anthropina* dependent on and derived from them. The organizational law must be discovered to interrelate the partial aspects and apprehend them as components of a meaningful structure, a *unitas multiplex*.

I. *ANTHROPINA* OF CREATIVITY

The point of departure is that man, as distinguished from the animals, has at his disposal neither specialized instincts nor organs highly specialized for particular operations. In contrast to animals, humans, according to an ancient observation, at first seem badly equipped – in fact, unfinished and incapable of survival. This lack of specialization is called the anthropine gap. It is, in a way, the transcendental basis of all the *anthropina*. Yet we hesitate to claim it as the first *anthropinon*, because lack of specialization is a purely negative concept. It originates from the comparison with the animal, and with the point of departure from the animal as standard; but if the person is regarded intrinsically, from his/her own center, then it is seen that lack of specialization is only the precondition for something with which it must be looked at together from the first, because the two things are composed with one another in mind and form, a unity – namely, creativity. What was regarded comparatively as a deficiency turns out to be a positive trait from the point of view of his own being.

Humanity is not a "way out of a dead-end street," a defective species that compensates for its deficiencies by *creativity* (1), but is creative and because of this creativity, has no use for specialization; since it would, on the contrary, hamper it. Through creativity, humanity acquires what the animal inherits. We design and shape our activity independently, and precisely this raises it above mere "behavior." Or, to cite the old Hermetic tradition, as expressed by Pico della Mirandola: "Other creatures were completed by God; but he left human beings incomplete, and instead, lent us a part of his own creative power. With it, we complete ourselves." "Artificiality is human nature," as Ferguson says. Above what humanity has by nature, there rises, Kant agrees, "what it makes of it as a freely acting creature." Humanity has no Platonic essence; "forced to be free," we choose ourselves, we "invent humanity," as Sartre concurs.

So it is false to understand humanity, as it were, as first an animal, to which mind [*Geist*] was then added by an ascetic refusal to follow the animalistic urge, as Scheler would have it. Our entire pragmatic structure is stratified differently. We need theory from the start. The road from the drive to the goal is not direct; because no instinct points it out, there is an "hiatus" between needs and fulfillments (Gehlen). It is therefore also not interrupted by the mind. Rather, from the first the mind shows the needs the way; indeed, it often even assigns their goal.

Lack of specialization and creativity extend to another dimension we have not yet mentioned. The animal's specialization also means that it can live only in a certain environment. Every animal species lives in a – respectively different – "ecological niche." Humanity, however, has *no such fixedly assigned surrounding world* (2), but maintains itself in every type of environment by constantly discovering new ways of survival. In forest or steppe, north or south, it gives a "creative answer" (Toynbee). If the living conditions change, then an animal species either perishes or it must develop biologically to adapt to the new conditions. By producing new economic and social forms, humanity, however, remains the same and still survives.

This announces two more *anthropina*. The first is *work* (3), which *Genesis* already singled out as a distinctive trait of our species and Marx again stressed. That humanity does not begin in a finished state, that individuals must make something of themselves, must make and unmake their habitat, means a primacy of the pragmatic – no matter how much it must rely on theory – not only in the life of the individual, but far more elementarily, by an historical and cultural standard, for humanity as a "generic being". Humanity first produces stable forms by living; in and through these forms it structures and programs itself.

Nonetheless, it would be wrong to put this factor of work or, in Gehlen's language, action, too early. Work and action exist only within fixed channels. It is typical of humanity that work is performed only within such channels, but that even the channels are invented. Work and action as such do not characterize humanity, but only because and insofar as they are creative.

The second *anthropinon* announced here is what can be called by the eighteenth century term, *perfectibility* (4). Humanity as we said, must complete itself. This can be done on different levels: stopping on relatively low or scanty forms of self-perfection; or producing more differentiated and richer ones. Thus, there is a scale of possible progress to better and higher things. Self-completion implies ideas of value, standards, goals. But regression also exists, a falling-short of an already perceived norm. There is also the danger of going astray in diabolical and destructive forms.

Creativity also means *freedom* (5). Self-creation does not mean only freedom of choice between existing possibilities. It does not mean only to do or not do something as the person that one is, but to make something self-conceived out of oneself. Philosophy has often not explored the full depth of the concept of freedom for fear of the possible aberrations. But the banner of freedom often connoted creativity too.

Creativity also means *higher consciousness* (6). For the vital processes do not unreel on the basis of innate reaction-patterns. They must be mentally anticipated, then realized; means must be found for ends. Therefore, the course of action takes place in the light of reflection.

Creativity also means *individuality* (7). Humanity, as we said, must, to attain full reality, go beyond what it first finds by nature – since it is incomplete – and add something new to it. This addition, since it is added in a respectively different way and different direction, is individual. It changes with the historical demands and free imaginative designs. A part of individuality is transmitted to the particular person by the great individuality of the nations and epoch to which he belongs. For they too find a respectively different way of being human. But within this framework, binding on a whole community, the individual too fills the anthropine gap in character and way of life in a respectively individual manner, whether by disposition or destiny, when this is permitted, as it was among the Greeks and since the Renaissance, by inclination and will.

II. *ANTHROPINA* OF CULTURALITY

Now we must go back once again to what we called at the beginning the "anthropine gap," which is filled, as we said, by creativity. But this information is still incomplete. This gap is filled by two things, of which creativity is only one. It is also filled by what former generations have produced and handed down: discoveries and evaluations, mores and techniques, customs and institutions. Creativity does not evaporate in a moment. Its findings and inventions that stand the test are solidified in the "objective spirit" of culture. As significent for man as the ability to produce culture is the obverse ability of receiving formerly created culture, moving within its channels, and being stamped by it. Creativity is followed by *culturality* (1), which, though historically always dependent on creativity, occupies a far broader area than in the immediate present.

Arnold Gehlen stressed that "institutions" exercised an unburdening function for the culture corresponding to the individual's use of habits and automatic actions. However, this characterizes the cultural objectivism only in its psychic effect. It is anthropologically more exact, that while creativity as such occurs at a particular point, enduring forms accumulate. Humanity lives among an abundance of such forms, and *would be much poorer if it had to rely only on spontaneous creativity* (2).

Creative power, since it always has available the previously created, may perfect what was previously created. It may turn its attention not only to something new, but to something more difficult. In this way, a qualitative refinement of objective forms may come about. But this refinement may also be transmitted to the subjects who move within the forms. Realizing himself through the medium of cultural structures, man betters and enhances himself (3).

Man does not create only one, single culture. There is no Platonically preexisting ideal pattern of morality, form of government, artistic style, etc.; for then he would not be creating culture, but merely finding or realizing it. Since his creating is a real act of something not predesignated, even ideally, nations and eras create a plurality of cultures. They fill the anthropine gap each in its own way. It is the essence of culture to be variable, i.e., to change in content according to time and place. Every concrete culture is, because created, historically different. This is not because, as rationalistic cultural Platonism would have it, each in a different way falls short of the eternal norm, but because there is no such norm and because every culture again reveals the inexhaustibility of creativity. Man's creativity and the historicity of cultures, which must be affirmed, were both discovered in the eighteenth century; they belong together and are two aspects of the same thing. Both were discovered at the moment when the rationalistic faith in eternal norms was being denied. What we said above of the individual, that because of his creativity, he is always a unique individual, is now repeated on the level of cultures as a whole in their historical specificity.

Man's culturality thus inevitably involves, as another ***anthropinon, historicity*** (4). The cultural being is also historical. With *power over* history, we shape history; *dependent on* history, we are shaped by it (cf. this author, on the five concepts of historicity, in: *Der Mensch als Schöpfer und Geschöpf der Kultur*, Munich, 1961, p. 77 ff.).

Often thinkers tend to take traits of man belonging to a particular historical period and anthropologize them, i.e., declare them traits of man as such. So Adam Smith anthropologized the acquisitive drive, and Heidegger, fear of death. Man never knows enough, however historically varied he may be and he therefore ascribes time-conditioned features to his permanent structure. Thus, the more the awareness of historicity increases, the more formal philosophical anthropology becomes. It separates "images of man," in which an epoch expresses and interprets itself, from the really perennial underlying structure. Behind apparent constants, it discovers variables that are determined historically in ever different ways. Anthropology can exist at all only after a certain level of historical awareness has been reached. Anthropology, as the delineation of the "permanent", constitutes the dialectical counterpole to historical awareness.

Because of this formalism of philosophical anthropology, the Marxist and Existentialist objection that anthropology staticizes, naturalizes man, does not apply. He is fixated by "images of man," but not by anthropology. Anthropology, rather, fathoms a more general, more elementary stratum. It can say only that the anthropine gap must be filled, and in ever different ways. But it will not confuse any of these historical fillings with the enduring structure of man. Man is not either unalterable nature or freely self-creating history. Both are strata in him. The second is superimposed on the first. The two things are part of his entirety. The permanent, contentless ***anthropinon*** of historicity calls for contentually flowing history.

Thinking is usually held captive by the ethical and educational task of establishing certain ideals before men. It moves on the level of "contents," not of formal structure. Plato designs the "best state." He presupposes the more general thesis that man is a being that must live in communities and organize them. He rejects the idea that there could be a plurality of systems equally justified in principle. That is the reason why anthropology, which deals only with the structural level, is rare, misunderstood, and rejected by many.

In a future static and dogmatic situation, anthropology could be the power that reminds civilization that man is the creative being, variable in his cultures.

The nations, as we said, produce the historically varying cultures; the individual person, however, who is born into such a culture, receives it. He does not inherit it biologically, like the animal its instincts, but he appropriates it through gradual assimilation. Every culture lives in the form of tradition. Only through tradition does it acquire relative duration. As culture is the objectivism of the created, so tradition is its prolongation. Preserving a store of knowledge, skills, and modes of behavior, humanity, cultural and historical, is also *traditional* (5).

This heritage is always passed down from the older to the younger generation and re-assimilated. In the process it usually undergoes unnoticeable and involuntary – or at least not consciously voluntary – modifications. Only at a higher level of freedom must traditions be legitimated before the forum of reason and, if required, altered. New discoveries and inventions are added to the stockpile of tradition, which thus constantly grows.

Traditionality presupposes *community* (6). Life in a community – which man shares with many animals – certainly has other functions, but here it acquires the additional, specifically human function that the community is the only place where tradition can be lived, transmitted, and received. Birds teach their young to fly, but in the animal most behavior typical of the species develops by itself as an inborn instinct. In man, however, the later-born must interiorize the ways of acting acquired by his predecessors; he must learn them. Therefore, man is more dependent on society than the animal. He is the most social being.

Two more *anthropina* must be added at the end of this section, though it may be asked whether they are separate *anthropina* or only the psychic equivalents of what we already summed up objectively as historicity and traditionality. They are, moreover, very closely interrelated.

The first is malleability (7). So that earlier creation can determine the later-born, individuals must be determinable; they must be plastic and impressionable. In the animals, there are "primary fixations," which however remain limited to a certain stage of development; human beings however have the ability to imitate, but it lasts throughout their lives. Whereas ducks raised among chickens follow their nature and, to the despair of the hen, go into the water, humans are so impressionable and dependent on their model that exposed children raised by wolves and bears take on the habits of these animals. Culturality is the anthropological counterpole to creativity. Malleability corresponds to productivity. The productive being is also the receptive one. That is no contradiction. The two things are mutually conditioned. Humans are concurrently the most individual and the most social beings all in one. Humans encompass a new dimension which they in turn expand.

Malleability means, in other terms *educability, capacity to learn* (8). But human beings not only *can* learn, they *must* learn. Every individual must receive the traditional heritage of his group, must master it and combine it adaptively with his own creativity, and then in turn pass it on to the next generation. As the cultural, not given by nature, is acquired creatively by mankind, so it is not passed on to posterity by biological heredity, as are animal instincts, but acquired by learning. For this structural-mental reason, humans are also biologically, as Adolf Portmann showed, the creatures with the longest period of childhood. Learning is in part required by the elders, but in part it also stems from spontaneous readiness. Humans are "more monkey-like than monkey" (Aristotle). The ability to learn continues to a degree and in different gradations even beyond childhood. This ability extends not only to the cultural heritage, but to discovering previously unknown parts and aspects of the world, and to learning from new experiences. Here the ability to learn leads to the "*anthropina* of knowledge."

III. *ANTHROPINA* OF THE INTERPLAY BETWEEN CREATIVITY AND CULTURALITY

The "anthropine gap" of missing instinctiveness is thus filled dually: by creativity and culturality, i.e., the reception of "objective mind" produced by former creativity, but then solidified and separated. All human existence is thus based on two polar but complementary powers. It is, like Leibniz's monads, "loaded with the past" and "full of the future." As so often, here too nature works by the principle of force and counterforce. From this polarity there results a further, individual *anthropinon*. For the two polar forces place humanity before *the task of reconciling them, of bringing them into a reasonable relationship, suited to the respective historical situations* (1).

In Hegel, every antithetical polarity has its synthesis. He sees the contradiction as residing only in our conceptualization, which has not yet reached the end of its dialectics. In contrast, Nicolai Hartmann recognizes "antinomies of being." Here, conceptual reconciliation would mean obscuring the objective "real repugnance." So the two fundamental *anthropina* are antagonistic to one another. They are not two separate facts, but each corresponds to an opposite demand. We must saturate ourselves with tradition; otherwise, we remain culturally poor barbarians. But we must not be so saturated with it that we are only *loci* and functions predetermined by objective mind. We must also develop the innovative and revolutionary potential in ourselves as well as freedom and independence. Otherwise, we fall pitifully behind the level of individualization attained by the Greeks and the Renaissance. We must satisfy both demands; in what blend is a matter of moderation. The balancing of these opposite claims remains a perpetual task, which is as much a matter of given conditions as of decision and never succeeds easily.

The power-ratio of the two is *always differently calibrated* (2) depending on the cultural sphere and historical circumstances. This could also, if desired, be listed as a separate *anthropinon*. Religion is usually conservative; innovations are more permissible in the arts. Historically, tradition is predominant in earlier times. "In the beginning was the quotation." It often represses creativity to such an extent that Toynbee spoke of a "fossilization" of primitive cultures. Generally creativity still stirs, though subconsciously and unnoticed, since social change takes place with corresponding slowness. It is not like in our time, when we see change occurring tangibly before our eyes. Only very late, among the Greeks and in the modern age, is creativity approved; it becomes independent and opposes tradition. Now, no longer the new, tradition is on the defensive. Historically, an increase of creativity, thus of individuality and freedom, seems to be taking place. But this increase is not guaranteed. After Antiquity, the Middle Ages again showed traits of rigid immobility. And today the objective potential of the creative seems to be exhausted. Historically, creativity seems to have reached an apex in the Middle Ages, whereas it lay fallow at the beginning and the end of civilizations, though for different reasons.

Cultural forms that were always handed down from the past, as was seen above, unburden the present creativity and raise it to a higher level. Since the more simple has already been found, it may turn to the more complex. But that is only the harmonious aspect of the present situation. Creativity does not always find a new way. Very often it lies fallow. What human beings today would have liked to create has already been co-opted by a predecessor, who has thus condemned us to mere imitation. Or, creativity could really go in a new direction, but it must not and cannot because the traditional forms have set themselves up as absolute and sufficient and permit no deviation or supplement. Culturality can also mean *hamstrung creativity* (3). It is a repression of the particularity and selfhood of the individual.

The cultural forms and the human subjects who live in them can also evolve away from each other. The cultural forms have their own inner evolutionary logic (Georg Simmel). They grow not only quantitatively beyond what the subjects can meaningfully receive. They also develop structurally in ever greater autonomy and rationality, regardless of whether they are still suited to the subjects for whom they really exist. On the other hand, new demands and evaluations also arise in the subjects in the course of time. They find the "classical" forms created by and for earlier and completely different subjects no longer acceptable. The self-love of the traditional forms and their firm roots in the community, however, make it hard to replace them with more suitable ones. For this twofold reason, *alienation* of humanity in its cultural forms results (4). As from the forms, so it is also alienated from fellow beings and from his/herself.

Marx traced alienation to a particular development within the economic sphere: namely, capitalism. But it is, as we saw, a much more broadly determined phenomenon. Therefore it cannot be undone by a single revolutionary act. It is already contained anthropinally as a possibility in the fact that both the objective forms and the subjects are dependent on history and therefore can become mutually inadequate. Alienation has different strength at different times. Or, rather, it is a matter of distinguishing here between its real strength and the degree to which it is felt. It may be actually oppressive, but human beings may not be conscious of it and therefore will be only indirectly effective. On the contrary, slight alienation can be registered with double strength by a sharpened sensibility.

In general, historically a transformation of cultural forms takes place and diminishes the tension between them and their supporting and experiencing subjects. At times, a revolutionary destruction of the forms occurs. Till now we defined historicity only as a variation of the manifold qualities of culture. Now, we must expand the definition to include one concept of *continuous change*, of which two forms are evolution and revolution (5).

Destruction can serve the purpose of preparing the ground for a new beginning, for a desired better, no longer restrictive, form. But it may also be motivated by a longing not so much for the new form as to liberate spontaneity which is repressed in and by all forms. All human life runs, with hardships and delays, through the cultural forms. We can never follow our direct impulse, but must be led by the forms. Instead of our present personal feeling being decisive, objective mind, which stems from the past and belongs to the community, works for us. This results in animosity towards the principle of form as such. In this animosity, one fundamental *anthropinon* denies the other, antinomially opposite one. In so doing, it turns against the human in its entirety. It seeks to break free of human existence, to shake off its burden, and to return to a pre-human behavioral structure. Therefore, it always has only limited success. But since human existence is based on two opposite principles, it is nonetheless typical for man to repeatedly try to live by one power instead of by a duality of powers. *Spontaneity rebels anarchistically and momentarily against its repression and relativism by tradition and institutions* (6). It wants to be autarchic and autonomous. But, conversely, *the established powers are incensed at the outbreak and revolt of spontaneity* (7). They relegate them to the area of the pathological and criminal.

IV. *ANTHROPINA* OF KNOWLEDGE

We must once again return to what we called at the beginning the "anthropine gap."

The animal, so we said, is specialized in its organs and instincts. It is accordingly also specialized in regard to its receptivity of world. Its senses are selective filters that only admit what is relevant to its vitality. For that by itself suffices to release the right instinct that is called for. Thus behavioral and perceptive structures are correlative.

Such a correlativity simply exists in humanity. When we previously defined the anthropine gap in terms of the missing equipment with organs connected to a single purpose and a lack of instincts, that was only half a description. The other half consists in the lack of specialization even of the receptiveness of man; or, in positive terms, in the *greater scope and depth of his perception and apprehension of the world* (1). As in the animal species, there is specialization in both. The two things intermesh and form a structure of meaning.

For that humans have no instinctual means; they can construct their actions independently and must always decide what they want to do. But this presumes that they know the world more comprehensively, in its own reality and its specific quality. For only then can they so shape their behavior as to prove themselves in it. The senses, as Feuerbach already knew, must not be coupled with specific needs in human beings. For the sake of humanity's specific form of actuality, his theoretical side must, from the first, display *a certain autonomy* from it (2). Only as relatively independent can it be of service to it. Contrary to Heidegger's thesis, the actual [*das Vorhandene*] retains priority over its potentially discoverable use [*das Zuhandene*]. But contrary to all pragmatism, those who with an old tradition, have humans as essentially theoretical beings, are right. But for their findings to then be of practical use – that remains anthropologically the original goal – knowledge must first approach things with a non-practical, non-utilitarian intent. Human, inventive action succeeds only within the horizon of "playful" *superfluity* of knowledge.

Classical anthropology nonetheless errs in setting the direction of independence falsely. It starts with man's greater power of knowing, with his subjective "reason" and states that because he had it, he was then able to create culture. But that is to put the cart before the horse. The logical sequence is rather that because man is the creative, independent, active, culture-making being, he therefore needs the stronger power of knowledge. But this must not be understood genetically and temporally, from below. Otherwise we fall into the type of naturalism that is the exact reversal of the "classical" idealistic model: lack of instincts; therefore, distress; therefore, as *ersatz*, knowledge; and, concomitantly, culture. In truth, creativity comes only subsequently, as its goal, and is not temporally prior. Both are mutually interdependent in a "living simultaneity." One could therefore say that the power of knowledge is a secondary fundamental *anthropinon*.

Herder spoke of the narrow "sphere" of the animals. Jakob von Uexküll proved empirically that they live in a specific surrounding world, [*Umwelt*], different from species to species. Starting with this, Scheler then antithetically set man's gift of reason in "*openness to the world*" (3). We stated above that man is not chained to a particular environment. "Openness to the world" is the gnoseological equivalent of this (cf. Franz Graber, *Die Weltoffenheit des Menschen. Eine Darstellung und Deutung der philosophischen Anthropologie.* Freiburg, Schweiz, 1974).

Erich Rothacker then introduced the modification that even man in his historically changing cultures also lives in respective worlds determined primarily by attitudes and styles of life. Depending on them, there results *a different selection* of what is relevant to life and, therefore, of what is known (4). But in contrast to the species-connected and inextricable worlds of the animals, these human and cultural worlds are only secondary worlds. Humans remain, in principle, open. Therefore, they can at all times transcend their surrounding world. Their history is theirs precisely through the lively intermeshing of objective world [*Welthaftigkeit*] and cultural world

[*Umwelthaftigkeit*].

Here, a further *anthropinon* of knowledge is added. The animal only *has* knowledge. He learns new things, but always within his inborn categoriality. Humans, however, are creatures that *acquire* knowledge (5). By always familiarizing themselves more closely with the world, they create new categories. These categories have become historical and historically variable. Knowledge thus becomes for him something mobile, something not simply given, but which is achieved and worked for.

There are psychic equivalents to this. Moving from what is known to what is not yet known, humans ask questions (Löwith). They do not wait until new things enter their field of vision from the outside, but explore them actively. *The exploratory trait*, which in animals is mostly limited to the childhood phase, becomes *habitual* for them (6).

Just as basically, or perhaps more so than through openness to the world – which is relative compared with his adherence to a cultural world and is only quantitatively greater – the human kind of knowledge is characterized by something else. In the animal, things have a signal-function as releasers of instincts. They are therefore not perceived as independent objects, isolated and separate. They are, rather, merely correlatives in the animal's vital stream, into which they are incorporated. Only in humans are they separated from vitality, and they become a strict opposite with qualities and laws belonging to them as such, independently of their vital significance for the knowing being, and on the basis of which their future action can be foreseen. In humanity, the subject of knowledge is *emancipated from its practical function "for something else"* (7). Therefore Plessner says that we live not only out of our center; things become a second center for us, from which we reorient ourselves. Together with things, our own body, our own subject becomes concrete for us. We experience ourselves, reflect upon ourselves, become self-aware.

Let us return to the passage where we said that humans make their own categories. The animal too subsumes under categories – under his innate schemata of perception. But these perception-schemata are, first of all, rigid and transcendentally hidden. They lie beyond the animal's awareness as such. In humans however, they are, firstly, movable and expandable; secondly, they occur on the objective side. Humans, by working at them, not only operate with them, but command and dispose of them. This work at the general, world-articulating concept, we can now add, is however not purely logical and system-immanent. It consists rather in checking the concepts against things, which have now become a self-valued counterpole that demands its right. The animal's perception-schemata are here always prior to the object, the human's conceptual world is constructed and enriched ***after the object*** (8).

The concepts, as we said, first appear in humans on the objective side. They do so in the form of linguistic fixations and objectivations. Humans convey a distinct materialization to their self-constituted forms with which they point at things and comprehend them. Next to things, they set up their own world of representations distinct from them. Therefore, humanity is *animal symbolicum* (9) (Ernst Cassirer). It is false, in the system of *anthropina*, to group language under the rubric of sociability. Language is based on the logico-cognitive ability to establish symbols. That society then uses them for its communicative, practical purpose, may be a factor in their formation, but it is objectively secondary.

The weight of Platonic philosophy through the ages has caused human reason to be defined as the capacity to generalize. But just as central is another determination that can be traced back to Democritus. The animal, as Wolfgang Koehler observed, perceives *gestalt*. It does not recognize a component in a different structural complex as the same. But man has analytical power. *He breaks forms down into elements* (10). He disintegrates and isolates the element. Therefore, he does not remain imprisoned in the once-given forms. He can even construct new forms out of the elements. Here, we are listing another condition of creativity.

The animal, by his instincts, behaves objectively purposefully, without consciously striving for purpose. The purposes are still set by nature and achieved naturally. Organs and instincts are as such chained to the goal; they are ordained directly toward it. Therefore, in the animal, present and future are not diastatically disjuncted, so that the present strives toward the future. The animal lives in an aggregate of present moments.

Humans set their own goals – and freely design ever different ones. Therefore their goals are conscious. They must also find means to achieve these goals. Human means are unspecific, not fixed to a single purpose. They can be the instrument for many purposes, including evil ones. Ambivalence and free application of means is a danger concomitant with man. Often, one means is not enough, but a whole chain of them is connected, and each one seen from below is itself a purpose. Human behavior is, as Max Weber described it, *purposeful behavior* (11).

If the goal is relatively short-range, we speak of purpose. If it lies far ahead, we speak of Utopias (Ernst Bloch). Structurally, however, they are the same. Humans imagine, anticipate a not visibly given better state and make it their objective to realize it. It is built into the structure of their behavior that they first form an image of what should be in their imagination and then transpose this image into action.

This better state is imagined to be a future one. The present state must be changed into this future image. This makes the present into something dynamic, géared to the future. Humans are the beings that are *aware of the future, design the future, live for the future* (12). Like Prometheus, we mentally enact future situations we could get into; we simulate them. We then determine our present behavior by the imagined consequences of our actions.

Setting objectives, activating means to realize those objectives, humans live not only out of themselves but on something else. The earlier listed *anthropinon* of work (I, 3) is here revealed even more clearly in the human structure.

Freie Universität Berlin,
Germany

[1] N.B. – In compliance with the author's wishes and in conformity with the format of his previous publications (e.g., 1969, *Philosophische Anthropologie: Menschliche Selbstdeutung in Geschichte und Gegenwart*, 3., überarbeitete und erweiterte Auflage, transl. David J. Parent, 1974, *Philosophical Anthropology*, Westminster Press, Philadelphia, Pa.), *Notes* and alphabetical *Bibliography* have been omitted. Professor Landmann prefers to signal the author of the expression or maxim in parentheses at the end of the sentence or phrase. The *numbers* in parentheses which appear at the end of a sentence or phrase throughout the text are not note numbers, but refer to the main points, i.e., *anthropina*, which appear seriatim beginning with number (1) of his four sections (signaled in Roman numerals). The final parentheses (I, 3) on this page, therefore, refers to Section I, and calls the reader's attention to 3, the *anthropinon* of *work*. – Ed.

PAUL OSKAR KRISTELLER

PHILOSOPHY AND MEDICINE IN MEDIEVAL AND RENAISSANCE ITALY

The efforts to establish a link between philosophy and medicine in which Hans Jonas has taken an active part and which find expression in the title and content of this volume are of rather recent origin. Earlier during this century, and for most of the nineteenth century, medicine and philosophy had been almost completely separated from each other. The two fields coexisted, to be sure, as different subjects within the common framework of the universities, but the teachers and students, the courses and subjects taught, were entirely different. With some notable exceptions such as, for example, Karl Jaspers or Kurt Goldstein, few philosophers have shown a specific interest in medicine and few physicians in philosophy. Even the philosophers of science have focused their attention on mathematics and physics and have taken little notice of medicine and of its specific problems.

The reasons for this state of affairs are not difficult to detect. Increasing specialization tended to separate the different sciences from each other and still more from other disciplines including philosophy. Moreover, important fields that had long been considered parts of philosophy such as physics, biology, and psychology had become more and more emancipated from philosophy and finally cut their ties with their mother discipline completely. Philosophy thus lost that part of its content which would have made its study most pertinent to physicians and other scientists. Finally, the positivism prevalent among most scientists encouraged the belief that the progress of empirical observation and research would not benefit but rather suffer from an acceptance of philosophical principles and assumptions that were considered purely speculative or even gratuitous.

This tendency has been reversed rather slowly during the last fifty years or so. The greater emphasis placed by mathematicians, physicists, biologists, and psychologists on the theoretical foundations and methods of their disciplines has also led them to a discussion of philosophical problems, if not to philosophy. The recurrent attempts to coordinate and unify the different sciences seem to point to a role which philosophy may have to play as coordinator, if not as a handmaid, of the sciences, and some philosophers such as Ernst Cassirer have envisaged, at least as a program, a comprehensive philosophy of culture that would encompass, in a broad and flexible system, not only the

S. F. Spicker (ed.), Organism, Medicine, and Metaphysics, 29–40.

different sciences but also all other branches of knowledge and all other areas of human experience [2].

At a moment when the links between philosophy and medicine are again being discussed and may even be strengthened, it might be useful and interesting to call attention to the close links between medicine and philosophy that existed for a rather long time in the history of Western civilization and that were especially characteristic of the Middle Ages and of the Renaissance in Italy. The subject is obviously broad and complex, and its adequate treatment would require much more space than we can give to it here and also much more research. I must limit myself here to a brief outline that is based on incidental observations that occurred to me during my work on the School of Salerno and Renaissance Aristotelianism [10, 11].

In classical antiquity, and especially among the Greeks, medicine was never considered as a part of philosophy but was practiced and taught from early times as a separate profession. Yet the links between medicine and philosophy seem to have been rather close. Hippocrates and his school originated in a region where natural philosophy had been flourishing for some time, and the writings of the Hippocratic Corpus in their problems, terminology, and doctrines show many traces of the early philosophical tradition. On the other hand, the methods and observations of Hippocrates had some impact on Plato and especially on Aristotle who was the son of a physician.[1] During the later centuries of antiquity, there were several rivaling schools of medicine that were not identical with the philosophical schools of the same period but had close relations with them. Sextus Empiricus, who is one of our chief sources for philosophical scepticism, was also a physician.[2] Galen, whose extensive writings summed up the medical knowledge of the preceding centuries and retained great authority down to early modern times, was both interested and learned in philosophy.[3] He is a well-informed source for many philosophical doctrines otherwise poorly attested and has been used as such by modern students of Greek philosophy [26, 17]. He also incorporated many philosophical theories in his medical teaching and thus encouraged his medical followers through his example to study the writings of the philosophers and to apply their terms, concepts, and doctrines to medical theory. Galen's chief philosophical source was Aristotle, and Galen's own authority was a powerful factor in establishing the authority of Aristotle among Arabic and Western physicians and other scholars in later centuries. Yet Galen studied and appropriated not only the doctrines of Aristotle but also those of Plato and of the Stoics, and thus his role as a transmitter of Platonic and Stoic doctrine, although less known and perhaps less influential, is far from negligible.[4]

When the Arabs translated and continued the work of their ancient Greek predecessors, medicine and philosophy, along with the mathematical disciplines, occupied a central place, and the leading Greek authors translated into Arabic included Hippocrates, Galen, and Aristotle [24, 27]. Their writings inspired an impressive body of Arabic literature on philosophy and on medicine. It is worthy of note that some of the most important and influential Arabic treatises on medicine were written by Avicenna and Averroes, the leading philosophers and Aristotelian commentators, and that other medical writers in Arabic such as the Jew, Isaac Israeli, showed a strong concern and familiarity with philosophical problems. This must mean that in the Arabic world, even more than among the Greeks, the same persons were professionally engaged in both medicine and philosophy, and this fact is important, not only in itself but also in its impact on the Latin West where the Latin translations from the Arabic, including the authors whom we have just mentioned, played a very significant role.[5]

Unlike the Greeks and Arabs, the ancient Romans made comparatively modest contributions to medicine, just as they did to the other sciences and to philosophy [21]. Roman medical literature includes only one author of note, Cornelius Celsus,[6] and very few Greek medical writings were translated into Latin during the ancient or early medieval period. After the knowledge of the Greek language had practically disappeared from Western Europe, with the exception of Southern Italy and of Sicily, medical knowledge, if we may judge it from the extant manuscripts and texts, declined to a rather primitive and elementary level [1]. Most Western medicine down to the eleventh century was either bookish lore transmitted as a part or appendix of the seven liberal arts in monastic or other clerical schools or pure practice based on observation without theory and transmitted through a kind of apprenticeship and without regular instruction by local lay healers who were usually unlearned and often illiterate.

The rise of Western medicine, which began towards the end of the eleventh century and continued without interruption during the following centuries, was closely connected with two important developments in the history of learning. A sizable body of medical literature was translated from Greek and Arabic, mainly in Southern Italy and in Spain, and this literature included a number of writings attributed to Hippocrates and Galen and to the leading Arabic physicians, and it contained a large amount of solid and detailed medical and philosophical doctrine that had previously not been available to the West [3, 7, 23]. Moreover, medical instruction became increasingly advanced and theoretical and was soon to be concentrated at some of the same universities which were the centers of instruction also for the other branches of later medieval learning such as theology, philosophy, and civil and canon law. The School of Salerno had originated during the tenth century as a center of medical practice and began during the eleventh century to produce a certain amount of literature on practical medicine. During the twelfth century, Salerno emerged as the leading center of medical instruction, theory, and literature, a position which it maintained through the first decades of the thirteenth century. During the twelfth century, medical instruction and literature began to flourish at the universities of Montpellier and of Paris that were to retain their position for many centuries. During the thirteenth century when the influence of Salerno declined, medical instruction began to develop at Naples and especially at Bologna. During the fourteenth century, medical instruction began to flourish at Padua, Pavia, and other Italian universities, and at the newly founded universities of central Europe [18].

The twelfth century was a crucial period, for the relation between medicine and philosophy no less than for many other developments in medieval Europe. The concept that an adequate training in medicine involved theory as well as practice, and that medical theory was based on natural philosophy or even a part of it, was widely defended and adopted. The medical doctor who had received theoretical as well as practical training came to be designated as *physicus*, a word that survives in the English term *physician* ([10], pp. 515–516). The term implied that the medical doctor, as distinct from the surgeon or practitioner, had studied theoretical medicine and also natural philosophy, usually on the basis of Aristotle and his commentators, a field that included many subjects that have since become separate sciences such as physics, biology, and psychology. The direct evidence that logic and natural philosophy were taught in separate courses at Salerno is rather scanty for the earlier period. However, the extant writings of the leading Salerno masters such as Bartholomaeus, Maurus, and Urso show beyond any doubt that they were familiar with the terminology and the teachings of Aristotelian philosophy, that they considered the theoretical part of medicine as related to philosophy or even as a subdivision of it, and that they dedicated a part of their thinking, writing, and teaching to philosophical problems.[7]

Outside of Salerno we find during the thirteenth century Petrus de Hibernia who taught at Naples and was active in both philosophy and medicine[8] and Peter of Spain, the later Pope John XXI, who was trained at Paris and was not only a respected medical writer but also the author of a textbook on logic that had a wide diffusion and influence down to the eighteenth century [8, 9, 12, 22]. We may also mention Peter of Abano who was active at Paris and later at Padua and who made important contributions both to medicine and philosophy ([14], pp. 1–74; [20]).

The strongest link between medicine and philosophy developed, however, not at Salerno which was archaic and untypical in its structure and whose influence, though strong, was purely literary after the middle of the thirteenth century, nor at Paris or the other Northern universities, but at Bologna and at the other Italian universities. The University of Bologna had originated during the twelfth century as a school of Roman law to which the study of canon law, of the notarial art, and of practical rhetoric (*dictamen*) were attached at a slightly later date. It was only during the second half of the thirteenth century, after the faculty of law had become fully organized, that the teaching of medicine and of other nonlegal subjects such as logic, natural philosophy, and also Roman poetry became important. Only at that time a faculty of arts and medicine, separate from the faculty of law, was formally established at Bologna ([20], pp. 105–115). This system of two faculties, one of law and one of arts and medicine, persisted at Bologna down to the French Revolution, and it was adopted after the late thirteenth century by all other universities in Italy or elsewhere that followed the model of Bologna. Paris, on the other hand, and the numerous universities that followed the model of Paris, had four faculties, namely, arts and philosophy, law, medicine, and theology. The faculty of theology, at Paris, Oxford, and other Northern universities, remained for a long time the leading faculty. The faculty of arts and philosophy was separate from the other three faculties and prepared its students for all the higher faculties, but especially for theology. Thus, we may say that at Paris and at the other Northern universities philosophy was studied and taught as a preparation for theology, and this explains why there are so many links between medieval Aristotelian philosophy and medieval Christian theology, as these subjects were taught and developed at Paris and the other Northern centers of learning. This does not mean, however, that there was no distinction between philosophy and theology, a view that is often held by non-specialists, and this is sometimes encouraged even by specialists who should know better. On the contrary, beginning with the second half of the thirteenth century, we find at Paris, Oxford, and elsewhere an increasing number of professional philosophers who never became interested or active in theology, whereas some of them became interested in medicine, as we have seen in the case of Peter of Spain and Peter of Abano.

The structure of Bologna and the other Italian universities differed from that of Paris in two significant ways. First of all, there was from the beginning and down to the French Revolution no separate faculty of theology.[9] Italian students who wanted to study theology went either to the schools of the monastic orders or to the Northern universities famous for their theological faculties. The Italian universities offered occasional courses on theology, and these courses became increasingly numerous over the centuries. Moreover, during the fourteenth century and later, most Italian universities requested and obtained the right of conferring degrees in theology. This right was exercised with the help of a few university teachers of theology and of the more numerous trained theologians who happened to reside at the local monastic schools. However, in Italy there were never separate university faculties of theology, and the comparatively few university lecturers of theology were members of the faculty of arts and medicine. This means that in Italy few university students of philosophy were preparing for a career in theology, as was the rule in the North, especially during the thirteenth century, but also afterwards.

The second feature that distinguishes the structure of Bologna and of the other Italian centers from those of the North is the fact that the arts and philosophy did not constitute a separate faculty but formed a single faculty together with medicine. Since medicine was a study leading to a respected and lucrative profession and was powerful enough to compete for influence and precedence with the older faculty of law,[10] it was obviously the most powerful element within the combined faculty of arts and medicine. The teachers of grammar and rhetoric were at first neither numerous nor influential, and the theologians counted even less. Only the philosophers, that is, the teachers of logic and natural philosophy, were of relative importance. Since the earliest professors of medicine at Bologna such as Taddeo Alderotti had shown a strong interest in philosophy [13], we are not surprised to encounter at least in the fourteenth century the pattern that the study and teaching of logic and natural philosophy was considered a prerequisite for the study and teaching of medicine. It became a normal procedure to acquire first a degree in arts and philosophy and then a degree in medicine and to begin a teaching career with logic and natural philosophy and to continue and end it with medicine. Philosophy and medicine thus became and remained for centuries complementary parts of the same professional study and career. Thus, the Italian Aristotelianism of the later Middle Ages and of the Renaissance was to a large extent a philosophy studied and taught by future physicians. The secular character of Italian Aristotelianism, which has often been commented upon, is directly related to this fact. For medicine had been, at least since the thirteenth century, if not before, a profession for laymen.[11]

I have been stressing an institutional development that had the result that many professional philosophers were also trained physicians, and that many professional physicians were also trained and sometimes productive philosophers. I do not wish to deny that institutions arise and develop on the basis of underlying ideas or that in the case of each individual thinker and scholar the institutional and professional pattern may be strengthened or modified by personal ideas and intellectual interests. However, I am inclined to attribute a good deal of importance to the professional pattern and to admit that the work of the individual student and thinker, however original, will develop within the framework determined by this pattern. The learning and thought of each period is characterized by a specific and also limited notion of the *globus intellectualis* that encompasses the different fields of knowledge as they are pursued in the schools and in the literature of the time and as their mutual relations are conceived. The work of the individual thinker and writer does not arise in a vacuum and will rarely transcend the given pattern, although it may modify and expand it.

The connection between philosophy and medicine during the Middle Ages and Renaissance which we have been trying to describe is in need of much further research. Such research will require also new methods of bibliographical and documentary investigation, for every new historical concept and problem must lead to a reexamination of the known facts and to a search for additional facts that were neglected before and may now become pertinent. The historian of philosophy must learn to find bio-bibliographical information for many Aristotelian philosophers of the Italian Renaissance, not in the reference books on the history of philosophy, but in the literature on the history of medicine. *Vice versa*, the historian of medicine who is accustomed to play down or ignore the philosophical background and interests of the physicians and medical scientists of the past must learn to pay attention to their place in the history of philosophy and to consult the literature and bibliography dealing with the history of philosophy. Moreover, both the historian of philosophy and the historian of medicine must not only learn from each other but also utilize the numerous documents that have been published by historians of the universities and that may throw much light on the medical and philosophical writers of the past by illustrating their careers as students and teachers.

I do not wish to imply that all philosophical thinkers of the Renaissance were trained in Aristotelian philosophy (though most of them were), or even less that they all were professors of philosophy (many of them were not). I only want to say that many Renaissance thinkers, and most Renaissance Aristotelians, were products of the university system that I have been trying to describe. Beginning with the fifteenth century, we find an increasing number of Aristotelian philosophers who never went into medicine, including Gaetano da Tiene, Pietro Pomponazzi, and Jacopo Zabarella. Yet we must be prepared to recognize that many physicians had a normal and professional interest in philosophy and many philosophers in medicine. When we study the faculty lists of Bologna which have been preserved and published almost completely for the fifteenth and sixteenth century, we find noted physicians such as Jacobus de Forlivio, Baverius de Bonittis, or Benedictus Victorius[12] who started their teaching career in logic and natural philosophy and noted philosophers such as Alessandro Achillini[13] and Girolamo Cardano[14] who concluded their career as professors of medicine. Cardano was in his time also a famous physician, and Achillini has been credited with an anatomical discovery. Outside of Bologna, we find at Padua Nicoletto Vernia who probably never taught medicine but expressed at least his loyalty to his faculty by writing a treatise in defense of medicine against law ([6], pp. 111–123). Agostino Nifo, a successful professor of philosophy and a prolific philosophical author, worked also as a practicing physician and left at least a few medical writings.[15] Bernardino Telesio, the natural philosopher trained at Padua, refers so often to medical sources that we may assume that he took at least some courses in the field [25]. Marsilio Ficino, the leading Florentine Platonist, was the son of a surgeon and began the study of Aristotelian philosophy and of medicine without obtaining the respective degrees. It is to this background and training, rather than to his Platonist interests, that we must credit his extensive familiarity with medical literature which is apparent in all of his writings and the interesting and influential medical treatises which he composed in his later years [4, 5]. If we go outside the circles of Aristotelian philosophy altogether, we may mention the humanist Niccolò Leoniceno who lectured at Ferrara and Bologna on both philosophy and medicine.[16] The contribution of humanism to medicine appears also in the rediscovery of the text of Cornelius Celsus[17] and in the numerous new translations of the Greek medical writers that were made during the fifteenth and sixteenth centuries, and that included several works of Hippocrates and Galen that had not been translated during the preceding centuries.[18] The earlier humanists had been uninterested in, or even hostile to, medicine. Petrarch wrote four invectives against the physicians, considering them as illegitimate rivals of his own favorite studies [19], and Coluccio Salutati, as other humanists after him, tended to side with the jurists in their rivalry with the physicians.

During the sixteenth century, we note the first signs that the tradition we have been trying to describe was coming to an end. Paracelsus, though strongly imbued with philosophical interests, questioned the entire medical tradition that goes back to Galen [15, 16]. Vesalius and other anatomists went in their discoveries beyond the limits of classical medicine, and William Harvey, a graduate of Padua, discovered the circulation of the blood and thus opened the way for later medical developments. The medical tradition that had its center in Italy through the sixteenth century and exercised its influence all over Europe, through the foreign students that came to Italy to obtain training and degrees and through the circulation of their lectures and writings in manuscript and print, began to decline, and with it the influence of philosophy, and especially of Aristotelian philosophy, also tended to disappear. This decline, which reached its climax in the nineteenth century and which coincided with some of the greatest triumphs of modern medicine, falls outside the scope of this paper.

It was my intention to show that there was a close link between medicine and philosophy during a rather long and interesting period of Western history. I do not wish to suggest that our own thought should be directly inspired by the precedents which I have been describing. We cannot ignore the changes and the progress that have occurred during the last few centuries in both medicine and philosophy, including their very conception and content. However, the precedent is appealing, and it might be adapted to a modern context. I really think that we should aim at a philosophy of medicine that would be an important and integral part of a philosophy of science and of a broader philosophy of culture. I am not equipped to contribute to this enterprise but earnestly hope that the task will be pursued by others.

Columbia University,
New York City, New York

NOTES

[1] Jaeger, W.: 1959, 'Die griechische Medizin als Paideia', in his *Paideia*, Vol. II, 3rd ed., Berlin, pp. 11–58. Cf. Edelstein, L.: 1931, *Περὶ ἀέρων und die Sammlung der hippokratischen Schriften*, Problemata IV, Berlin, pp. 117–122; 1967, *Ancient Medicine*, Baltimore, pp. 349–366.

[2] Sextus Empiricus: 1935–49, *Works*, R. L. Bury (ed. and trans.), 4 vols., London. Empiricus suggests that he was a member of the Empirical school of medicine, but he shows a preference for the Methodist school (*Pyrrhonian Dissertations*, Book I, Sect. 236 ff.). Cf. Ueberweg, F.: 1926, *Die Philosophie des Altertums*, 12th ed., K. Praechter, Berlin, p. 583. Edelstein, L.: 1967, *Ancient Medicine*, Baltimore, pp. 197 ff.

[3] Galenus, C.: 1821–33, *Opera omnia*, C. G. Kuehn (ed.), 20 vols., Leipzig. Cf. Temkin, O.: 1937, *Galenism*, Ithaca.

[4] Galenus, C.: 1874, *De placitis Hippocratis et Platonis*, I. Mueller (ed.), Leipzig; reprinted in Amsterdam, 1975. This work is also an important source for the Stoic philosophers Chrysippus and Posidonius. Cf. bib. 26.

[5] Avicenna: *Canon*, first printed in Latin before 1473 (*Gesamtkatalog der Wiegendrucke*, 3114) and often thereafter; Averroes: *Colliget*, first printed in Latin in 1482 (*Gesamtkatalog der Wiegendrucke*, 3107): Isaac (Israeli): *Opera omnia*, printed in Lyons, 1515. All these works circulated in manuscripts for centuries before they were printed, and were used as textbooks of medical instruction in the West.

[6] Celsus, A. C.: 1935–38, *De medicina*, W. G. Spencer (ed. and trans.), 3 vols., London. According to Stahl (pp. 96–98), Celsus was a good writer but depended for his medical doctrine on Greek sources.

[7] For Bartholomaeus and his commentaries, see Kristeller, P. O.: 1957, 'Nuove fonti per la medicina salernitana del secolo XII', *Rassegna storica salernitana*, **18**, 61–75; Kristeller, P. O.: 1959, 'Beitrag der Schule von Salerno zur Entwicklung der scholastischen Wissenschaft im 12. Jahrhundert', *Artes Liberales von der antiken Bildung zur Wissenschaft des Mittelalters*, J. Koch (ed.), Studien und Texte zur Geistesgeschichte des Mittelalters V, Leyden and Cologne, pp. 84–90. I shall deal with the subject more fully in an article to appear in *Italia Medioevale e Umanistica.*

[8] Crowe, M. B.: 1967, 'Peter of Ireland: Aquinas's Teacher of the Artes Liberales', *Arts libéraux et philosophie au Moyen Age, Actes du Quatrième Congrès International de Philosophie Médiévale*, Montreal, pp. 617–626. To the philosophical works discussed by Crowe, we should add a medical writing preserved in Erfurt, Ms. Ampl. Fol. 335, ff. 119v–160.

[9] The existence of a separate faculty of theology is asserted by Sorbelli ([20], pp. 129–148) and implied by the titles of two important works: Ehrle, F.: 1932, *I più antichi statuti della Facoltà teologica dell' Università di Bologna*, Bologna; Brotto, G. and G. Zonta: 1922, *La facoltà teologica dell'Università di Padova*, Padua. However, the claim is refuted by the documents: Dallari, U.: 1888–1924, *I Rotuli dei lettori legisti e artisti dello Studio Bolognese dal 1384 al 1799*, 4 vols., Bologna; *Acta graduum academicorum Gymnasii Patavini (1406–50)*, G. Zonta and G. Brotto (eds.), 1922, Padua; *Acta graduum academicorum (1501–50)*, E. Martellozzo Forin (ed.), 3 vols., 1969–71, Padua. The error is caused by the ambiguity of the term 'faculty' and by the failure to distinguish between the teaching faculty and the *collegium doctorum*, a kind of permanent examining commission that includes many members who do not teach and excludes many who do.

[10] There was a whole literature on the superiority of law or medicine. Most of it has been edited by Garin, E.: 1947, *Coluccio Salutati, De nobilitate legum et medicinae*, Florence; Garin, E.: 1947, *La disputa delle arti nel Quattrocento*, Florence. The earliest text known to me that pertains to this subject is a *Questio de comparatione scientie medicine ad scientiam legalem*, presented by Julianus de Preamptis at Bologna in 1342 as an introductory lecture (Vatican Library, ms. Vat. lat. 2418, f. 227).

[11] For the contribution of Aristotelian philosophy, especially at Padua, to the theory of the scientific method, and especially to the conception of analysis and synthesis, see Randall, J. H., Jr.: 1961, 'The Development of Scientific Method in the School of Padua', *The School of Padua and the Emergence of Modern Science*, Padua, pp. 13–68. As Randall rightly observes, the distinction between analysis and synthesis was known to the Paduan philosophers not only from Aristotle but also from Galen whose *Ars medica*

begins with this distinction (C. G. Kuehn ed., I, pp. 305–306). This text was the subject of numerous commentaries ever since the twelfth century. Cf. Niedling, J.: 1924, *Die mittelalterlichen und fruehneuzeitlichen Kommentare zur Techne des Galenos*, diss. Leipzig, Paderborn. The alliance between medicine and philosophy thus bore important fruits in the history of logic and epistemology. Cf. Gilbert, N. W.: 1960, *Renaissance Concepts of Method*, New York. It should be noted that Galen adds to analysis and synthesis a third method which he calls the dissolution or division of a definition and which he may have derived from the method of division used in Plato's *Phaedrus*, *Sophist*, and *Politicus*. Cf. De Lacy, P.: 1966, 'Plato and the Method of the Arts', *The Classical Tradition, Literary and Historical Studies in Honor of Harry Caplan*, Ithaca, pp. 123–132 (I am indebted for this reference to Professor Richard Patterson). For the sixteenth century, and especially for Galileo, the discussion of analysis and synthesis must be related to a third source, in addition to Aristotle and Galen, that is to Pappus and the other Greek mathematicians. Cf. Gilbert, N. W.: 1963, 'Galileo and the School of Padua', *Journal of the History of Philosophy* 1, 223–231. For Galen's influence on Renaissance methodology, see also Gilbert, N. W.: 1960, *Renaissance Concepts of Method*, New York, pp. 13–24, 98–107, 164–165.

[12] Jacobus de Forlivio taught natural philosophy (1384–86), moral philosophy, and medicine (1392–93 and 1395–96) and again logic and natural philosophy (1398–99). Baverius de Bonittis taught logic (1429–32), philosophy and logic (1432–33), philosophy (1433–34), and medicine (1438–81). Benedictus Victorius taught logic (1503–05), philosophy (1505–10), and medicine (1510–61). See Dallari for all these data. Cf. also Siraisi, N. G.: 1973, *Arts and Sciences at Padua*, Toronto.

[13] Achillini taught logic (1484–87), philosophy (1487–1500), medicine and philosophy (1500–10). His writings include a work on anatomy. See Nardi, B.: 1960, *Dizionario Biografico degli Italiani*, Vol. I, pp. 144–145; Matsen, H. S.: 1975, *Alessandro Achillini (1463–1512) and His Doctrine of Universals and Transcendentals*, Bucknell University Press, Lewisburg, Penn.

[14] Cardanus was professor of medicine at Bologna from 1562 to 1571.

[15] Mahoney, E. P.: 1966, *The Early Psychology of Agostino Nifo*, thesis, Columbia University. Several medical writings by Nifo are found in Rome, Biblioteca Lancisiana, ms. 221. Cf. Schmitt, C. B.: 1970, 'A Survey of Some of the Manuscripts of the Biblioteca Lancisiana in Rome', *Medical History*, Vol. XIV, pp. 289–294, especially p. 291.

[16] At Ferrara, he taught first mathematics, beginning in 1473, and later moral philosophy. Cf. Borsetti, F.: 1735, *Historia almi Ferrariae Gymnasii*, Vol. I, Ferrara, pp. 60–64; Visconti, A.: 1950, *La storia dell'università di Ferrara*, Bologna, p. 45. At Bologna he taught both philosophy (on the basis of the Greek text) and medicine during the academic year 1508–09 (Dallari, Vol. I, pp. 201–202). Cf. Vitaliani, D.: 1892, *Della vita e delle opere di Nicolò Leoniceno Vicentino*, Verona.

[17] Celsus was rediscovered in Siena in 1426. Cf. Sabbadini, R.: 1914, *Le scoperte dei codici latini e greci ne'secoli XIV e XV*, Vol. II, Florence, p. 207; the first printing was edited by Fontius, B.: 1478, Florence, *GW*, 6456.

[18] For the printed versions of Galen, see Durling, R. J.: 1961, 'A Chronological Census of Renaissance Editions and Translations of Galen', *Journal of the Warburg and Courtauld Institutes* 24, 230–305. For Hippocrates, cf. Kibre, P.: 'Hippocratic Writings in the Middle Ages', *Bulletin of the History of Medicine* 18, 371–412.

BIBLIOGRAPHY

1. Beccaria, A.: 1956, *I codici di medicina del periodo presalernitano*, Rome.
2. Cassirer, E.: 1944, *An Essay on Man*, New Haven.
3. Dunlop, D. M.: 1958, *Arabic Science in the West*, Karachi.
4. Ficino, M.: 1481, *Consiglio contro la pestilenza*, Florence.
5. Ficino, M.: 1489, *De vita libri tres*, Florence.
6. Garin, E. (ed.): 1947, *La disputa delle arti nel Quattrocento*, Florence.
7. Haskins, C. H.: 1924, *Studies in the History of Mediaeval Science*, Cambridge, Mass,; reprinted in 1960, New York.
8. Hispanus, P.: 1947, *Summulae Logicales*, I. M. Bochenski (ed.), Turin.
9. Hispanus, P.: 1972, *Tractatus* (called afterwards *Summulae Logicales*), L. N. de Rijk (ed.), Assen.
10. Kristeller, P. O.: 1956, *Studies in Renaissance Thought and Letters*, Rome.
11. Kristeller, P. O.: 1965, 'Renaissance Aristotelianism', *Greek–Roman and Byzantine Studies*, Vol. VI, pp. 157–174.
12. Mullally, J. P.: 1945, *The Summulae Logicales of Peter of Spain*, Notre Dame.
13. Nardi, B.: 1949, 'L'averrosimo bolognese nel secolo XIII e Taddeo Alderotto', *Rivista di storia della filosofia*, Vol. IV, pp. 11–22.
14. Nardi, B.: 1958, *Saggi sull'Aristotelismo Padovano dal secolo XIV al XVI*, Florence.
15. Pagel, W.: 1958, *Paracelsus. An Introduction to Philosophical Medicine in the Era of the Renaissance*, Basel, New York.
16. Pagel, W.: 1962, *Das medizinische Weltbild des Paracelsus, Seine Zusammenhaenge mit Neuplatonismus und Gnosis*, Wiesbaden.
17. Posidonius: 1972, *The Fragments*, L. Edelstein and I. G. Kidd (eds.), Cambridge.
18. Rashdall, H.: 1936, *The Universities of Europe in the Middle Ages*, F. M. Powicke and A. M. Emden (eds.), 3 vols., Oxford.
19. Ricci, P. G. (ed.): 1950, F. Petrarca, *Invective contra medicum*, Rome.
20. Sorbelli, A.: 1944, *Storia dell'Università di Bologna* I, Bologna.
21. Stahl, W. H.: 1962, *Roman Science*, Madison.
22. Stapper, R.: 1898, *Papst Johannes XXI*, Kirchengeschichtliche Studien, Vol. IV, No. 4, Muenster.
23. Steinschneider, M.: 1956, *Die europaeischen Uebersetzungen aus dem Arabischen bis Mitte des 17. Jahrhunderts*, Graz.
24. Steinschneider, M.: 1960, *Die arabischen Uebersetzungen aus dem Griechischen*, Graz.
25. Telesius, B.: 1910–1923, *De rerum natura*, V. Spampanato (ed.), 3 vols., Modena; reprinted by Vasoli, C. (ed.): 1971, Hildesheim.
26. von Arnim, H. (ed.): 1905–1924, *Stoicorum veterum fragmenta*, 4 vols., Berlin-Leipzig.
27. Walzer, R.: 1963, *Greek into Arabic*, Oxford.

OTTO E. GUTTENTAG

CARE OF THE HEALTHY AND THE SICK FROM THE ATTENDING PHYSICIAN'S PERSPECTIVE: ENVISIONED AND ACTUAL (1977)[1]

The purpose of this paper is to formulate a conceptual framework that will be useful to everyone concerned with discussing the problem of health and health care. The concepts of health (good health and ill health) and health care are rarely studied from the standpoint of the attending physician and within the framework of ultimate actuality – that is, the ultimate, irreducible categories, or 'existentials,' of our experience.

By ultimate actuality, I do not mean the world of theoretical-reductive empiricism but the world of an empiricism "that links itself most intimately to the object and thus becomes the theory proper" [4]. It is the world sought by the phenomenologists and described in Husserl's emphasis: *back to the things themselves* ([7], p. 96) or in Paul Tillich's phrase of "receiving knowledge" ([15], p. 98). It is the world in which we live and which confronts us in the endless variety of its ultimate concreteness, particularly with regard to the dichotomy between the living and the nonliving and to the interactions of human beings with each other. By ultimate, irreducible categories of our experience I mean such concepts as myself-ness and other-ness, including the differences between Martin Buber's I–thou and I–it relations, autonomy and bondage, and individuation and community, and not the concepts of the special disciplines such as drives and volition in psychology or gravity and temperature in the mechanical sciences.

The current literature usually deals with the concepts of health and health care predominantly in the terms of some special discipline, e.g., sociology or psychology. Concepts of health and care of health are examined, for instance, with concern for the "social construction of illness" or the professionalization of medical care ([2], pp. viii, 186) or with regard to death and care of the dying in terms of anxiety, loneliness, and isolation [9].

Thanks to certain recent analyses by the phenomenologists and philosophical anthropologists, we now clearly recognize the difference between the cognitive, rationalistically detached, objectifying approach to comprehension of our human existence and the participating, existential, subjectivistic approach. Tillich succinctly points out ([15], p. 98) that understanding of man requires the unity of both approaches.

The attending physician is the key figure in the entire Western medical

S. F. Spicker (ed.), Organism, Medicine, and Metaphysics, 41–56.

counseling effort, the person who brings the concept of health and health care into sharp and complete focus. Therefore, it is appropriate to elucidate the concepts of health and health care from this standpoint and not only from that of an investigating onlooker who is unencumbered by the responsibilities of personal advice and action (e.g., of a sociologist free to adopt or not to adopt medicine's own concepts) or of someone seeking counsel.

Physicians are pragmatists whose decisions perforce are concerned with the immediate here and now. The unparalleled recent and continuing advances in medical research have provided no reason to explore theoretical premises. Sociologists are often similarly pragmatic. They quite convincingly point out significant deficiencies of current medical care and complain (not unjustifiably) about the lack of imaginative, cooperative effort by professional organizations to correct these deficiencies ([2], pp. 31–33). However, they treat rather cavalierly the ultimate actuality of health and health care, which is the realm of the attending physician, e.g., the problems of the comprehensibility of life, of health, and of the relation between life and health and their resolutions.

This cavalier attitude is unfortunate, because it ignores a most important aspect of health and health care. It also forestalls any discussion among physicians as well as between physicians and representatives of other disciplines. Interestingly, the lack of the discussion of the concepts of health and health care in terms of the ultimate, irreducible categories of our experience and from the standpoint of the attending physician is less detrimental in relation to actual care of patients. Medical historians report that physicians who differ greatly in their interpretation of humans and their health may differ little in actual patient care.

Finally, and perhaps most importantly, one cannot reliably evaluate the assets and liabilities of medicine of a given time and place in their full compass unless one has first tried to recognize clearly the axioms of medicine in their trans-historical and trans-geographical (i.e., ultimate) actuality. Of course, no investigator can escape entirely the contingencies of his time and culture and, indeed, of his own person. However, merely accepting these contingencies, perhaps even blindly, is quite different from trying to transcend them to the best of one's ability.

As in all analyses of the type presented here, the results are subject only to experiential, not experimental proof. In other words, it is range, not repetition of a single set of observations, that is the essential methodologic, empirical factor.

It should be emphasized, at the very beginning, that the immediate practical value of the reflections of this paper is negligible. Obviously, no specific empirical research projects or rules or regulations can be directly derived from them. The value of the reflections lies in the attempt to provide the most adequate theoretical basis or background for such undertakings.

I. THE PRIMORDIAL YET ULTIMATE SITUATION OF MEDICAL CARE

The range of phenomena and situations to which the attributes of health and health care have been attached is extensive. For example, in sociologic terms that relate to *health*, identical psychologic or bodily manifestations may be judged differently in different societies. Pinta, a patchy red coloring of the skin produced by spirochetes, is considered a disease in western cultures. Among many South American tribes, however, the red coloring is so common that the few men who do not show it are regarded as aberrant to the point of being excluded from marriage ([1], p. 143). In sociologic terms that relate to health *care*, different societies also pay different attention to health care and its availability. Diversities range from complete *laissez-faire* to legislative insistence on licensure of health care personnel, on keeping statistics concerning health and disease, on support of health research and health care research, and from no provisions for general availability of health care to what is called socialized medicine.

The diversity among individuals who give care range from the motivations of monks and nuns, who may occasionally and for purely spiritual reasons take care of the healthy and the sick, to those of ambulance drivers, who may drive their vehicles exclusively to make a living. Within this range, many, including nurses and physicians, may have chosen their disciplines for both reasons. From the standpoint of the object of medical care, the patient, individual relations to society may range from those persons consciously feigning illness to those on hunger strikes. Operational concepts in the actual giving of care range from the rituals of the shaman to the over-the-counter sale of drugs by a pharmacist or (to mention two current extremes) from the activities of the psychotherapist to those of the proctologist. Finally, there is the ever present and all pervading dualism of restorative and preventive care of health.

In view of these diversities, what then is the common irreducible given underlying the expression 'health care?' The answer may be formulated as follows: It is the positive, active response by someone who is convinced of being able to respond effectively to another's overt or hidden request: "Help me in the care of *my* health." However else the situation may be defined, someone calls to have health needs fulfilled, and someone responds to this call.

It should be clear that the feigner of illness, the person on a hunger strike, the institutionalized restriction of individual choices, and the delivery of preventive medicine do not represent the primordial and ultimate situation of health care. Intentional feigners of illness, persons on hunger strike, and health care dictators do not face the situation of health care in its genuine immediacy, they exploit it. As to the dualism of restorative *vs.* preventive health care, preventive medicine flows from, and is not ontologically equal to, restorative health care. Finally, whatever the relationship may be between *my* (personal) health and what is called public health, and whatever health care laws society may make, public health refers to the mean of many personal healths, and public care is identical with a legislative mandate for many persons.

Briefly, then, dyadic solidarity or fellowship of personal health assistance sought and given is the primordial and ultimate situation of medical care.

II. ON MAN AND HIS HEALTH AND ON THE PATIENT-PHYSICIAN ENCOUNTER

The point of origin of all health care and the point to which all health care returns is the positive, active response by someone to someone else's open or hidden request: "Help me in the care of my health." It will facilitate the analysis of this characterization, if we translate it into the more objectified phrase: "Medicine deals with the care of health of human beings by human beings." An analysis of this expression in regard to the patient–physician relationship requires that we: (a) compare the inter-human solidarity of the physician–patient encounter with the human care of nonhuman beings, e.g., of machines or animals; (b) elucidate the nature of human beings, of their health and of the interrelationships between the two; (c) bring into focus, on the basis of this elucidation, the mode of interpersonal relationship between patient and physician; and (d) compare this mode with those modes of interpersonal relationships employed in filling non-medical needs (for example, those of a buyer of merchandise and its seller or of a client and a trial lawyer) in order to pinpoint, thereby, what is meant by medical excellence.

What is the significance of the phrase "of human beings, by human beings?" The answer is readily given: Patients and physicians are *fellow* beings, whereas maintenance engineers and machines, veterinarians and animals are not. Further, engineers may know everything about the objects of their care because machines are man-made. But how much do physicians know about patients? Certainly no more than they know about themselves, for they are fellow beings or, more precisely, fellow human beings. Thus, the question arises: What do we human beings know about ourselves? Expressed more generally: Who are we? What do we mean when we say we have lost our health or are in possession of it?

To answer the first question, "Who are we human beings?" we shall first paraphrase a characterization by Tillich: We experience ourselves as multiphasic physical beings of finite freedom ([15], p.165, [18]). Clumsy as the phrase may be in its experiential immediacy, it is broad enough to avoid the objections of any interpreter of the nature of man who is of good will, yet it is precise enough to bring the concept of health and the problems of health care into proper ontological focus. It is also brief and succinct enough to pinpoint the significance of this insight for the specific purpose of this paper. The words 'physical' and 'multiphasic' need no detailed elaboration. They refer to the experience of ourselves as beings of mass and extension, of storage and transmission of energy, in other words as machine-like beings, and to our experience of the sequential developmental character of our nature, i.e., the various stages of our being, implying an ultimate fulfillment, "Man is a *Zeitgestalt*" ([19], p. 57).

The word 'freedom' in the context of this paper refers to a conscious experience of oneself as someone, not merely something: as an 'I,' not as an 'it,' as a self, an irreducible center of autonomy or spontaneity, as a being of unfathomable privacy and inwardness, and as a moral and spritual being with the corollary insight that trust and confidence lie at the basis of every human encounter ([10], pp. 8–18).[2] The word *conscious* is stressed, for being conscious (not merely aware), reflective, responsible is the characteristic that makes man judicable. It denotes the difference between man and the rest of the biological world, i.e., all those objects of our experience that are not adequately described as its, things, machines but adequately described only when described as subjects, centers of autonomy, in other words as living.

The word 'finite' refers to our experience of the limitation of our being, of our knowledge both theoretical and practical, and especially of what we call death, the total, inevitable, and unpredictable loss of the self, with the implications these phenomena have for our spiritual and moral self-fulfillment. Two elaborations must be made here. First, death as fulfillment must be differentiated from actual death [6]. Death as fulfillment is a regulative principle. Actual death lacks this aspect of fulfillment. It is always untimely and 'absurd' and prevents possibilities of further development that always appear to be present. Second, because of the awareness of our finitude, the freedom of our morality is constantly challenged by our temporality. "Self intrudes into every ideal when thought gives place to action" ([12], p. 13).

Finally, some remarks about the word "of" in the phrase, physical beings of finite freedom: In the context of this phrase, the word is as usual vague. It implies a dualism, but does not articulate the ambiguity of that dualism. Is this a composite dualism as, for instance, in René Descartes's separation of mind and body, or is it a polar dualism, that is, is it personal versus physical poles, similar, for instance, to the polar structure of the earth or of a magnet? The difference between the two meanings of dualism is profound. In composite dualism, the parts precede the whole and must be integrated to create the whole. In polar dualism or duality, the whole precedes two antithetic propensities that are differentiated; the whole is structured. It is, of course, polar dualism to which the "of" refers, for an "I" may be structured but is not divisible; our body is ultimately always *our* or *my* or *your* body. As Gabriel Marcel has so succinctly expressed it: "I am my body" ([11], pp. 332–333).

How then does the concept of health, its possession and its loss, fit into this characterization of our nature?

Two lines of thought might be followed: (a) Health must be viewed not only in terms of good health and ill health but also in terms of other aspects of human nature; and (b) Marcel's dictum "I am my body" must be considered in relation to ill health.

Man not only is more or less healthy; he also is more or less wise and intelligent and more or less moral (regardless of any given standards for what is wise, intelligent, or moral). Obviously, a healthy Hitler does not fall into the same cateogry of human existence as a healthy Gandhi or a healthy "man in the street." Health refers to the most basic (pre-rational and pre-moral), inescapable, and universal, i.e., biologic conditions of human existence ([16], p. 42).

Good health is the optimal state of these conditions in terms of fulfillment of our finite freedom (aside from social restrictions); ill health is their sub-optimal state. Health is measured in terms of potential, typical life span and in terms of the range of affirming one's self in the face of the infinite challenge (biological and social) to our self-fulfillment.

Marcel's dictum "I am my body" refers unqualifiedly only to the healthy body. I kiss someone I love – I, not a pair of lips that I carry and can dispose of as I can dispose of a pair of gloves. This dictum does not refer unqualifiedly to my body, me, when I am sick. Therefore, the term "ultimately" was inserted above – when it was stated: Our body is always *our* or *my* or *your* body. When I am sick, I may remove myself from parts of my body: That lip tumor of mine is not me – I hardly identify with it at all. It is very close to the physical pole of my nature. Indeed, I can remove myself to a surprisingly great degree from many parts of my body. (Similarly, I can integrate foreign bodies into my body, me, as in organ transplantation.) This potential for separation from parts of the (my) body is exceedingly variable from person to person. The little finger of the left hand may be *me* to a typist or a violinist but perhaps not to a sprinter or the conductor of an orchestra. It is research, not nature, that determines the commodity aspects of health, i.e., which parts of my body, of me, are disposable. But however much this disposability may vary, I am, ultimately, an "incarnate being" ([11], p. 269). It is against this background of the nature of human beings and their health that we will review the nature and structure of the patient–physician relationship in terms of ultimate actuality and in terms of medical excellence.

Medicine does not focus on human intellectual and moral formulations. It does not differentiate between law abiding and criminal patients, between moral and immoral ones. The physician considers these aspects of human nature only as they affect either the basic prerequisites (health) or indicate their reduced state (illness). The physician is, to paraphrase Owsei Temkin's succinct formulation, the fellow being who assists in the care of the natural side of human beings [14].

Humans are beings of finite freedom. What should this elicit in characterizing the patient–physician relationship? To illustrate, we shall compare the abilities required of a physician with those required of a trial lawyer. Both, physicians and lawyers, are counselling (service) experts ([2], p. 73) in fields highly relevant to central values of society [13]: posited law and health. But, although both physicians and lawyers may be considered equals in terms of existential effort, of solidarity with their callers, what a difference of fields in terms of consulting concern and human nature! Law, the field of the lawyer's concern, refers essentially to human freedom, to decisions and acts of humans as humans, to humans as creative humans; health, the field of the attending physician's concern, refers ultimately to human finitude, to the incomprehensible bondage of our nature, to our inescapable creatureliness. As the language expresses it: we *commit* crimes but we *fall* ill.

This difference does not mean that the lawyer can neglect seeing the lawbreaker in the duality of human nature, or that the physician can neglect viewing this duality in the sick. The asocial act may have been triggered by the most basic, inescapable, and universal conditions of the lawbreaker's existence, or ill health may have been triggered by some willful act. The difference does mean, however, that the manifestation of human freedom, objective knowledge, plays quite a different rôle in the two fields. The characterizations of various crimes are fixed. Crimes can be classified quite rigidly, and punishment is closely tied to the crimes; but the physician, with all the knowledge at his command and all the carefulness of his proceedings, may see a minor wound become the portal of death, is aware of the fluid and abstract character of what is called a single disease, knows the loose relationship between diagnosis and therapy. Lawyers (themselves included) focus on the rationalism of human freedom; they center on the act. Physicians (themselves included) always have to give equal weight to human finitude; they focus on a creature [5, 8]. The effort, the "expertise," of the physician is never exhausted in rationalistic insight. The effort to participate in the patient's situation existentially must also always include the effort to come as close as possible to grasping the patient's ultimate elusiveness. The demands for such effort may vary widely, of course, depending on the individual patient and the particular therapy required. It may be added that the more physicians advance in insight, the more they "know," the less they need translate diagnoses into disease labels.

To illustrate the difference between the medical and judicial approach, we may briefly discuss the obligation of telling the truth to the patient about his medical condition. This obligation poses a problem for the physician. A corresponding problem hardly exists for the lawyer.

It is obvious that the patient as a fellow human being has every right to know the truth about his biologic condition. Yet, two other considerations enter into the situation: the mutual influences of the somatic and the personal poles (the dualism of our nature) and our anxiety concerning death, i.e., actual death. As mentioned above, the experience of ourselves as developing selves carries the connotation of fulfillment, of death as "natural" death. However, the determining causes of any actual death have no ultimate necessity. Depending on the fulfillment of our lives, any knowledge of approaching death makes us more or less anxious, because it implies the inevitable and total denial of fulfillment (my fulfillment). The same, to a lesser degree, holds true when one becomes aware of some irreversible somatic functioning. Telling an unwelcome truth to the unprepared patient is as ill-conceived as trying to hide the truth from the prepared. In the words of Thomas Addis: "Honesty with patients requires thought and discipline and effort."[3]

Broadly then: At a given time, silence may become a therapeutic tool, an instrument of the help for which the patient called. At this point, two considerations must be added: First, the physician as a person of creaturely imperfection may, of course, simply err in understanding the uniqueness of the patient's freedom and may misjudge when "the time is ripe;" second (and this, from the standpoint of fellowship, is more detrimental), the physician, in narrow egotism of the self that intrudes into every ideal when thought gives place to action, may either "spring" the truth on the patient, in order to avoid the moral responsibility implied in fellowship, or may remain silent in self-righteous paternalism. No objective procedure is possible to test in advance the physician's adequacy in this personal area of the patient–physician encounter. His "expertise" can only be subjectively tested and sensed by "reasons of the heart," by trying to evaluate the physician's effort to participate in the patient's being in all its dimensions. But however risky this area of the patient–physician encounter is, however different the risk may be in different modes of ill being, it can never be ignored without detriment to the patient.

Viewing the patient in the complexity of our human nature, in the polarity of our dualism, von Gebsattel elucidated the medical situation by dividing it into three stages ([3], p. 371). In slight modification of his original characterization, they may be called:

(1) The stage of attentive fellowship: the primary stage of receptive participation. Alert listening (history taking in patients' or witnesses' words but in the professional framework).

(2) The stage of detachment: the objective, scientific stage, nosography, and the concept of *a* disease.

(3) The stage of personal communication: the final subjective stage of reciprocal participation. The ethical element in the patient–physician relationship.

True, the ratio among the three stages may vary and each single stage may be more or less pronounced. Nevertheless, as in a musical theme with variations, the nature and structure of the whole situation is always present.

The points made in this section can be restated as follows: The patient–physician relationship is fiducial. It is a matter of trust for two reasons: first, for the reasons inherent in any service function *per se*, i.e., the caller's trust in the responder's effort to try his best (the patient's trust in the integrity of the physician); and second, for reasons of the object of the service: restoring or maintaining a patient's health, *my health*, the most basic, inescapable, and universal condition of *my very existence.*

Such service demands technological knowledge of the objectified facets of human nature as well as participation in those that are logically incomprehensible to both patient and physician. The expertise of technological knowledge can and, of course, must be tested objectively. The test for expertise in grasping the incomprehensible, creaturely facets of our biological nature can only be the patient's decision to trust the physician in grasping these facets and accept him or to distrust the physician and reject him.

III. THE CURRENT SCENE

What is the value of all that has been said so far in terms of providing and maintaining optimal, or improving sub-optimal, medical care in this country today? The question is particularly appropriate because, as was stated in the very beginning, the immediate practical value of the reflections of this paper is negligible.

In order to answer the above question, it is useful to survey briefly the current medical scene as it presents itself to an open-minded observer, meaning one who does not ignore the viewpoint of the attending physician. To such an observer, the situation presents several paradoxes. On the one hand, it is difficult to visualize that anyone who views the recent and significant advances of medical care would prefer to live in an earlier time or in a setting where the interpretation of health is less differentiated from spiritual or moral concepts. On the other hand, and equally obvious, one finds great and widespread dissatisfaction with the current status of medical care. This dissatisfaction is present both in the physician–patient relationship itself and in the broader sociologic context, that is, the level of commitment of our society to provide optimal availability of optimal medical care.

In the patient–physician relationship, dissatisfaction refers, from the patient's point of view, to what is called the "management" attitude of physicians, their viewing patients not as fellow human beings in distress but rather as malfunctioning machines or as other inferior beings. On the side of the physician, it expresses itself in the "using" of physicians, openly and not merely by simulating ill health, for non-medical services, for economic or social benefits. Dissatisfaction with the patient–physician relationship also arises in the complaints of the evergrowing number of workers in the field of medical care who are not physicians, but who are needed and who feel unduly restricted in the scope of their responsibilities by physicians as well as in the counterclaims of physicians that the sovereignty of physicians must be maintained.

In the social area, the sociologist's dissatisfaction concerns the difficulty of obtaining medical care for the poor and not-so-poor as well as the difficulty of obtaining optimal medical care regardless of personal wealth because of the insufficient number of diagnostic and therapeutic facilities. On the other hand, public agencies and private organizations find that the complaints are exaggerated; they follow a different scale of priorities in appropriating funds. The dissatisfaction in this area also finds expression in some sociologists' complaints that the physicians' expertise and power is inadequately checked by society. The physicians, in turn, insist that such dissatisfaction is ill-founded, and that, even if it were well taken, such complaints need a justification other than the sociologists' view of medical expertise.

How, then, does the attending physician, who reflects on the current medical situation in terms of ultimate actuality, evaluate the many charges and countercharges of dissatisfaction?

Aware of their basic frame of reference, reflective physicians will side with those who unequivocally reject the "management" attitude of attending physicians because it is a violation of the principle of fellowship. They will also side with those who try to improve the current delivery of health care, for its inadequacies are impediments to the attending physicians' response to the call for help. In the context of this paper, no further analysis is required of these dissatisfactions. A glance at medical-educational and medical-political literature will show that serious efforts are underway to remedy these situations, although there are, of course, honest as well as self-serving differences of opinion.

The weight of tradition and the superior position of being the counselor and not the counselee tends to buttress a self-centered attitude on the part of the physician. Therefore, we need to examine carefully both the data and the complaints of the non-physician members of the medical team and their interpretation as supplied by the sociologists in order to interpret present-day medical care as it relates to the ultimate actualities of the patient–physician relationship.

The conflict concerning the structure of the medical team will be examined first. Obviously, the self-centered "I" may intrude into the claims of non-physicians just as it may into the claims of physicians. But such bias does not clarify the issue. The issue is: Can an "I" who calls for my help be satisfied with a divided response? Can an "I" accept or reject different advice at the same time? At first glance, the request of the non-physician members of the team seems understandable and, perhaps, well-founded. Today's medical knowledge requires specialization, and our society's structure often necessitates evaluation by non-physicians of such factors as the patient's financial status and chances of reintegration into the social process. However, we experience ourselves not as composite beings but as differentiated beings, as individuals. Hence, the responsibility for following up an advised action must be shouldered by one person, diversified as the advice of the medical team originally may have been. The physician, on whose shoulders the ultimate responsibility for the recommendations and actions of the medical team lies, is the person most qualified to serve as the speaker for the team.

The physician must, of course, not conceal from the patient the input from other members of the team or prohibit their meeting, nor may the physician in self-centeredness neglect to ask for such input. The physician may, with the consent of the patient, delegate responsibility to any member of the team, but such action is quite different from shifting responsibilities according to different facets of the patient's being. Briefly then: In terms of the ultimate actuality of the medical situation, the leadership of the medical team, and, of course, the primary fellowship with the patient rests with the physician.

We must also deal with the assertions of some sociologists that physicians as a group are allowed too much freedom by society to supervise the organization of their own professional performance, their expertise, and their general conduct. Reference is made to lack of professional self-regulation in protecting those who seek help from falling prey to intraprofessional and extraprofessional economic pressures, to the efforts at limiting the practices of competitors such as chiropractors, and to inadequate awareness of communal responsibility.

The yielding of individual physicians to intra- and extra-professional pressures for economic reasons would seem to be more a general problem of society than specifically a medical one. The admonition to resolve it addresses itself primarily to the society in which the profession finds itself, not only to the profession itself.

However, on the basis of examples given by Freidson ([2], p. 149) and in terms of the responsibilities of physicians as a group, there can be no doubt that much is to be desired in the field of communal responsibility, for example, in the profession's efforts to protect the community from obviously incompetent physicians. Similarly, in the field of professional supervision of its expertise, it is unacceptable to reject mainstream therapy on the other than carefully evaluated empirical grounds. The egotistic self intrudes into an organizational ideal no less than into a personal one when thought gives place to action. Indeed, the most important contribution of the medical sociologist to the improvement of medical care lies in pointing to such intrusion. Bierstedt's statement, "The profession of sociology has something to say to the profession of medicine," is quite appropriate and needs careful attention ([2], p. ix).

But the reverse is equally true. As the onlooker, detached and objectifying, is the best critic of the intrusion of self-serving egotism ([2], p. xix), so too is the participant the best person to evaluate the intricacy of the mission. It is at this junction that some clarification is necessary from the standpoint of the attending physician and in terms of the ultimate actuality of medical care. For Freidson, for instance, health and disease are essentially sociologic concepts and to a much lesser degree "neutral scientific" ones ([2], p. 208), as he chooses to characterize the alternative attributed to physicians. For the attending physician, this characterization of the physician's attitude is ill-conceived. Physicians do not deny that humans are social beings, that included in definitions of health and disease may be appropriately sociologic categories, and that, therefore, society has the right to enact health laws. However, physicians are not free to belittle, as Freidson is, and does, the existence of human beings as biologic beings. Each human being is also a single self, a single center of autonomy, finite and socially contingent though this autonomy may be. Indeed, the awareness of the incomprehensibility of actual finiteness, the anguish about the absurdity of actual personal death, about the inevitable and absolute loss of self, and of further self- realization are ineradicable from the inner life of each human being. No less ineradicable is the awareness of the possibility and the actuality and anguish of suffering from more or less comprehensible, occasionally self-limited, occasionally only artificially reversible limitations of the biologic prerequisites of self-fulfillment, that is, of the actuality of falling or being ill. In other words, attending physicians are always obligated to view anew the concepts of health and disease in terms of biologic-sociologic polarity and to present, if need be, this polarity to their patients.

Because the self-realization of man's biologic nature is obviously not man's only choice for self-realization, the attending physician will certainly explore with the patient – the one who called for care of *my health* – the intricacy of the situation at hand. It very well may be that they come to a parting of the ways for one or more of a variety of reasons: different attitudes about what gives meaning to their individual lives, intrusion of narrow egotism on either side, or they may be forced to stay together, because of the structure of their society.

Perhaps, in view of the ever-increasing technologic sophistication with its ensuing institutionalization and rising costs, a new type of physician will confront the attending physician, namely, the "fiduciary physician of the cost carrier," who must focus on the patient in sociologic terms, primarily as an asset or liability to society. But in any case, it seems clear that the biologic–

anthropologic approach, that is, the recognition of man as a single physical being of finite freedom, plays a more decisive rôle in medical care than many non-physicians, onlookers, and patients alike seem to acknowledge.

IV. SUMMARY

How can the preceding analysis be summarized in order to be useful to those attempting to improve medical care?

Perhaps the briefest and, at the same time, most comprehensive answer is as follows: Providers of medical care fail or will fail in their mission, if they ignore the experiences that underlie the words "life" and "living," and the rôle that the thorough awareness of these experiences plays in the performance of optimal medical care. More specifically, there are three points:

(1) The words "life" and "living" refer to the phenomena of selfness and subjectivity that we experience in ourselves and that, in our observations, we use to differentiate between living and non-living physical entities of our experience, between the biologic and the inanimate world. They also refer to the specific temporality of biologic existence, to the irreversible, non-preventable, ultimately incomprehensible, finite flow of life and to the temporal and modal unpredictability of actual death. Finally, they refer to the experience of the polar, not composite, dualism of the physical and subjective aspects of our nature and, thereby, to the presence of such dualism in all beings we call living. Attending physicians, then, focus upon a modal range of a human being's biologic nature: from optimal to sub-optimal status, from being in good health to being in ill health, from being healthy to being ill.

(2) We are unique biologic beings, conscious of our finite autonomy. We strive to give meaning to our lives, to "lead" them; we are able to choose between the endpoints of self-sacrifice and self-centered egotism. We are anxious about life's incomprehensibility, about the lack of ultimate self-fulfillment and the absurdity of any actual death, about "falling ill" and other fated contingencies. This awareness of our finite autonomy may become the cause of illness for one or both of two reasons. We may be unable to resolve inner conflicts and may develop subjective as well as physical manifestations of illness. Conversely, our awareness of physical manifestations of illness may invite other physical or subjective constrictions.

By responding to a fellow human being's call to "help me in the care of my health," attending physicians do not change their original focus upon a human being as a biologic being; that is, they do not focus on humans as rational or moral beings *per se*. They focus on human intellectual and moral conceptualizations only as they affect the biologic prerequisites for these conceptualizations (the caller's health) or as they indicate their reduced state (being ill). Attending physicians are concerned, however, with human beings in the awareness of their biologic creatureliness, with human beings in their existential anxieties and the attendant sequelae. Because the self intrudes into every idea, when thought gives place to action, physicians have both risked their lives for the sake of patients and relinquished human fellowship by manipulating patients.

(3) In times and cultures dominated in general by calculating reason, acknowledgment of our biologic nature and ultimate creatureliness seems difficult. Certainly, "Nobody can say where the limits of human power lie" ([17], p. 78). Yet it is equally certain that we cannot permanently suppress the awareness of the biologic aspect of our nature, the aspect of our spontaneity as well as of the ultimate incomprehensibility of sickness and death. Technical expertise is never enough, and patient–physician solidarity lies at the basis and at the summit of the patient–physician relationship.

There is little doubt that the social organization of medical care needs to be changed. But these changes are for society to decide, not for physicians as physicians. Aside from participating as members of society in the formulation and execution of new proposals, the task of physicians as physicians is to articulate the full compass of the basic and ultimate medical situation. This task is indispensable, for any program of medical care that does not fully perceive the phenomenon of life is doomed to failure.

University of California School of Medicine,
San Francisco, California

NOTES

[1] For Professor Hans Joans, defender of life in times of biotechnologic and sociocratic aggression.

[2] The experience emphasized here refers to that immediate subjective experience of being an *I* that is simultaneously ultimate, i.e., nonreducible to any other subjective experience of self.

In view of the current interest in split-brain exploration, it may be pointed out that the subjectivity, the I-experience, emphasized here, underlies – is the necessary premise of or presupposition for – any discourse in the field of split-brain research. It is not a controversial part of such discourse. Forgetting this leads to utter confusion. In the words of Goethe: "The worst that can happen to physics as well as to some other scholarly disciplines is to take the derivative for the original and to try to explain the original in terms of the derivative, since the original cannot be derived from the derivative. This creates endless confusion, a chaos of words and a constant effort to look for and find excuses whenever truth appears and tends to become powerful." Viz., Goethe, J. W. von: *Jubilaeumsausgabe*, J. G. Cottasche Buchhandlung Nachfolger, Stuttgart/Berlin, Vol. 40, Zur Farbenlehre V, Nachbarliche Verhaltnisse, para. 718.

[3] Addis, T.: 1948, *Glomerular Nephritis: Diagnosis and Treatment*, The Macmillan Company, New York, p. 296. For thoughts similar to those expressed in this paragraph, see Jaspers, K.: 1963, *General Psychopathology*, J. Hoenig and M. W. Hamilton (trans. from German 7th ed.), Manchester University Press, Manchester, England, pp. 796–797.

BIBLIOGRAPHY

1. Ackernecht, E. H.: 1947, 'The Role of Medical History in Medical Education', *Bulletin of the History of Medicine* **21**, 135–145.
2. Freidson, E.: 1972, *Profession of Medicine*, Dodd, Mead and Company, New York.
3. Gebsattle, V. E. von: 1954, *Prolegomena einer medizinischen Anthropologie*, Springer Verlag, Berlin.
4. Goethe, J. W. von: 1900, *Goethes Werke*, herausgegeben von K. Heinemann, vol. 24; *Maximen und Reflexionen*, Naturwissenschaft 876, Bibliographisches Institute, Leipzig und Wien.
5. Guttentag, O. E.: 1956–57, 'Prison vs. Closed Ward', *Kentucky Law Journal* **45**, 251–254.
6. Guttentag, O. E.: 1959, 'The Meaning of Death in Medical Theory', *Stanford Medical Bulletin* **17**, 165–170.
7. Husserl, E.: 1965, 'Philosophy as a Rigorous Science', *Edmund Husserl: Phenomenology and the Crisis of Philosophy*, Q. Lauer (trans.), Harper and Row (Harper Torchbooks), New York.
8. Kadisch, S. H.: 1968, 'The Decline of Innocence', *Cambridge Law Journal* **26(2)**, 273–290.
9. Kubler-Ross, E.: 1969, *On Death and Dying*, The Macmillan Company, London.
10. Løgstrup, K. E.: 1971, *The Ethical Demand*, T. I. Jensen (trans. from 8th ed.). Foreword by J. M. Gustafson, Fortress Press.
11. Marcel, G.: 1952, *Metaphysical Journal*, B. Wall (trans.), Henry Regnery Company, Chicago.
12. Niebuhr, R.: 1938, *Beyond Tragedy, Essays on the Christian Interpretation of History*, Charles Scribner's Sons, New York.
13. Rueschemeyer, D.: 1964, 'Doctors and Lawyers: A Comment on the Theory of the Professions', *The Canadian Review of Sociology and Anthropology*, pp. 17–30.
14. Temkin, O.: 1949, 'Medicine and the Problem of Moral Responsibility', *Bulletin of the History of Medicine* **23**, 1–20.
15. Tillich, P.: 1951, *Systematic Theology*, Vol. I, The University of Chicago Press, Chicago.
16. Tillich, P.: 1952, *The Courage to Be*, Yale University Press, New Haven.
17. Tillich, P.: 1960, *Love, Power, and Justice*, Oxford University Press, New York.
18. Tillich, P.: 1962, 'What is Basic in Human Nature', *The American Journal of Psychoanalysis* **22**, 115–121.
19. Uexküll, J. von: 1928, *Theoretische Biologie*, Verlag von Julius Springer, Berlin.

ERIC J. CASSELL

THE CONFLICT BETWEEN THE DESIRE TO KNOW AND THE NEED TO CARE FOR THE PATIENT

There was a symposium about the moratorium on recombinant DNA research. Someone said that the research would start up again, that it had to. Hans Jonas asked the simple question – "why?". A scientist answered that if we did not do it, someone else would. Therefore we had to. Again Hans Jonas asked "why?". Others may move you to examine your dearest unquestioned assumptions by the persuasiveness and complexity of their reasoning; but such is the force of Hans Jonas' intellect and moral authority that his simple question – "why?" can be more provoking of reflection and inner questioning than longer and more complicated arguments. Certainly, he has had this effect on me and this essay is a small acknowledgement of what I have learned from him. I would not think as I do were it not for him.

In the art of medicine, knowledge about the body is used in the service of the sick. That knowledge varies from a vast theoretical structure of human biology through ever more practical and less generalizable facts used in the care of the ill. The profession of medicine would appear at times to love its knowledge more than its practice. Thus, the modern ideal of medicine is the research center not the clinic, and young physicians are more often taught by researchers than practitioners. (Although that is less true of surgeons.) Yet, doctors are, ultimately, the caretakers of the sick and their knowledge is meant to be used for the good of their patients. It seems to me, therefore, that medicine and physicians may be in relation to their patients as science and scientists are in relation to nature – if Hans Jonas is correct, that man is caretaker of his world. I believe he is correct and that therefore the relation of the doctor to knowledge in medicine provides an opportunity to explore the problem of the double agent – caretaker versus knower.

In an attempt to explore this conflict in medicine, it seems helpful to describe a case that lends itself to an examination of the dilemma. A thirty-three years old woman who had always been healthy was finishing her doctoral dissertation in sociology, when a nagging pain in the upper right side of her abdomen forced her to see her doctor. The pain, after coming on suddenly, became progressively worse for two weeks. In addition, she was aware of loss of appetite, weight loss and weakness. Her lack of desire to do anything distressed her greatly, because it prevented her from working, despite the

S. F. Spicker (ed.), Organism, Medicine, and Metaphysics, 57–72.

approaching deadline. Finally, she developed a fever. She was obviously ill when her doctor examined her. The abdomen in the region of the liver was quite tender. The doctor explained that the story of the illness, the examination and office laboratory tests all seemed to indicate that she had had an attack of gall bladder disease with gallstones. He told her that she had probably developed an inflammation or infection of the bile ducts of the liver (cholangitis). She was advised to enter the hospital where tests could be done which should prove the diagnosis.

Within the first few days, the blood tests and X-rays suggested by the diagnosis had been done. They had caused the young woman some, but not much, discomfort. The tests were, however, all normal. There was no apparent gall-bladder disease and the blood tests showed no evidence of liver infection. Although she was aware of the possibility that the diagnosis would not be substantiated, she was disconsolate because her weakness and fever persisted, and she continued to lose weight. The pain and tenderness over her liver were less, but still distressing. Other tests, although not pointing to a diagnosis, continued to show that her illness was serious.

The doctors' (for by now interns and residents were active in her care) concern for her future heightened. Increasingly, it was suggested (although not to her) that malignant disease, probably lymphoma, was the cause of her illness. The pressure to make a diagnosis increased as the days passed and illness deepened. Although apparently obvious, it seems reasonable to inquire into the source of the pressure.

The case was mystifying. There could be no question of the seriousness of the illness. To doctors that means an active disease process which threatens death or which would significantly impair or interfere with normal life; a situation in which the disease or its effects dominate the patient's life. It is in the nature of the physician–patient relationship that what threatens the life of the patient in some sense threatens the physician. It seems reasonable that the threat to the physician inherent in the threat to the patient, is one source of the sense of responsibility the physician feels for the patient.[1] The responsibility of the doctor for the patient is central to the conflict between the desire to know and the need to care for the patient.

The evidence in the case suggested infection except for one small but important test which was normal (white blood count). But infection of what? The pain and tenderness in the upper right side of the abdomen pointed to an infection of the liver or something near the liver. The adjacent kidney was normal and no abscess could be found (negative scans and sonogram). The very absence of any concrete evidence for a specific disease entity in this sick young woman did not calm the physicians' fears, but rather increased them.

The most obvious reason for pressing on toward the diagnosis – the medical equivalent of knowledge – was to make the patient better. In fact, however, as the diagnosis became more and more obscure, the probability diminished that a treatable disease would ever be found. Instead, experience suggests that such patients usually either get well without specific treatment or are ultimately found to have a malignancy. But further, the need (in terms of the patient's physical well being) to press for a diagnosis recedes, because the necessity for ***urgent*** treatment becomes extremely unlikely. In terms simply of the patient's body, it could be argued that the best course might be to wait and allow further events to unfold. In other words, to allow the diagnosis or lack of necessity for it, the patient having become well, to express itself through time.

I. THE PATIENT'S NEED TO KNOW

Such a course seemed most difficult for both the patient and her doctors. The young woman had a need to know what was the matter with her. But what is the nature of the knowledge that the patient desired? Clearly, it is not abstract but rather very concrete. The focus of her desire to know is primarily the patient herself rather than that which is making her sick. What the illness is doing to her is her concern. The young woman's knowledge of her disease is experiental and whatever she knows of patients in similar positions is regarded in relation to her own situation. It is, if you will, a kind of "I" knowledge, a knowing in which the central position from which perception will come can only be herself, her well-being, pain or suffering. The label that will ultimately be attached to her illness will be (to her) symbolic of her suffering rather than of the thing apart from her. And so it must be of all knowledge to the degree that the knowing of anything has an experiential component. But where others can to one degree or another move their knowledge and manipulate it as a thing apart from both themselves and their experience of it, such ability to abstract is denied the sick. So the limits on this young woman's desire to know are set by the price in pain and suffering that further knowledge may cost. She has competing needs. She is caretaker of her body and must protect its integrity and her dignity not only from the disease and from her doctors but from her own desire to know.

In fact, however, the patient did know *what* was the matter: she had pain, weakness, no appetite, and she knew these things better than anyone. But what she did not know was why. The symptoms were events that had occurred in her body and she could not put them from her mind. Generally, it seems impossible for anyone to perceive an event without pursuing cause. An example is the lengths to which someone will go to identify the source of a strange sound: as long as the sound is present, it cannot be put from mind. A story of Sholom Aleichem makes the point. A man has lost his money, and disconsolate, asks everyone whether they have found it. Several days later, his friend finds him cheerful again and asks whether he is happy because he found the money. "No," says the man, "but I found the hole in my pocket." The desire to know (desire seems too small a word, perhaps drive would be better) is not merely to know what, but to know why. More than this is involved.

When the desire to know is stimulated by the occurrence of an event, learning what and why does not put the mind to rest; also required is some idea of what will happen. Further, it seems to me that events are always perceived as occurring, because of the operation of something. Some object, whether it be government, a tin whistle or a disease is thought to be present and actively operating. Thus, the why, or antecedent conditions, and that which will be, or consequences (as we understand them), are always connected to our conception of an object. For the young woman of our case to know what is the matter, it is not sufficient (in this era) for her to know simply that she has fever, pain, lassitude and so on. Because although each of these is also a thing in itself, she will perceive them as characteristics of some, more inclusive, object. To know the more inclusive object is also to know causes and outcomes. It would seem, then, that to perceive an event is to desire to know the object of which the event is a characteristic. And to know the object is to have some idea of its origins and its functions or outcome. The conception of an object in this view is a dynamic thing. It always comes from something and it always does something. Remember, in this regard, the toy of a few years back that was a small black box with randomly flashing lights. The humor was in the fact that the box *did* nothing. So it is that our patient needs to know what is the matter and she has the expectation that what is the matter is an objectified process. And the process has a name – the diagnosis. Clearly, however, the names serve purposes even when the knower of the name does not thus also possess a clear knowledge of the cause and outcome. Much medical humor comes from the lack of content of some medical terms. A man has pain in the tailbone and is satisfied to hear the doctor say he has coccyalgia or coccydynia – both of which merely mean painful tailbone. The name itself has some power.

As a resident at Bellevue, I had a patient who was persistently dissatisfied with our lengthy explanations of his condition. My intern satisfied this patient by telling him that his problem was that he had a gastric stomach. Why was the patient now satisfied? He could not have been relieved, because in knowing the name he knew the cause, characteristics and outcome. The humor (for the intern) derived from the fact that the patient was happy knowing that he had a "stomach" stomach. If he knew that to have a gastric stomach was simply to have a stomach stomach, we cannot conceive of his being satisfied, but rather insulted. What function, then, did the name serve for that patient? For one thing, he now knew that his physicians knew the name of the disease and thus all that is implied in the conception for which the name is label. As the doctor may be perceived (for some patients and in some situations) as the instrumental arm of the patient, if the doctor knows, the patient is provided the same comfort as if he knows himself. But more, the very fact of a name signifies that the thing is known. Labelling in itself signifies the move from unknown to known, providing all the comfort and reduction of uncertainty that goes with a known object. Further, the label or signifier can then be manipulated as though the object itself was being manipulated but without concern for the actual content of the conception so signified. As evidence for this, think of the many meaningless or at best imprecise signifying words that function well, precisely because we do not look too closely at their content. As we shall see, the fact that physicians can name something may also imply they can do something. To name the beasts of the field implies the power to dominate them.

In any case, we know that the sick young woman wants to know what is the matter as well as causes and outcomes. And that she, with the rest of us, views causes in terms of processes or objects whose characteristics are signified by the name.

She also needs to know the diagnosis, the name of the thing that is making her sick, for social reasons. She has to tell her mother and the rest of the family why she is in the hospital. Not to be able to say what is wrong would cast doubt on the quality of her medical care, the competence of her physicians and the excellence of the hospital. For her mother, as for others, the absence of a name allows worry to seek its own name. Her parents have to tell others and it is not sufficient to say that their daughter has fever, weight loss and abdominal pain. Those informed would only half believe that the doctors did not know. They would also wonder whether someone was not telling the whole truth, and in avoiding the name was concealing a more terrible truth as in newspaper obituaries that say, "he died after a long illness." If the facts are being concealed, those being told might wonder who is doing the hiding, the mother, the daughter or the doctors? It does not overdo things to point out how much interpersonal relationships depend on the sharing of knowledge and how much meaning is read into the absence of a name. The patient also needs to know, so that she can tell her employer why she cannot work and why her dissertation will not be finished when anticipated. She needs the diagnosis to justify her hospitalization and even her assumptions of the sick role. The name avoids ambiguity in such situations. But here, the important content of the conception labelled by the disease name deals with such things as length of hospitalization, post-hospital convalescence, ultimate health, and ability to function. In other words, the content of the conception labelled by the same name may or probably will be very different depending on the function of the knowledge represented in the conception.

II. THE PHYSICIANS' NEED TO KNOW

The physicians also have a pressing need to know what is the matter. Their need stems from their relationship to the patient, from matters of influence and authority, from social or institutional pressures as well as from an intellectual desire for true knowledge.

As mentioned earlier, the threat to the patient from her illness is to some degree felt by the physician as a threat to him. As she is not whole without a diagnosis, neither is he. In part, this is because he cannot know what his attitude is to be towards the patient and her disease. If it is curable, then the connection between them may need to be strengthened to help make her better. If it is fatal, then preparation must be made for separation or at least handling those special aspects of the dying. That such changes in attitude may not be conscious does not diminish their reality. Of more concern here is that the physician will, in learning about the disease, be acquiring experiental as well as abstract knowledge. Further, as he experiences some threat to himself in the danger to her – and it is not cynical to realize that some of that threat is to his reputation, as well as arising from a vital if poorly defined human bond – then some of the knowledge must be, in common with the patient's personal knowledge, focused upon himself and related to himself. But unlike the patient, we expect him to be able to de-center and to abstract his information into true knowledge. The fact that abstraction is not always possible, or even that it is not always desirable in medicine, is beside the point. It remains true that the knowledge that he brings to the patient is tested by his personal need to make her better. And perhaps equally by his need to test the knowledge itself. Thus, not only does the patient need him, and he the patient, but his knowledge needs the patient in order to be tested, to become "real" and thus to become part of him. In this latter regard, we understand the constant desire of medical students to take responsibility for the care of patients. Until their knowledge of medicine is thus applied, they are not a part of medicine and medicine is not truly a part of them.

III. KNOWLEDGE AS POWER

The doctors have a need to know based on the desire for influence and authority. There are those who believe the physician maintains his control and power over the patient by controlling the knowledge and its communication. But further, status is conferred on the physician who makes the diagnosis in difficult cases. Especially for the younger physician, a demonstration of knowledge is seen as the sceptre that confers power and authority. This function of knowledge is often emphasized in the academic setting of the university medical center, where the faculty member with the greatest store of esoteric disease information seems to be most admired. It is not uncommon for physicians in training to try and upstage one another by quoting the latest medical journal. Somewhat more pernicious is the attempt to be the first to record the correct diagnosis on the chart of a patient with an obscure disease, even though such mention may bend the rules of etiquette.

Because the young woman's illness continued unabated and still undiagnosed, her physician called a respected and experienced senior associate in consultation. On the telephone to the consultant, her doctor described the case in detail. Overhearing the conversation, the consultant's younger but very knowledgeable associate was struck by the similarity to a case he had recently seen in which the diagnosis had turned out to be Fitzhugh–Curtis Syndrome (an inflammation around the liver (perihepatitis) occurring in the course of gonorrheal infection of female pelvic organs (salpingitis)). Passing the hospital floor on which the patient was situated, the younger associate mentioned the diagnosis to the resident. When the specialist arrived to make his consultation, he found that the resident had already written a note on the chart suggesting the diagnosis and crediting it to the young associate. The specialist examined her and her records in his usual careful manner and then wrote his opinion which concurred in the suggested diagnosis and recommended treatment.

There was no evidence that the young woman had gonorrhea (such overt evidence may be lacking in Fitzhugh–Curtis Syndrome), but the diagnosis offered the chance to act. Increasingly the sciences of medicine have become interventionist sciences. Pedro Lain Entralgo relates that 100 years ago, the great Austrian physician Joseph Skoda gave a brilliant lecture at the bedside on the nature and origin of a patient's disease. When his assistant asked what should be done for the patient, the question was brushed aside as of no interest. The object was not to do, but to know. It was believed that only such a view of its role could elevate medicine. Now, however, knowledge has become the power to act in medicine as in the other sciences. We are all aware of the evolution of man's relationship to nature that has occurred hand in hand with the belief that knowledge provides the power to act against nature. It may be well to stand in humble tranquility before nature's majesty revealed in the distant stars, but it is humiliating helplessness to stand before the same nature revealed as the inexorable progress of a disease. Thus, for her doctors to know the young woman's disease was to have the power to act. It is true that physicians have always done things to their patients. What distinguishes the present era is that so much action is not only based in theory, but is also effective. Theory has always been present, but effectiveness is recent.

In the case of the young woman, cause being discovered, action soon followed and she was started on high doses of intravenous antibiotics. Everybody was relieved at the lifting of ambiguity. Attributing the illness to gonorrhea created certain strains for the patient, but such negative aspects would be a small price to pay for a cure.

IV. INSTITUTIONAL PRESSURE FOR KNOWLEDGE

Having a diagnosis relieved other institutional pressures on her doctors. It is not an exaggeration to say that the entire institution of medicine revolves around the system of diagnostic categorization we call disease. Although a detailed discussion of institutional pressures is not pertinent to this essay, any analysis of the physician's need to know what is the matter cannot entirely neglect these pressures. Indeed they are so prominent that one could be led to believe that institutional forces are primary in the push towards diagnosis. Institutional in the narrower view; all insurance forms, requests for X-rays and laboratory tests, hospital admission forms and virtually all the other bureaucratic paraphernalia of medicine require diagnostic entries. Institutional in the wider view; diagnosis and diagnostic categories provide the basis of conversation among physicians and other staff members about patients or about their work. In other words, the group structure of medicine and the system of individual interactions revolve around disease categories. I do not believe institutional pressures are primary, but rather, that they are derivative from the other forces being described. It cannot be denied that institutional forces are both stabilizing and essentially conservative, holding the profession within the view of the world based on disease categories long after new concepts have emerged that will provide the basis for the institutional pressures of the future.

In the light of all these pressures, it is not surprising that everybody was pleased that the diagnosis was made. Unfortunately, several days of intravenous antibiotics produced no improvement in the woman's condition. The bacterial cultures that might have confirmed the diagnosis were also negative, while the evidence of serious illness continued.

V. RESPONSIBILITY VERSUS THE DESIRE TO KNOW

By this time, the patient had been in the hospital three weeks. Her fever had continued, her weight decreased, and she had become progressively more anemic. It appeared inevitable to most of the staff that she should have exploratory abdominal surgery. The evidence that the disease process was within the abdomen seemed unequivocal. Her physician felt that little was to be gained by delaying. It is important to reiterate that while the belief that the diagnosis could be made by surgery was probably justified, the chance that she would be directly benefited by making a diagnosis was, by this time, even less probable than earlier in the case. It was still most likely that she would either improve spontaneously, the diagnosis manifest itself without harm coming to her, because of the delay, and/or the disease would prove to be untreatably fatal. That being the situation, the search for a diagnosis would now become for her a source of a danger in itself. In other words, she was now being directly threatened by the various pressures to know. Different physicians had somewhat different views of the matter. The surgeon saw no point in waiting, indeed he felt she should have been operated on ten days earlier. The students, interns and residents generally believed the operation should be carried out promptly. Her own doctor felt that surgery should proceed but was concerned by the risk and by the patient's manifest fear of surgery. It was, after all, a matter of weighing risk against benefit. The benefit was that a diagnosis would be made, but the risk was more direct. As there are institutional pressures in medicine to know the diagnosis, there are also very strong institutional pressures that protect the patient from undue risk and reinforce the responsibilities of the physician and all of medicine for the patient. The risk of death in exploratory surgery in patients like this is probably less than 1 percent. If the risk were to approach 3 percent, I doubt whether anyone would have felt surgery justified in this patient. Throughout medicine, patients are exposed to risks in order that knowledge be gained and a spectrum exists of risks that are considered acceptable. It seems fair to say that when the primary desire is knowledge, as in research settings, the risk to the patient that may be acceptable to the physician is higher than when the primary goal is patient care.

While I have assumed that there exist two needs within the physician – the need to care for the patient and the need to know – more must be said to substantiate these needs, and to show that they may exist in tension within the profession of medicine and are ideally, internalized within each physician. Earlier, I discussed some of the roots of the doctor's responsibility for his patient. Another might say that it is unnecessary to prove that doctors have responsibility for their patients; the profession of medicine arose out of the need to care for the sick, indeed is in bondage to the sick for its very existence. While some may argue the nature of professions in general, in medicine (as in divinity and law) responsibility is inherent in the definition. Also inherent in the definition of these professions is a body of knowledge developed in the service of responsibility. But it is the case that knowledge comes to have a life of its own. As the mysteries of the body have been exposed throughout the history of medicine by the operation of disease, systems of thought have arisen to solve the problems posed by the mystery. It is the nature of any system of abstract or formal thought not to be content with mystery, but to continue operating on any problem until understanding results. Mystery is a threat to the adequacy of the system of thought itself. But, as I noted earlier, where knowledge is inadequate, so too may the owner of that knowledge find himself inadequate, so inextricably bound are the person and his knowledge. Mystery is the point where the responsibility to patient and the desire to know may come in conflict. The ancient but respected aphorism of medicine "above all do no harm" makes clear that, out of fear of harm to the patient, the body cannot be invaded merely to resolve mystery. That conflict was exposed in this case.

VI. THE FORCE OF TECHNOLOGY

At this point another force entered that seems, at least in part, responsive to the tension between the desire to know and responsibility for the patient – technical novelty.

A bone marrow examination was done in the hope of revealing the cause of the illness which would be possible if the disease were lymphoma. The results were not helpful, but the hematologist suggested that instead of surgery, laparoscopy should be done. The laparoscope is a recently developed instrument which, when introduced through a small incision in the abdomen, allows visualization of the abdominal contents. It has been very helpful in examining the pelvic organs of women and in recent years has been used more widely for examining other abdominal contents. It is not necessary here to examine the lure of the novel, although it does seem to be basic to mankind and to be related to the desire to know. In medicine, at least, new techniques and technology frequently seem to offer the double reward of reduced uncertainty and reduced risk. That experience seldom bears this out may account for the reluctance of older physicians to leap to new techniques. Yet, the lure is there for young and old. Another consultation was sought from an experienced clinician who believed some kind of abdominal exploration was necessary and recommended laparoscopy as the least risky.

To the patient, it was a reprieve from surgery and she accepted the idea gratefully. Her physician was doubtful that it would be useful, but he hoped that it would reveal the diagnosis and he was relieved that he would not have to bear the responsibility for the surgery.

The laparoscopy was done (and also a liver biopsy through the instrument) and no disease was found. Unfortunately the limitations of the instrument precluded visualization of that area around the liver which the patient's symptoms suggested might be the site of disease. The laparoscope also could not examine the intestines, and other areas where disease might be hidden, but the laparoscopy ended the active search for a diagnosis. No one was willing to expose the patient to further risk. It was almost as if the laparoscopy, having served the purpose of searching for the sake of searching had made further probing unnecessary. The physicians were now willing to fall back on the probabilities, cited above, that she would either get well or would not be treatable and that the disease would ultimately show itself.

Before we discuss the nature of the thing everybody was looking for, I should describe what ultimately happened. After all the weeks of hospitalization, two things occurred. First, the young woman began to improve. Her fever subsided, her appetite began to return and pain almost disappeared. Second, a resident, reviewing the entire case, discovered that although the stomach and large intestines had been X-rayed, the small intestines had not. Promptly done, those X-rays revealed the pattern (X-rays are only pictures of shadows) of the disease, regional enteritis, involving the end of the small intestine. That particular inflammatory disease of the small intestine does, uncommonly produce an illness similar to that of the young woman. Thus, although the X-rays were only suggestive and the extent of visualized disease was small compared to the severity of her illness, the diagnosis was accepted with both relief at its essentially benign nature and a wait-and-see attitude. The first consultant, whose special interest is inflammatory bowel disease, was chargrined that he had been diverted from making the diagnosis because of the more dramatic Fitzhugh–Curtis syndrome.

VII. CONCLUSION

The case of the young woman in whom a search for a diagnosis proceeded to a point where the search seemed to represent a greater threat than her disease has allowed me to examine the sources of the pressure to know. Whether expressed as social, emotional, intellectual, institutional, power oriented, or technological, these forces seem to reveal a desire to know and an admiration for knowledge whose only check in medicine is the conflict of that desire with responsibility for the well being of a patient. The evidence suggests that it was the push to know from *within* individuals – socialized, institutionalized and reified, rather than primarily a pull from the object about which knowledge was desired – that motivated the search of the physicians. The word object seems to be the key. Knowledge itself becomes the objective increasingly detached from that which is to be known, until at length the knowledge itself becomes the object. Throughout this case and in general, knowledge is dealt with as though it is a thing. The scientist told to suspend his research, because it may be dangerous replies that "if we don't find *it* someone else will." As though there is an *it* in nature that is calling and to which man responds. But that is not the case, from man's response the call is constituted. On the other hand, it cannot be denied that there was something inside the patient that made her sick. In fact, the thing ultimately made shadows on an X-ray and, one presumes, could have been seen directly, if the small intestine had been exposed.[2] The thing has a name, regional enteritis, (also called granulomatous colitis or Crohn's disease), and defining characteristics that distinguish it from other inflammatory bowel diseases. In that sense, regional enteritis seems to be almost as unique an object as is, say, an oak tree.

I would like now to explore in greater detail the nature of that object which the doctors had sought. In common with the patient, when the physicians were seeking an explanation for the event represented by the woman's illness, their explanation also had to satisfy the demand for antecedent conditions, characteristics and post-cedent conditions (outcome). It is clear that the characteristics of the physicians' conception labelled, regional enteritis, are quite different from the patient's conception with the same label. As noted earlier, the patient's view of the disease is primarily experiential and self-related. What she means by antecedent conditions – where did the thing come from – is most probably "what did I do?" or, "what in me caused me to get this disease?" By outcome she means, "what will happen to me?". The physicians are looking for a more generalized knowledge. Under what circumstances do individuals acquire regional enteritis? What is the usual age, sex, predisposing situation, or even personality make-up of patients with the disease. The doctors' description of outcome will also be more general. In a certain percentage of patients, healing is spontaneous. The frequency of recurrences, how it spreads, when surgery is necessary and what are the responses to treatment are also assigned probabilities. These are statistical descriptions. These probabilistic statements are attached to a pathologic description of the disease. The intestinal wall has a certain appearance to the eye and the microscopic appearance of the cells has certain characteristics. All these things together form an idealized conception of the disease to which this particular woman's disease is only an approximation. In a certain sense, the object the doctors sought was not real but rather a classifactory device. An occurrence existed in her small intestine and the classification is only good to the extent that it can include this occurrence. Further, each doctor's conception of regional enteritis is only useful to the degree that it can accommodate individual variations in the disease (will not *exclude* bona fide cases of the disease nor *include* situations which are not truly regional enteritis). That means that each doctor's conception is useful to the extent that experience fleshes out the conception and moves its content away from the generalized ideal towards the particular. The problem to be solved is uncertainty. The more nearly the instance is approached by the conception, the less the uncertainty. The less the match between the instance and the guiding conception, the greater the uncertainty, as in the case of this young woman. One of the things that frequently startles patients, when they become seriously ill, is the enormous amount of uncertainty with which physicians work. In instances where things go as expected, the uncertainty is not revealed; however, the more questions raised by the circumstances, the more the uncertainty is manifest. We are not prepared for uncertainty, because we come to confuse the conception of an object with the object itself.

Thus, in their search for a diagnosis, for the real thing in this patient, the physicians are guided by a conception of the object and that conception is contained within themselves, but that conception is not the object itself. That conception serves the function (among many others) of guiding perception and of reducing the stimuli from the outside world (and from other parts of thought) that must be dealt with in order to solve the problem introduced by the event of the woman's illness. But it must be the case that if the conception guides perception and indicates what information must be considered in solving the problem, then the same conception must indicate what information is not relevant. To put it metaphorically, as the conception clears the vision for some things, it must blind the vision to others.

To return to the original problem, the hunt for knowledge as represented by the search for a diagnosis in the sick young woman was a result of a complex push from within (the basic desire to know and reduce uncertainty, the social, institutional, power oriented and technological aspects of the desire) and a pull from outside. But that real object called knowledge (the diagnosis) is not an object at all. From its place in the continuum of nature, it is formed up into its status as object, as discrete entity, by the search for it, and no matter how tenuous its original claim, its status as object is retained and even grows by the fact of being named and having that name (which *is* an object) communicated from person to person. Another might say that it achieved its status as entity, because of its operation in this patient simply because the illness presented an event to be solved. But, other eyes in other worlds (historically or culturally) looking to solve the uncertainty raised by the same event might see a different entity. Physicians may object that I have confused *a-patient-with-regional-enteritis* with the pathologic entity known as regional enteritis. The way we see this young woman, her symptoms and the findings of examinations are the expression of that pathologic entity in this particular patient – dependent for the way it expresses itself in this particular woman on variables as disparate as the state of her nutrition and her unconscious conflicts. The pathologic entity, regional enteritis, it might be argued, is much more discrete, better defined, and less subject to variation. The confusion of the two – the patient with the disease and the disease entity – is quite common. Indeed, I think the patient-with-the-disease is the more real object than the pathologic entity itself. Regional enteritis is an occurrence, but it acquires its form as object – its boundaries and definitions as an individual disease in the course of the search for it – in this patient and in all the preceding patients back to its first description, all of which gave the conceptualized entity its history and status as object.

If it were truly an object as discrete as (say) a pine tree, there would not be so much current argument about the classification of the inflammatory bowel diseases. My own guess is that in this instance and in the other increasing areas of medicine where diagnostic categories are being called into doubt, the question is not about this or that definition of a disease, but definitions of disease in general as they have been inherited from the past.

Present day re-examination of the definitions of disease should be seen, I believe, more as attempts to put in question our whole classification of disease than as attempts to make "better" definitions. Whatever classification emerges, however, will be essentially a product of human thought. But as we begin to see the sick in a more holistic manner, purely cellular or biochemical definitions, to say nothing of definitions comprised solely of arbitrary statements of abnormality will begin to become increasingly inadequate. The conceptions represented by those disease names will begin to come in conflict with other, newer understandings that guide our views of the sick.

More is involved. "Regional enteritis" did not make this patient sick but rather a number of basic biologic mechanisms involved in her disease. (It would be common to say basic biological mechanisms that went wrong – pathophysiology, but the word 'wrong' is also definitional.) The beauty and robustness of twentieth century medicine lies in the search for the fundamental mechanisms of disease. Increasingly, what matters in medicine is not a knowledge of sophisticated criteria for disease classification (as was the case in nineteenth century medicine) but knowing how the body works in health and disease. Such information certainly leads more truly to knowledge in the universal sense than does knowing a diagnosis. Since the body does exist, no argument can be made that the mechanisms uncovered by science are mere artifice created by the search. But the direction of the search, what knowledge we value, what technology we use to find it, how we relate each part to another, under what categories we organize it, our priorities (in the modern argot) are clearly more derivative from the search itself than the phenomenon pursued. If all of this is true for diseases and their biologic mechanisms, it is much more true in research on DNA recombinants where the knowledge truly does not even exist before it is found.

Knowledge is not an object; knowledge takes the form of object by the act of seeking for it and by the fact of being communicated. Using the case of a woman severely ill with an undiagnosed disease, I have tried to show the various sources, within us and outside us, of the desire to know. But clearly, we have this drive to know inherent in our existence. The only check on the force to know that otherwise would have caused her to be taken apart as one opens a stopped watch, was the sense of responsibility felt towards her; by herself, by her physicians because of their bond to her, and by medicine in its institutionalization of that bond.

If knowledge is not a thing waiting only to be found; if it is even in part a desire of man, then it is subject to moral restraint like any other drive or desire. If this case is even in part a metaphor for man and his world, then the author of moral restraint on the desire to know is man's responsibility for mankind and the world.

"All men by nature desire to know. An indication of this is the delight we take in our senses; for even apart from their usefulness they are loved for themselves; and above all others the sense of sight."[3] The love of knowledge exists in us like the love of seeing. But it is possible to control our desires, as all men learn out of the love of humanity, that some places we look and from some things we turn our eyes.

Cornell University Medical College,
New York City

NOTES

[1] A physician is authenticated at least in part by his ability to help his patients. His role as physician cannot be seen wholly separate from other aspects of his person. Where his medicine is helpless, the instrumental element of himself is blunted. Where his knowledge does not explain and cannot predict a part of himself is threatened. But further, as we know, his acts, if they are wrong, directly threaten the very existence of his patient. Thus there is a bond between the doctor and the patient which can be better understood, if one sees that the physician cannot be a physician without a patient and to have a patient is to have the ability to act. The patient is there, because she believes the doctor will act for her benefit. Thus to act, and he himself is part of his actions, the doctor must accept responsibility for the effect of his actions on the patient. Otherwise, he would be rejecting the patient's trust and dealing with the patient as only an object, a means by which he exercises himself. Since neither his role nor even his knowledge exist apart from his person, what threatens the patient must at least in part be a threat to him.

[2] For the sake of completeness I should point out two things. First, because so little disease was visualized on the X-rays and because what was seen was open to interpretation the diagnosis was by no means certain, when the woman left the hospital; and second, the diagnosis might have been missed even if she had undergone exploratory surgery.

[3] Aristotle: *Metaphysics*, Book A, 980a, 20–24. *The Works of Aristotle*, trans. W. D. Ross (Oxford: Clarendon Press, 1963).

STUART F. SPICKER

THE EXECUTION OF EUTHANASIA: THE RIGHT OF THE DYING TO A RE-FORMED HEALTH CARE CONTEXT

I. INTRODUCTION

At risk of overstatement a physician and historian of medicine, Chester R. Burns, the past President of the Society for Health and Human Values, remarked that "Contemporary man is transforming death from a taboo to an obscenity" ([5], p. 3). As he rightly observes, there are books, articles, symposia, editorials, societies, panel discussions, political caucuses expressing ideologies and hospices. Distinguishing as we should the *phenomenon* of death from the *concept* and language in which the concept reveals itself, it is the latter which is in danger of becoming indecent, impure, and even obscene. The phenomenon of death was never taboo. However, only the explicit bringing to consciousness of the latent affect associated with the phenomenon was forbidden. Indeed, in many quarters the topic is still very much taboo, in spite of the fact that many of us move in circles where the phenomenon of death is "discussed." We should not, after all, assume that our particular social nexus is typical. In truth it is more correct to say that authentic and open talk about death is still taboo *and*, especially in academic circles, at times even obscene. The reality, in short, is expressed by the conjunctive.

To avoid the Charybdis of taboo and the Scylla of obscenity we would, perhaps, have had to have lived during the Middle Ages. As a Dutch phenomenological psychiatrist has observed in his regrettably little known *The Changing Nature of Man*, "in the middle ages, death was accessible and therefore communicable. Death was there. The sick were sitting at the roadsides, and their loud presence confronted every passerby with a *memento mori.*" Van den Berg continues, "Epidemics raged over Europe and allowed no one any doubt as to the reality of an unexpected and abrupt end. The diseased man died in his own home; if in his dying hour he was conscious, and bade farewell to those dear to him, he did not omit the children" ([24], pp. 89–90).

An adequate elaboration of the reconstruction of history makes it even more evident that the contemporary hospital and hospice care settings, in the United Kingdom and the United States, are condemned historically to history. They are the products of the cultural and anti-cultural forces which have

S. F. Spicker (ed.), Organism, Medicine, and Metaphysics, 73–94.

shaped Western civilization, such as it is. Specifically, St. Christopher's and St. Luke's Hospices in Britain and now, in the United States, the hospices in Santa Barbara and Marin County, California, Paoli, Pennsylvania and the Hospice for the Greater New Haven Region (as well as units within hospitals) serve as the most recent expression and modification of the clinic, whose birth took place during the last quarter of the eighteenth century. During that period the ancient question: "What is the matter with you?" was being replaced, as Michel Foucault reminds us, with the question: "Where does it hurt?" ([11], p. xviii). By 1770 the patient's signs and symptoms signified disease, not illness, and the language of the clinic – the discourse about disease – was co-temporaneous with the abandonment of the patient, not a spatio-temporal abandonment, to be sure (the patient was, after all, "confined" to his bed), but a communicative abandonment, since we can only consummate our love for the Other by cherishing the language through which we speak and listen to the Other, in this case the Other as patient, as person in need of and crying out for our care.

How far have we come since the eighteenth century? Dr. Henry O. Heinemann writing in a recent issue of *Man and Medicine* (an unfortunate title) remarks that "the emotional needs of man afflicted with incurable illness cannot be met in large urban medical centers as they exist today" ([13], p. 281). Furthermore, "In our present society encounters between patients and physicians are incidental" ([13], p. 283). In sum, "the fatally ill remain the stepchildren of the modern hospital" ([13], p. 284). But whence the "modern hospital"? The story begins, in modern form, in 1770 and has a complex two-hundred-year development. It begins with the earliest form of the clinic, prior to the advent of the teaching hospital; prior to the introduction of scientific methods to the care of the sick and dying; prior to the subtle and sophisticated nosology of disease which is now undergoing even further refinement and standardization. The earliest form of the clinic witnessed what we might call "practical medicine", which included home practice where doctor and patient met, where observation of a certain prescientific form took place, in a care setting innocent of theory. The doctor passed on his learning – empirical, not experimental – to his disciples, most often without complex linguistic impediments. One might call it, for want of a better expression, the traditional ideology of the *corpus Hippocraticum*. Let us turn to *Ancient Medicine*, written by members of the school of Hippocrates of Cos. Book XXI opens with the following passage:

"In convalescence from illness, and also in protracted illnesses, many disturbances occur, some spontaneously and from things casually administered.

I am aware that most physicians, like laymen, if the patient has done anything unusual near the day of the disturbance – taken a bath or a walk, or eaten strange food, these things being all beneficial – nevertheless assign the cause to one of them, and while ignorant of the real cause, stop what may have been of the greatest value . . . " ([14], Vol. I, pp. 55, 57). Ignorance of real causes compelled the Greek physician to remain confined to procedures, attempting by the accumulation of *particular* experiences to determine how *particulars*, like dietary regimen, affect the *particular* patient. Hence Hippocratic medicine, though based on observation and a collection of experiences, was not truly scientific. Some regimen found to be efficacious may well have been adopted since clearly successful in practice. The important point is that *the reason* of its efficacy was unknown. The notion of *empirical* medicine eventually connoted quackery identified with the practice of charlatans, positive outcomes being the result of mere accident.

I stress this methodological aspect of the ancient clinic not to disparage the empirical tradition in favor of the scientific one on which contemporary medical practice is now so "securely" based, especially in the post-Flexnerian era, but to emphasize the point that the ancient clinic, until the late 1770's, was as Foucault remarks: "a universal relationship of mankind with itself" ([11], p. 55). The question for us then, is whether the ancient clinic as described here is to be reinvented again, rediscovered, or even recreated. The earliest form of the clinic, one should note, included teaching, but in a limited fashion, as master instructs pupil or disciple, based upon empirical method and human memory. Succinctly put by Foucault in his monumental *Birth of the Clinic*,[1] "In no sense was the clinic to *discover* by means of the gaze; it merely duplicated the art of demonstrating (*démontrer*) by showing (*montrer*)." That is, the traditional clinic, unlike its later form, the hospital, was through and through empirical. But the hospital was organized to sustain experimentation with patients as subjects. Foucault cites *Observations sur les hôpitaux*, published in Paris in 1777, in which the author, J. Aiken, remarks that "Hospital patients are, for several reasons, the most suitable subjects for an experimental course" ([11], p. 83; p. 87 Note 57). Carried to chauvinist extreme, the maternity clinic at Copenhagen functioned as follows: Those women who claimed to be unmarried were admitted. The Directoire, J. B. Demangeon, remarked that "It seems that nothing better could be imagined, for it is precisely that class of women whose feelings of modesty are likely to be the least delicate." Thus these women would be of the greatest possible use to honorable families. The maternity patient, judged to be in no state to exercise beneficence, would "at least contribute to the training of good doctors and repay their benefactors with interest" ([11], p. 85).

The birth of the hospital, then, replacing the traditional clinic, immediately raised an ethical issue: by what right may one transform a person as patient into an object of clinical evaluation, especially a person whose poverty often compels him or her to seek the assistance of a hospital? What justification, if any, is there for transforming the doctor-patient healing relation into a researcher-subject relation, where knowledge (scientific to be sure) is the end or telos? Could we not raise these same questions for those planning the contemporary hospice? To what extent, if any, is it justified to include in the hospice care setting a research component? Does the rebirth of the clinic not signal, if for no other reason than historical virtue, the need to refuse to utilize the hospice patient as a research subject? Can the care context envisioned by the hospice planners be preserved if hospice too adopts the research program insinuated into the contemporary hospital? Recall that medicine as practiced through the end of the eighteenth century was not founded on research or scientific language, but only on a "gaming language" (*une langue de jeu*). Medicine perhaps had to forego the truth and, as in playing games, even serious ones, could only hope to win more than it lost. The ancient clinic was able to teach medical practice. By the turn of the century, that is by 1800, the clinic underwent a radical restructuring, a transformation of the very language of the physician, which of necessity was a transformation of medical perception as such. The culmination almost two hundred years later is embodied in the contemporary temples of Aesculapius, possessors of the "truth" which paradoxically leads chronically ill patients to have false expectations. The infirm, chronically debilitated, dying, terminally ill patients seeking help in a modern hospital, though often misguided into expecting cure for all ailments, cannot even obtain comfort and reassurance from fears which accompany their illness, writes Dr. Heinemann ([13], p. 282). The physician and health professional, well educated in scientific medicine, are somehow ill-equipped to care for the incurably ill. This same physician is somehow untrained to provide what Dr. Heinemann calls the "emotional support" to the patient. Such judgments lend credence to the need for the hospice movement, of course, which is the most recent attempt to establish separate institutions where patients with fatal illnesses can receive the most appropriate care in a setting more social than medical – more like the milieu of one's home, perhaps – the home being that locus of personal space in which exists the family expressing its love, concern and a common desire for care if not cure. The modern hospital is, of course, an invention whose tacit motif is the innovation of a newly differentiated and distinct space, one which paradoxically protects and preserves the patient's disease. The hospital which, through language, has created disease (not illness) itself creates a new social space which is designed to abolish it, yet which (unintentionally) generates a new "disease" – alienation, anxiety, inauthentic dying, and in many cases (let us not forget) poverty. So it seems we must create another new space. If we cannot go home again and if we cannot accept the space of the contemporary teaching and research hospital (even if it has a *vacant* bed), then we must create an alternative and not one suitable for only the dying patient.

II. EUTHANASIA

It would be out of place here, of course, to presume to suggest additional models which might serve to engender humane care settings suitable not only for the dying patient but for *any* patient, suggestions which would of necessity, at the very least, challenge the institution known as the contemporary American hospital. Suffice to say that there are sufficient reasons to accept the generalization offered by Dr. Arthur S. Abramson, Chairman of the Department of Rehabilitation Medicine of the Albert Einstein College of Medicine:[2] "The hospital as an institution has outlived its folkloric reputation as a house of comfort for the healing of the sick and for the graceful and dignified dying of the hopeless" ([3], p. 403). "The truth of the matter is," he continues, "that in the past the hospital was not a very effective instrument in the matter of healing of the sick and is not yet a comfortable place to die" ([3], p. 403). Given the generalization rather frequently expressed in the literature that "the hospital stay . . . may actually do harm by unwittingly reducing the individual's capacity for successful social reintegration," ([1], p. 213; [18], pp. 33–36) in combination with the current spectre of the technological imperative which has recently been superimposed on the traditional medical ethic – "The doctor has a duty to preserve, spare and prolong human life wherever and whenever he can. . .and must fight against death under all circumstances" ([25], p. 19) with whatever is available in the medical armamentarium – is it any wonder that the medical and philosophical literature proliferate madly with items and arguments seeking to morally justify passive and/or active euthanasia? It should really not surprise us.

The current burgeoning of interest in the moral implications of modern medical techniques which enable the physician to prolong life (or dying) has led many persons to question the present illegality of euthanasia, the form of the question being philosophical. That is, if it is morally permissible (and perhaps morally obligatory) in at least one case, to end the life of a person *in extremis* and by so doing to act morally (perhaps illegally), then the law should be modified, amended, significantly altered in some way; for, after all, ought not the law to reflect the most moral position determinable by rational persons? But the matter is extremely complex and serious since euthanasia is the killing of a person who voluntarily and *in compos mentis* requests an assisted suicide. The assisting partly expressly intends to terminate the life whose quality is such that the person making the request himself adjudges it not worth living.

Voices are heard coming from many directions (and they are not hallucinatory) calling for the legalizing of euthanasia on the basis of the claim that it is sometimes morally permissible and, more rarely, morally obligatory to so act. I am not thinking simply of the formally organized Euthansia Educational Council of New York or the voice of the distinguished Cantabrigian professor of English law, Glanville Williams [26] ; or of the physicians who here and there in the literature warn us of our inhumanity in forbidding euthanasia in any form. (Responding to a questionnaire prepared by Robert H. Williams of the University of Washington, 87% of the physicians polled voted in favor of so-called 'passive' or 'negative' euthanasia [the planned omission of therapies that would probably prolong life] and 80% of this group indicated that they had used it in practice. Responses regarding 'positive' or 'active' euthanasia [the use of techniques that are hoped will promote death sooner than otherwise] were affirmative in 15% of questionees.[3]) I am more directly concerned with the writings of professional humanists – specifically the philosophers. What precisely, one might ask, is it that they are doing? Since one of the central functions of philosophical inquiry is clarification of concepts, what concepts are in need of clarification and analysis? What is the general intent of those philosophers (and theologians) who have very recently expended considerable energy in presenting carefully articulated arguments supporting euthanasia? It may be useful to appreciate these efforts, since they are often in tension not only with the obdurate and recalcitrant law prohibiting euthanasia in any form under any and all circumstances in all States, but more importantly perhaps for our purposes, in tension and conflict with health professionals who advocate the legal status quo. More pointedly, the tension may well be felt among hospital and hospice personnel. For where, precisely, do hospice physicians stand when confronted directly with the questions: "Do you advocate passive and/or active euthanasia? Do you believe it is morally justifiable to *actively* intervene and end the life of a competent person who has *requested* your assistance in ending his life? If you believe it is never morally permissible to end the life of a patient *in extremis*, do you believe it is ever morally permissible to let the patient die, that is, to terminate treatment or therapy with the express intent of shortening the life of the patient? If so, do you accept the charge that you advocate negative or passive euthanasia? And if so, what reasons would you give for your position?"

As a reader of Cicely Saunders, the founder and physician director of London's St. Christopher's Hospice, I confess I cannot determine very clearly what her position is in this regard. Even so, it is worth citing her article "Living with Dying" which appeared a year ago in *Man and Medicine*. Doctor Saunders' reflections in this article are ambiguous, though her fully clarified position may not be ambivalent at all. And I surely make no accusations against those "in the trenches" and "on the firing line" who in fact care for the dying on a daily basis. Philosophers, like most of us, rarely ever set foot in hospitals or dying wards, and we certainly have no American hospice extant at this time to provide an invitation for self-education. These matters aside, however, Dr. Saunders writes: "It's important that we recognize the point at which our aim should be comfort rather than cure or the maintenance of a travesty of life" ([21], p. 228). A few lines above she writes: "Certainly they [physicians] should take *no deliberate step* to hasten the process of dying" ([21], p. 228). This remark suggests no ambiguity whatsoever – Dr. Saunders is opposed in all cases to positive or active euthanasia. She continues: "We should consider *withdrawing* steroid therapy for cerebral tumors when we are merely maintaining a travesty of living." Further on she writes: "Some of the requests for the legalization of euthanasia stem from the misguided idea that legislation is the only way of preventing these interferences" ([21], p. 235). By "interferences" she means certain kinds of extensive surgery, for example, tracheostomy and gastrostomy, and these, she agrees, can be understood as nothing but "meddlesome medicine." The question raised in this reader's mind is whether Dr. Saunders is, in fact, an advocate of 'passive' or 'negative' euthanasia, since she does not wish to maintain what she calls a "travesty of living"? Positively put, does Dr. Saunders, speaking perhaps only for herself and not for her disciples, advocate at times and under special conditions passive euthanasia? For our purposes it is not crucial that we receive an answer from Dr. Saunders in London or Dr. Sylvia Lack in New Haven. What is important, as I suggested above, is to appreciate what some philosophers and theologians have to say such that any response to the formulation of the moral issues ingredient in euthanasia can signify anything intelligible at all.

Let us turn to the philosophers: One concept requiring clarification is 'mercy'. Claudia Card of the University of Wisconsin at Madison, in her article "On Mercy", raises two general questions ([7], p. 182). (1) "How is mercy related to justice in the practice of punishing? (2) What are the moral bases of a case for mercy?" She eventually argues for the truth of the proposition that in the context of the "ethics of mercy" it is not the case that mercy is a form of benevolence, but a form of charity. This analysis is, from Professor Card's standpoint, propaedeutic to any deliberations on the justifiability of mercy killing: "What, if anything, justifies being merciful not to someone who was an offender but to someone who requests that his life be ended by the hand of another kindly disposed toward him?" In another article, published three years earlier, Professor T. Goodrich offers his reflections on the morality of killing. Once again the topic is not quite euthanasia. Rather, Professor Goodrich is concerned with clarification. He writes: "At first sight there doesn't seem to be any problem about killing. Most people would say that it is wrong to kill and that's all there is to it. . . . But there are several issues involving the morality of killing where ordinary men, secular and religious alike, make judgements or evince perplexity which reveals that common sense morality is less clear about killing than it at first appears." He continues, "I want to elucidate these issues and to show that the usual ways of settling them are fallacious. At the same time I want to show how attention to these issues reveals the inadequacies of many philosophical theories about morals" ([21], p. 127). At one point in his analysis Goodrich argues: ". . . consider '*X* would benefit by dying (i.e. ceasing to exist)'. This implies that we can say something like '*X* is benefiting by ceasing to exist', and this implies that *X* exists through ceasing to exist, which is absurd [self-contradictory]. This is the solution to the puzzle whether it is a kindness to an individual to end his existence. If we do something for the good of an individual, this presupposes that the individual exists throughout the process of being done good to. Now, questions of immortality apart, dying is not just an event which happens to an individual, like any other. An individual does not exist throughout the process of dying in the same way as he exists through other processes. Dying is just ceasing to exist, and one cannot exist through the process of ceasing to exist. So how can we say that it is for the good of an individual to die?"([12], p. 136). Hence Goodrich argues for the conclusion that, as Professor Marvin Kohl in *The Morality of Killing* points out, "there is something logically odd in the idea that it may be a kindness to an animal or human being to end its existence" ([15], p. 82). Notwithstanding Professor Goodrich's claim that he intends to clarify, Professor Kohl counter claims that Goodrich's "analysis is mistaken." Dying, Kohl argues, is not simply ceasing to exist as Goodrich suggests. Furthermore the process of dying, which might be said to take place in stages, is surely distinct from the terminal stage of the process. Simply put, dying is not *being dead*. Finally, Kohl continues, "just as there is nothing logically odd about saying that we can help a dying patient by making him more comfortable, there is nothing odd about saying that we can help a person by hastening the process of dying" ([15], p. 82). Thus it is not self-contradictory or absurd to claim that it is to the patient's benefit to die, to cease to exist. To make the matter still more complicated (I'm afraid) Kohl's response does not quite touch Goodrich's argument. For Goodrich never argued that we could never help a person by hastening the *process* of dying. Rather, Goodrich warned against asserting both that '*X* exists' and 'it is not the case that *X* exists' in the same argument, since surely any and all sentences follow from a set of sentences that contain two contradictory premises. The question is whether we are thrown back on older philosophical worries which emanate from the predicate 'is' or 'exists.' Surely Goodrich is wrong in his claim that "Dying is just ceasing to exist." Rather, *being dead* is just ceasing to exist. Furthermore it surely presents no logical difficulties to say that one exists through the *process* of ceasing to exist, as long as we agree that 'ceasing' only means 'becoming more and more disfunctional, including the dysfunctions which pertain to the organic and biological basis of our existence.' So we might plausibly conclude that it is best that an individual be assisted to die painlessly throughout the process of becoming more and more dysfunctional, though we might prefer to resist saying it is *good* to be dead. In short, a reading of Goodrich and Kohl may still leave us dissatisfied regarding what is meant by 'dying' and 'being dead' and the relation between them.

Another important distinction which continues to appear in euthanasia debates is denoted by the terms 'active' and 'passive' euthanasia, as special cases of acting and refraining. P. J. Fitzgerald, of the University of Leeds, writing in *Analysis* asks: "Is it any worse to cause a man's death by doing something than by doing nothing? Or is it immaterial whether the death in question results from an act or an omission?" ([10], p. 133). To answer these questions the author refers to the now common decision problem: Given the mother-child context, the obstetrician is faced with the choice of killing the unborn child to save the mother or refraining from operating so that the mother will die and child survive. That is, *either* (1) operate, kill the child and save the mother or (2) not operate and the mother dies while the child is saved. The question will resolve to: "Is there any morally significant difference between acting and not-acting, where both will result in death? Fitzgerald concludes that there *is* a morally significant difference in some cases. The problem, of course, is that this is not the typical case of euthanasia, but surely an important preliminary discussion for such cases. What we now require is the distinction between 'killing' and 'letting die' in the specific context of euthanasia. Daniel Dinello of the University of Wisconsin, writing in the same journal, *Analysis*, notwithstanding his title "Killing and Letting Die" does not take up the euthanasia case. But once again his analysis is instructive: "The killing/letting die distinction drawn in terms of the number of moves an agent could make clearly can have no moral significance. It is not obvious, though," he concludes, "that the distinction as I have now drawn it could have no moral significance" ([9], p. 85). Taking up the specific case of euthanasia, Professor James Rachels of the University of Miami addresses, quite directly, the distinction between active and passive euthanasia ([19], pp. 78–80).[4] Rachels reiterates the typical argument offered by health professionals to justify passive euthanasia but never active euthanasia. Arguments, often proffered by physicians and others, usually conclude that there *is* an important moral difference between active and passive euthanasia, such that even though the latter is sometimes permissible, the former is always absolutely forbidden. Rachels spends the remainder of his article arguing that there is no moral difference between active and passive euthanasia. Thus, for Rachels, if one has no objection to passive or negative euthanasia, then one can have no objection whatsoever to active euthanasia either. Rachels works to show that in a typical healer/patient setting we are morally free to choose which form of euthanasia to employ, depending on the needs of the particular case at hand. Hence it may well turn out, on Rachels' view, that active euthanasia is morally preferable to passive euthanasia, rather than the reverse. This conclusion, we can see, already suggests that the absolute prohibition against active euthanasia in all cases (Dr. Saunders' view) is irrational, though surely the expression of cultural bias and irrational reflection; so much the worse, then, for culture, habit, squeamishness, and unsound medical practice. Hence Rachels indicts the claim that there *is* a morally significant difference between actively ending the life of a patient *in extremis* and passive withholding of treatment, "a dogma", one which should be recognized as irrational and therefore rejected by medical professionals. Furthermore, it is implied that the law and public policy presently reflect an indefensible moral position. Hence the law ought to be changed in order to permit the physician to carry out humane care. Finally, for our last illustration, I call your attention to a very brilliantly argued paper by Professor Natalie Abrams of New York

University Medical Center. In her "Active and Passive Euthanasia: The Relevance of the Killing/Letting Die Distinction,"[5] Abrams argues that the moral significance of the acting/refraining distinction is different for different cases. Her astute analysis suggests that for certain cases there *is* a morally significant difference between acting (killing) and refraining (letting die); for other cases, on the other hand, there is *no* morally significant difference between killing and letting die, the more restrictive case of acting and refraining. Although I cannot reiterate it here, Abrams' argument rests on the central point that examples usually used to discuss killing and letting die are *inapplicable* to situations of euthanasia. Hence Abrams challenges Rachels' contention that there is no morally significant difference between acting and refraining in the euthanasia situation. To the contrary, she suggests that there *is* a morally significant distinction to be made in cases where outcomes are *positive.* Since it is *ex hypothesi* true that death is *preferred* and a 'good' (in any truly euthanasia situation) it follows that all euthanasia cases are positive in outcome. After all, if the death of the patient is not a more desirable and preferable outcome than the suffering or the endured pain, how could we begin to discuss euthanasia at all? Hence we must understand, as a given, that death is not undesirable or a *negative* outcome. Were it so judged, then we would be acting against the patient's interests which is self-contradictory given our ordinary understanding of the euthanasia context.

For Abrams, then, it does indeed matter whether a person kills a patient or allows him to die in situations of "mercy killing." Abrams thus concludes that active euthanasia is preferable to passive euthanasia. That is, the physician is profoundly morally responsible if he or she intervenes actively and in so acting ends the life of the patient. Further, the doctor should want to be responsible for the death of his patient, since this is, in this case, judged to be *best* for the patient, and is the patient's voluntary and informed choice. To repeat, if death is not the best end given the situation, no form of euthanasia is morally permissible. But that by hypothesis is not here an issue. Hence Abrams concludes that if euthanasia is ever morally justifiable, active intervention is the more moral act of an agent, the passive withdrawing or withholding of treatment being a less responsible act on the part of health practitioners. Put positively, if only passive euthanasia is permissible, (in which case death is allowed to occur when death is thought the best for the patient) the physician is in fact prevented from being directly responsible for the death of his patient. Is that to act irresponsibly in some cases? The physician's mandate is not simply to alleviate pain and suffering. More importantly, what if the physician cannot relieve that suffering and pain of the patient *in extremis* (which is a precondition for most arguments that claim to justify some form of euthanasia)? Let us return to the prerequisite conditions for discussing the morality of euthanasia in the first place.

III. PALLIATIVE TREATMENT

For a case to be one which meets the necessary criteria for euthanasia, certain conditions must be fulfilled. For example, the patient we are discussing must be undergoing excruciating, chronic (not acute) pain and must be suffering. In addition, the patient must be "terminal," i.e., one in whom death seems inevitable to the clinician, in days or weeks ([6], p. 76); that it is beyond a reasonable doubt on the part of the clinician that the patient has to die; that the patient, being fully informed of his condition, proffers his wish to have his life ended by an assistant, and does so declare when *in compos mentis* thus fully consenting to his own will; that the patient believes his death is a *good* and prefers an easy death; that the person assisting has no other motive but the patient's own good, and so on.

Hospice and other institutions have developed methods of palliative care and treatment, whereby relief from chronic pain is not far from complete. *Stedman's Medical Dictionary* informs us that 'palliative' signifies "mitigating" or "reducing the severity of. . ." or denotes "a method of treatment of a disease or its symptoms." The term 'palliative' has its roots in 'palliate' – namely "dressed in a pallium or cloak." Implied here is the relief of suffering and/or pain or the distress of symptoms of a disease, slightly. That is, one obtains partial relief and protection. Keep in mind that a typical (though not necessary) condition of the euthanasia case is the presence of chronic pain which gives rise to the request for assisted suicide or euthanasia. Typically, it is the physician's goal to preserve and protect life, not to deliberately end it. But the physician may empathize with the dying patient's request, and feel that the request is entirely justified rather than have the patient face a few more days of agony and suffering.

Now when philosophers cite a case of euthanasia they tend to suggest a cancer patient who is currently undergoing horrible, intractable, chronic pain (say) cancer of the throat. Given the three options: (1) continue the treatment with a few more days of agony, (2) discontinue treatment with a few more hours of agony, and (3) administer a lethal injection and the patient dies at once, one might ask what the moral thing is to do. This is a typical formulation of the options that emerge after analysis. Given these options, philosophers tend to advocate the second and third options (Rachels and

Abrams, respectively). The cessation of suffering and intractable pain seem an adequate ground for merciful, kind and helpful intervention.

But what of the real health care context? Are the options really so limited in fact? Marvin Kohl is careful to point out that such cases are not "as common as some advocates of euthanasia would have us believe" ([15], p. 75). [I should point out, as an aside, that Kohl's case too is one of "disseminated carcinoma metastasis," but this malady is not the only type which appears at the Hospice door, though the National Cancer Institute may only reimburse Hospice for such a patient "problem." But that is another matter and best saved for another place.] The important point for our purposes is the method of symptom control which is promulgated in hospice care settings. The symptom list includes everything from lack of sleep to general discomfort, acute pain, and chronic pain. Observers of the hospice care setting frequently note that physicians who come to realize that they cannot cure a patient tend to forget that an entire program of palliative treatment regimen is available for these patients' "side effects". The euthanasia case of the type encountered in common hospital settings includes intractable, chronic pain and suffering, probably the worst torture to be endured by the terminal cancer patient. Terminal, chronic cancer pain is devoid of meaning, viewed as senseless (in the strict sense), pointless and needless by the patient, and usually gets worse rather than better; it frequently occupies the patient's entire attention, thus isolating the self from the world. Acute pain as is endured when delivering a child, having a toothache or recovering from surgery, is shot through with meaning for the person and has an end-point; the patient knows that it will end. "But terminal cancer pain is meaningless," says Dr. Sylvia Lack. It is a constant reminder of the disease, with no foreseeable end in sight, and impossible to predict when it will end. All the patient sees is a continuation of the pain. When that happens life no longer seems worth living. "Some even seriously consider suicide" ([4], p. 8), Dr. Lack adds. All we need add is the notion of "assisted suicide" and we have, of course, the context of euthanasia. But what bearing does chronic pain control have on that patient request? Dr. William Lamers replies: "We have learned to give doses of pain-relieving drugs on a regular basis, every four hours around the clock, even if the patient is asleep and we have to wake him up. In this way hospice physicians avoid producing anxiety in patients, the anxiety that comes when the patient expects pain. And anxiety only increases pain" ([4], pp. 8–9). In this matter of pain control extremely subtle distinctions should be drawn. First, the distinction between acute and chronic pain; second, in spite of public misunderstanding, Dr. Twycross observes that "it is theoretically possible to relieve the pain in every case. Success depends on the doctor having an adequate concept of the nature of pain, an understanding of the correct use of analgesics. . ." ([23], p. 212). Third,[6] chronic pain requires preventive therapy not so-called PRN dosages.[7] Fourth, not all pain suffered by cancer patients is caused by the cancer problem. Fifth, the psychological significance of pain is still not fully understood, but it is clear that diversional activity does much more than just pass the time; pain is experienced as far worse when it occupies the patient's entire life space and attention. Sixth, addiction is not a problem since one must distinguish psychological from physiological addiction, at least in a generally rough way. According to Dr. Twycross, ". . . psychological addiction – a compulsion or overpowering drive to take a drug to experience its psychological effects – does not occur when pain control is part of a pattern of total care." Finally, the physician/patient relationship itself often serves to allow reduction of drug dosage. Paradoxically, perhaps, as the physician comes to be less anxious about the dying of the dying patient "the need for drugs becomes less" ([23], p. 214), which suggests that drug dosage and regimens are a function of physician anxiety and not simply determined by the so-called objective signs offered by the patient.

At this point it should be possible for us to understand that the tension that obtains between clinicians who attend the terminally ill and the lay public (which includes the philosophers who write on euthanasia but who do not care for the terminally ill themselves) is founded on mutual misunderstanding. For a long time I was perplexed as to the reticence of astute, competent and compassionate physicians to advocate euthanasia in either its active or passive versions. It was not that I could not get them to admit that active euthanasia could be morally justified (for I could generate thought experiments or even go to real albeit rare cases and set them up so as to constitute needless and endless suffering with no assistance available) it was that I was somehow talking past them. Why did these physicians not see the importance of the new movement by other equally compassionate and competent persons to legalize euthanasia, to free the physician to exercise by his action his responsibility to the patient? The answer rests by and large with the actual health care context, not always as it should be but as it *might be*, and the empirical experience of the physicians who spend their professional life with individual dying patients.

IV. THE MORAL IMPERATIVE OF THE HOSPICE CARE CONTEXT

When philosophers teach "medical ethics" there is a tendency to illustrate a problem and to initiate student discussion by citing a case like the following: "An 80-year-old man is admitted to the *hospital* with recurrence of metastatic carcinoma of the bowel. He has *intractable pain* requiring daily *increases* in the dose of *narcotics and shorter intervals between doses*. He has lost 40 lbs., is unable to hold his body even in a sitting position, cannot swallow, requiring I.V. feedings and is *only at rest when completely unconscious from drugs. He begs to have his suffering ended.* An intern gives him a dose of morphine which depresses his respiratory center and he dies." Clearly a case of active or positive euthanasia! We might therefore focus on the legal repercussions for the intern. The philosophy student will no doubt be brought to the edge of the analysis of the concept of mercy; soon the student will be expected to discuss the morality of killing; the moral difference, if any, between acting and refraining; then on to the moral difference, if any, of the killing/letting die distinction. He or she may even move on to ontology and the problem with the terms 'exist', 'is' and 'to be' only to learn that 'dying' is not identical to 'being dead', *salva veritate*, and that 'existence' is not a predicate (and thus to discover that the days and nights spent with Kant were not completely for naught, though I can't think of anyone more uninterested in infirmity than Kant). All of this should receive our moral approbation, to be sure. After all, it was not until very recently that philosophers, teachers of philosophy and their students began to talk seriously of death and dying and to formulate truly moral issues around the bedside of the patient and to begin to converse with medical personnel. What was once taboo is also in some circles obscenity. But discussions need not border on the obscene if language is not reduced to mere talk and we are willing to set foot inside the clinic or the hospital or the home of the dying. The student might be tempted to forego, temporarily at least, his or her newly acquired habit of examining language and analyzing concepts and, perhaps, follow a key precept of more phenomenologically oriented philosophers. Could it be that the *context* in which we find the dying patient lying in bed has significant bearing on the philosophical analysis to follow? Consider the case described above. First the 80-year-old man was admitted to a typical American *hospital*. Surely the particular setting has a particular history, a specifically designed space, the result of some very special assumptions and values as Foucault has taught us. Furthermore this *setting, context* or *milieu* is an odd therapeutic community, one in which the chronically ill in need of long-term rehabilitation for illness and disability and the terminally ill (never to return home in most cases) do not usually receive adequate care. Dr. Abramson speaks: "The physicians' convenience is largely in mind when hospitals are designed but it is rare indeed that the same consideration is given to the complex needs of the patients. It is taken for granted that the medical profession knows what is best. What may be effective and appropriate in a hospital which prides itself on short stay may not result in the most effective or appropriate care for those with disabling residua of disease requiring long-term institutional management. Once designed for brief intensive treatment they cannot now fulfill our objectives for the chronically ill and disabled" ([1], p. 213). And, of course, *a fortiori*, for the terminally ill and dying.

The hospital environment itself, then, must be taken into account when confronting the issues ingredient in euthanasia. We must keep in mind that "The typical hospitalized patient, as has been well documented elsewhere," writes the late psychologist Bernard Kutner, "temporarily relinquishes his socio-biological autonomy. . ." ([17], p. 9). The milieu in which care and therapy is given must now come under scrutiny. For "Twelve months following discharge, we find 2-1/2 times the mortality among patients arbitrarily assigned to the conventional treatment program as opposed to those assigned to the 'therapeutic community' " ([17], p. 14). The reference here to an experimental group constituting the "therapeutic community" is based on a study conducted by Drs. Kutner and Abramson at Albert Einstein College of Medicine in New York City. A program designed as milieu therapy (or therapeutic community) was initiated for patients in need of rehabilitation. The study was designed to look inside the hospital environment and to identify those factors which lead to success or failure in the patient's adjustment to normal living. It was assumed that the hospital environment provides few stimuli which resemble the challenges which the patient must once again face in the real world when discharged from the hospital. (Rehabilitation is defined as "the re-establishment of the individual to a viable set of functionally useful roles in life, by whatever means are available, without stressing him beyond his limited physical capacities") ([16], p. 518; [20]). The experimental model of the therapeutic community created indeed a new institutional space, though one designed for the rehabilitation of the patient and not for care of the dying patient. The reason for mentioning it here is to alert us to a model which in outline is strikingly congruent with the general hospice model, one which principally utilizes palliative treatment, since the disabled patient under the care of physiatrists cannot expect "cure" in most cases and will have to live with some degree of disability once having returned to the home environment and the milieu of personal space. The problem for the rehabilitation patient is that "long term studies have shown that gains made in the hospital or clinic too frequently become eroded later" ([2], p. 59).

Dr. Abramson asks: "Can the environment itself be utilized as an instrument of therapy?" ([2], p. 60). The general results indicate, according to Dr. Abramson, that "the human environment of institutional care influences the long-term results of rehabilitation management" ([2], p. 65). Furthermore, a moral point is made as an appeal: "We have seen enough of the accrued benefits to be certain that we can no longer afford generic treatment of the disabled which leaves a legacy of potential failure, because of the absence of social realism and dynamic relationships" ([2], p. 65). The therapeutic community, and the evolution of millieu therapy has its historical antecedent in psychiatric models. Dr. Abramson adopted this model for the patient in need of long-term rehabilitation. The principles which govern the therapeutic community could well be included in the set of principles governing the hospice care setting: "(1) the preservation of the patient's right to self-determination, (2) the maintenance and strengthening of family and community ties, (3) the active participation of families in the process of treatment, (4) development of a strong and coordinated team with positive and healthy attitudes toward each other and the patients, (5) creation of a community of patients on the ward with an active program of mutual exchange and assistance, (6) deliberate efforts to render semipermeable the boundaries between hospital and community, (7) modification of the conventional therapeutic program to suit the patients' needs on a twenty-four hour basis and (8) elimination of unnecessary self-perpetuating hospital regulations" ([20], p. 257). In short, the application of these principles can be formulated as empirical maxims for the care of the physically impaired patient in need of long-term rehabilitation for chronic problems. But more importantly for us, these principles serve to sketch the hospice care setting as well.

It is time that we no longer sequester terminally ill patients in dead-end institutions. They are not simply to receive the leftovers and residue of acute models embodied in the typical American hospital. The rebirth of the clinic is needed now without further bureaucratic procrastination. The new hospice must be totally functional rather than mechanical or routine. Most importantly, it must serve as a subtle transition from home milieu in continuity with the patient's life, to guard that life in the process of dying.

V. CONCLUSION

A Physicians' Attitude Survey published in May, 1974 in *Medical Opinion* opens with the following sentence: "The medical mandate for euthanasia is far stronger than was previously known. . . . One-in-ten physicians would bring on death by active means, with the bulk of the remainder allowing passive 'neglect' to take place" ([22], p. 31). The central point here is not resolved by additional arguments which justify passive euthanasia. The moral spotlight is to be concentrated and re-focused on the institutional care *setting* itself, in all its particulars and ramifications. Here I, of course, mean the typical American hospital, already shown to be in essential conflict with the proper needs of the terminally ill and the chronically ill patient. The reason that the survey reveals that "More respondents would direct someone to end his/her suffering than would ask that the entire medical armamentarium be employed," ([22], p. 31) is the ubiquituous failure of the contemporary institution known as the hospital. Physicians and other health professionals are now obliged to examine the moral and value laden principles underlying humane care and superimpose that reflection on the contemporary American hospital care setting. Once this is done we will see that we have far too long ignored the social setting, the milieu, the full context in which the physician, other health care professionals and the patient meet. We will even realize that Physicians Attitude Surveys are devoid of context thus leaving a bald statistical litany. We will see that medical ethics courses, when lifting out the "essentials" of a complex contextual situation to make a "case" for discussion, fail to underscore the real health care context. Remember the 80-year-old man with recurring metastatic carcinoma? It was said that the patient required not only increases in the dose of narcotics, but "shorter intervals between doses." This "given" is sharply countered by evidence drawn from hospice physicians. Furthermore, we were told that our patient was only at rest when "completely unconscious." The hospice physicians have shown that this need not ever be the case. Most important of all, we were told that "the patient begs to have his suffering ended." The hospice physicians inform us that this suffering can be controlled and eliminated to a great extent, given the proper *context* which alleviates psychological anxiety, the proper medical regimen and trained professionals to administer it thus controlling not only chronic pain but all symptoms producing discomfort. It is no accident that the Physicians Attitude Survey indicates that psychiatrists "were more than twice as likely to say 'end my suffering' than other clinicians" ([22], p. 32). My guess is that most philosophers would respond in a similar fashion. The general debate in the medical and philosophical literature, then, seems to focus on the merit of active and/or passive euthanasia, with some physicians joining the philosophers, one physician noting that "I feel it is ridiculous to make a distinction between 'passive' and 'active' euthanasia. In certain situations it is more appropriate to be active than passive" ([22], p. 34). What we are suggesting is the new moral imperative: examine the hospital context closely.

The specialists who work least with the dying patient, tend to support negative euthanasia. This suggests the importance of listening not only to those who minister to the terminally ill but listening to them in their particular care setting. Joan Craven and Florence S. Wald remind us that "The depression, worry, and anxiety that accompany terminal illness are often more debilitating than cancer itself" ([8], p. 1819). The contemporary American hospital encourages patient loneliness, isolation and fear of abandonment. Non-health professionals might do well to suspend judgment regarding the presence of chronic pain and suffering in the hospice care setting. Their elimination in fact and not only in principle makes the application of philosophical argumentation on euthanasia by and large moot, though there will always be a few exceptions among those persons who request assisted suicide for libertarian reasons like the right to determine one's own destiny, including the right not to continue living in this world if the psychological suffering is unbearable.

In conclusion, the moral point of view compels us to respond to the societal imperative which presently awaits our acknowledgment and action, urging us as it now does to place our resources (given our *embarras des richesses*) in the service of a reborn clinic, like Hospice, a health care context which if fully successful, will make it unnecessary, undesirable and rare for the terminally ill patient to non-involuntarily request that his responsible physician, in whose care he remains, exert his or her agency against his life. The physician, after all, and not just for reasons of nostalgia and historical sentimentality, did take an oath. We perhaps are duty bound to help the doctor live up to the spirit if not the word of that tradition, established on the island of Cos. There it was written: "I will use treatment to help the sick according to my ability and judgment, but never with a view to injury and wrong-doing. Neither will I administer a poison to anybody when asked to do so, nor will I suggest such a course" ([14], Vol. I, p. 299).

Our collective failure to foster the most humane care settings for the terminally ill, compel physician and layman alike to revisit Hippocrates. The moral point of view, I suggest, requires that we relocate the debate. While we permit the philosophers to do the work necessary to provide a lucid clarification of concepts, we must also re-form ourselves in the political arena and join forces on behalf of the terminally ill (and perhaps all patients) by demanding that our health care institutions undergo drastic and dynamic changes. That may require nothing less than the rebirth of the clinic.

University of Connecticut School of Medicine,
Farmington, Connecticut

NOTES

1 The translator indicates that the French *la clinique* signifies for Foucault both clinical medicine and the teaching hospital. More importantly, *la clinique* is ambiguous in the English translation; that is, it appropriately refers to the ancient clinic through 1770, but the term is also used where the connotation is the modern *l'hôpital*, an institutional transition which takes place from 1770 until the early 1800s.
2 A division of Yeshiva University, New York City.
3 R. H. Williams' colleagues employed a slightly modified version of the same questionnaire and extended the study to staff physicians, first and fourth year medical students and others. Cf. An editorial "Euthanasia", *Journal of the American Medical Association* **218**, No. 2 (October 11, 1971), 249:

Percentages of responses favoring
the practice of euthanasia*

	Negative Euthanasia	Positive Euthanasia
Physicians (staff)	59	27
1st-yr medical students	69	46
4th-yr medical students	90	46

*36% of graduate nurses and 50% of second-year nursing students favored changes permitting positive euthanasia.

4 For an abbreviated version of the original paper "Active and Passive Euthanasia," disseminated at the Council for Philosophical Studies "Summer Seminar on Medical Ethics," Haverford College, Pa., 1974, see [19].
5 Forthcoming in *Philosophy*.
6 Dr. Twycross remarks, "Accordingly 'four-hourly as required' has no place in the treatment of persistent pain." Cf. Dr. Twycross' award winning entry, "A Plea for 'Eu Thanatos' " published by the Voluntary Euthanasia Society and distributed in Documentation in Medical Ethics (December 2, 1973). The author is grateful to Sylvia Lack, M.D., for calling Dr. Twycross' paper to his attention and for providing a copy of the essay.
7 'PRN' (*pro re nata*) 'when the condition is born'. We might suggest as an alternative abbreviation 'ARN' (*ante re nata*) 'before the condition is born'.

BIBLIOGRAPHY

1. Abramson, A. S.: 1965, 'The Therapeutic Role of the Hospital Community: Editorial, *Archives of Physical Medicine and Rehabilitation* **46**, 213–215.
2. Abramson, A. S.: 1968, 'The Human Community in the Rehabilitation Process', *Archives of Physical Medicine and Rehabilitation* **49**, 59–65.
3. Abramson, A. S.: 1976, 'Health, Healing and Rehabilitation', *Connecticut Medicine* **40(6)**, 403–6.
4. Alsofrom, J.: 1977 'The "Hospice" Way of Dying', *American Medical News* **20**(8), 8–9.
5. Berman, R.: 1977 'The Favor of the Gods', (*Annual Oration of the Society for Health and Human Values*, November 11, 1976, San Francisco, California), Society for Health and Human Values, Philadelphia, Foreword 3; 5–16.
6. Brown, N. K. 1970, *et al.*· 'The Preservation of Life', *Journal of the American Medical Association* **211**(1), 76–81.
7. Card, C.: 1972 'On Mercy', *Philosophical Review* **81**, 182–207.
8. Craven, J. and Wald, F. S.: 1975 'Hospice Care for Dying Patients', *American Journal of Nursing* **75**(10), 1816–22.
9. Dinello, D.: 1971, 'On Killing and Letting Die', *Analysis* **31**(3), 83–86.
10. Fitzgerald, P. J.: 1967, 'Acting and Refraining', *Analysis* **27**(4), 133–39.
11. Foucault, M.: 1973, *The Birth of the Clinic: An Archaeology of Medical Perception*, trans. by A. M. Sheridan Smith, Random House, New York. The original text: 1963, *Naissance de la clinique*, Presses Universitaires de France, Paris.
12. Goodrich, T.: 1969, 'The Morality of Killing', *Philosophy* **44**(168), 127–139.
13. Heinemann, H. O.: 1976, 'Incurable Illness and the Hospital in the Twentieth Century', *Man and Medicine* **I**(4), 281–285.
14. Hippocrates: 1962, *Ancient Medicine*, trans. by W. H. S. Jones, Loeb Classical Library, Harvard University Press, Cambridge.
15. Kohl, M.: 1974, *The Morality of Killing*, Humanities Press, New York.
16. Kutner, B.: 1970, 'Social Barriers to Cardiac Rehabilitation', *New York State Journal of Medicine* **70**(4), 517–21.
17. Kutner, B.: 1970, 'The Hospital Environment', *Social Science and Medicine* **4** (Supp. 1), 9–14.
18. Netsky, M. G.: 1977, 'Dying in a System "Good Care": Case Report and Analysis,' *Connecticut Medicine* **41**(1), 33–36.
19. Rachels, J.: 1975, 'Active and Passive Euthanasia', *New England Journal of Medicine* **292**, 78–80.
20. Racker, F. W. *et al.*: 1963, 'The Therapeutic Community: An Approach to Medical Rehabilitation', *Archives of Physical Medicine and Rehabilitation* **44**, 257–61.
21. Saunders, C.: 1976, 'Living with Dying', *Man and Medicine* **I**(3), 227–242.
22. Scott, B. T.: 1974, 'Doctors and Dying: Is Euthanasia Now Becoming Accepted?' *Medical Opinion*, 31–34.
23. Twycross, R. G.· 1975, 'Diseases of the Central Nervous System: Relief of Terminal Pain', *British Medical Journal* **4**, 212–214.
24. van den Berg, J. H.: 1964, *The Changing Nature of Man*, Delta Books, New York.
25. van den Berg, J. H.. 1969, *Medische Macht En Medische Ethiek*, G. F. Callenbach, N.V., Nijkerk, Holland.

26. Williams, G.: 1975, 'Euthanasia and the Physician', in M. Kohl (ed.) *Beneficent Euthanasia*, Prometheus Books, New York, 145–168.

SECTION II

PHILOSOPHY OF ORGANISM

LEON R. KASS

TELEOLOGY AND DARWIN'S *THE ORIGIN OF SPECIES*: BEYOND CHANCE AND NECESSITY?[1]

I. INTRODUCTION

Few books have turned men's minds more than Charles Darwin's *The Origin of Species by Means of Natural Selection or The Preservation of Favoured Races in the Struggle For Life*. Yet where the turning will or should come to rest is still uncertain, for the full significance of Darwin's ideas remain a subject of inquiry and controversy. Granting that Darwin was right about the fact of evolution – granting, that is, that all of nature flows, that nature is subject to "history" – it is by no means yet settled what difference this insight should make for our thinking about nature, about man's nature and his place in nature, about the good life for man, or about God. In this essay, I will consider one of the disputed questions, one which may also be central for several of the others: the question of teleology, of the presence of ends or purposes in nature.

Let me introduce the dispute by means of two sets of quotations. John Dewey praised Darwin for having done away with the need to think about nature in terms of final causes. Dewey claimed that, thanks to Darwin's discovery,

> philosophy forswears inquiry after absolute origins and absolute finalities in order to explore specific values and the specific conditions that generate them ([6], p. 13).

Hans Jonas, reporting on the consequences of Darwinism for philosophy, appears to agree:

> It was the Darwinian theory of evolution, with its combination of chance variation and natural selection, which completed the extrusion of teleology from nature. Having become redundant even in the story of life, purpose retreated wholly into subjectivity ([8], p. 44).

Yet in 1874, Darwin's friend and colleague, Asa Gray, the great American botanist and evolutionist, wrote:

> Let us recognize Darwin's great service to Natural Science in bringing back to it teleology; so that, instead of Morphology *versus* Teleology, we shall have Morphology wedded to Teleology ([7], p. 237).

S. F. Spicker (ed.), Organism, Medicine, and Metaphysics, 97–120.

Darwin replied to Gray:

> What you say about Teleology pleases me especially, and I do not think any one else has ever noticed the point ([4], Vol. III, p. 189).

Who is right, Dewey and Jonas, or Gray and Darwin? The answer, of course, is that both are right. Should anyone protest that both *cannot* be right, we shall have to concede that he too is right.

Let us consider the question afresh, and in these four stages. First, beginning with what is first for us – on the one hand, living plants and animals, on the other hand, the confused term "teleology" – let us establish provisionally the place of teleological notions in the understanding of a living organism. Second, turning to nature as a whole, let us consider the major teachings of Darwin's *The Origin of Species* and their implications for teleology. Third, let us see whether, with the help of a *closer* scrutiny of *The Origin*, we can discern any directions or tendencies in the course of evolution. Finally, with the aid of some free speculation, let us try to explore beyond the limits of the theory as presented by Darwin. The argument, in many places merely suggestive, is presented more as an outline for future inquiry than as the fruit of past research. Nevertheless, by the end, I hope to have indicated both the need for, and the possibilities and limits of, a sound teleological view of organism and of nature as a whole.

II. ORGANISMS AS PURPOSIVE BEINGS: ELEMENTS OF TELEOLOGICAL ANALYSIS

Do plants and animals have purposes? Do they come-to-be and do they function for "the sake of something?" To say "yes" makes one a teleologist. But the genus "teleologist" has many species, and some are more fit than others.

I begin with several cruder kinds of teleology. While these will ultimately be of little interest for us, I mention them for two reasons: First, these are among the simple-minded teleological notions that have brought a bad name to any and all teleological thinking. Second, these notions were dominant in teleological thought prior to Darwin, and, as we shall see, played a major rôle in the genesis of his theory.

Many of these cruder views take their bearings from the fact that plants and animals live not only by and for themselves, but also and always in relation to other plants and animals. One animal's substance is another animal's food: Men are food for worms, worms are food for ducks, ducks are food for men. Mutual service, rather than predation, is sometimes the dominant relation:

the bumble-bees are the indispensible fertilizers of clover, and the nectar of clover becomes food for the bees. Plants fix and use carbon dioxide released by animal respiration, and give off oxygen consumed in turn by animals. Also, organisms are structured as if designed for some of these external relations; the beak of a cardinal is a perfect nut-cracker for sunflower seeds, the beak of a woodpecker a perfect drill to insect treasures stored under bark.

These very evident uses of one organism and its productions by another, and these very evident elements of structure that permit such exploitation, gave rise to the view that all of nature is one grand system, an organic whole, in which and for which every living thing plays its proper and necessary part. Each animal and each plant has a wordly purpose, outside of itself, relative to and for others. In a common anthropocentric version of this external – or relative – teleology, all of nature is seen as being for the use and benefit of man. These views are usually grounded on a view of nature as *designed*, as *created*, by an intelligent designer or creator, who, in most such opinions, is also thought of as perfect and beneficent in his making. Sometimes it is even implied that a conscious and intelligent purposive being is *always* actively at work in the world, suitably arranging and ordering all these means and ends.

But these notions of external teleology, and of design and conscious purposive planning, are all secondary notions. The primary home of teleological thought is the *internal* and *immanent* purposiveness of individual organisms, in their generation, their structure, their activities. It was almost certainly his viewing the activities of living things in and for themselves that called forth Aristotle's view of nature as directed toward ends, and which led him to consider the "that for the sake of which" as an indispensable part of an understanding of natural phenomena. Let us review the evidence.

First, consider the *generation of living things.* Each organism comes to be not at random, but in an orderly manner, starting from some relatively undifferentiated but nevertheless specific seed or zygote produced by parents of the same species, and developing, unfolding itself from within, in successive stages that tend toward and reach a limit, itself, the fully formed organism. There is a natural end to the process of development which defines the previous motion, through its various stages, *as* a development, and the *specific* character of the end determines whether it is the development of a dolphin or a daffodil. Each stage depends on the preceding one out of which it emerges, and prepares for the succeeding one into which it passes. The emergence of the differentiating parts of the whole is coordinated, each part being related always to every other part but also, prospectively, to the mature form of itself and of the whole. The adult which emerges from the process of self-development and growth is no mere outcome, but a completion, an end, a whole. To use Aristotle's terminology, living beings come to be by a process that has a natural end or *telos*; the *telos* of coming-to-be is the *eidos* (i.e., species or "looks") of the particular organism in question. To use a more up-to-date formulation – which adds only to the *how* and not to the *what* of generation – a chicken comes to be as the result of numerous biochemical reactions, coded for, specified, regulated, directed, and ordered by the blueprint for making a chicken, which blueprint is "in" the chicken DNA contained in the nucleus of a fertilized chicken egg.

Second, consider any *fully formed mature organism or adult.*

(a) It is an organic whole, an articulated whole, composed of parts. It is a structure, not a heap, and to be a structure means to have a function; to understand a structure is to understand *what it is* in terms of *what it does.* And function, I need not remind you, is a kind of teleological notion; function is not a material or a mechanism, nor does it have extension; rather, it is the end of the extended material structure and its mechanism. (We note, in passing, that machines too are structures, not heaps; thus, one cannot understand even a machine mechanistically, i.e., by knowing only the how of its workings. To understand a machine *as* a machine is to understand its work, to know it as a functioning whole. The Lilliputians might have taught to take apart and reassemble Gulliver's watch, and might have learned everything about the workings of its gears and springs, without ever learning that it was an instrument for telling time rather than the God he worshipped.) Usually, the search for a mechanism begins only *after* one has identified the function, the activity, the work - usually one asks "*how* does it work?" only after one knows the "what" and the "what-for" of the "it."

(b) The parts of an organism have specific functions, which define their nature as parts: the bone marrow, for making red blood cells; the lungs for exchange of oxygen and carbon dioxide; the heart for pumping the blood. Even at a biochemical level, every molecule can be characterized in terms of its function: hemoglobin is a protein that binds and carries oxygen, hexokinase is a protein that phosphorylates glucose, DNA polymerase is a protein that copies the existing DNA molecules, faithfully reproducing the sequence of bases so that the encoded messages are not scrambled when the cells multiply and divide. The parts, both macroscopic and microscopic, contribute to the maintenance and functioning of the other parts, and make possible the maintenance and functioning of the whole.

(c) A mature organism shows itself as a whole, maintains itself as a whole, and functions as a whole in characteristic ways *above and beyond merely maintaining itself.* The mockingbird delights in its own imitative sounds, the adult cardinal hops from ground to low branch and squawks for the young bird to follow suit, a coyote howls at the moon, the otter turns identical underwater somersault after somersault for hours on end, the beaver cuts down trees, the dog sniffs the ground for traces of his pals, the young deer engages in ceremonial duels, the lizard sunbathes on a rock, the penguin struts and parades, the peacock shows off his plumage. These looks and ways serve to define the animal; its activities, taken together, are most of all what the animal *is* and what the animal *is for.*

(d) In addition to the self-producing, self-organizing, self-maintaining, self-preserving, and self-fulfilling characteristics of organisms, there is that most remarkable power of self-healing. A plant-cutting will re-grow the missing roots, each half of a divided planarian will regenerate the missing half, and more generally, nearly all living things heal wounds or breaks and tend to restore wholeness. This tendency toward wholeness also encompasses, in some cases, the rejection of additions to the whole, as in the immune system of higher animals which combats the entry of alien elements, whether infectious agents, tumors, or grafted tissue.

Third, living things display a *directedness*, an inner "striving" toward a goal, both in their coming-to-be and in many of their activities. A seedling sprouting beneath a large rock will bend and grow around the rock to reach the light. A young bird will continue to struggle to coordinate wing and tail motions until it finally learns to fly. A beaver will make many trips to build a dam, or a bird a nest, or a spider a web. And for many animals there is an elaborate pattern of behaviour leading up to mating. *In none of these cases is the activity planned or conscious or intended, yet it is just the same a directed and inwardly determined activity to an end and for a purpose.*

Fourth, many organisms function not only as individuals but also as *members of some larger group* toward ends that reside with the group: the specialized contributions of the various kinds of honey-bees to the work of the hive is only the best-known example. Even the so-called unsocial or solitary animals are social animals in respect of reproduction. Much of their appearance and activity is for the sake of attracting a mate, and this for the sake of reproducing their kind.

Before proceeding further, let us take stock of the ingredients of the notion of internal teleology developed to this point: (1) Orderly coming-to-be, reaching an internally-determined end or completion; (2) An organic and active whole, a unity of structure and function, with parts contributing to the maintenance and functioning of the whole; (3) Directedness, activity pointing toward a goal; (4) Serving and preserving one's kind. Implied in all of these notions is a fifth element, the notion of "good" or "well" or "fit" or "successful" or "perfect." Was the animal perfectly formed? Did the part function well? Was the end successfully reached? Was the organism fit for reproduction, i.e., fertile? When considering the functioning of an animal, we ask not only whether it functions in a characteristic way, but whether it functions well or normally as such and such an animal. For example, we regard a child born blind or with a congenital heart deformity as imperfect, not-fully-whole, not completely formed. These notions of "good" and "well" are, in a way, implied by the notion of "complete," "whole," and "functioning;" on the other hand, it may be useful to distinguish "having a function" and "performing it well." This holds for machines, no less than for organisms, indeed for anything which works toward an end. *The end is a standard as well as a goal.* Teleological analysis will be concerned both to identify the end and to evaluate how well or badly it is achieved.

The questions of completeness and of the well- or ill-functioning of a particular animal relative to a standard for that kind of animal invite a further question about completeness and well-functioning simply. If there is a completeness or fitness or excellence of moss, of barnacles, of grasshoppers, and of chimpanzees, is there no completeness or fitness or excellence over-all? I shall return to this question in Part V.

What I have presented in this section are some of the rather obvious reasons why we must regard living things as purposive beings, as beings that cannot even be looked at, much less properly described or fully understood, without teleological notions. I am well aware that many scientists believe this view is vitalistic – which it certainly is not – or unscientific, and that some philosophers of science claim that all teleological descriptions can be reformulated in non-teleological terms without loss of meaning. I know that, drunk on the truly amazing successes of modern biology – including the discoveries of Mendelian genetics; the chemical structure of the gene; the detailed biochemical mechanisms of gene replication, transcription, and translation; the achieved enzymatic synthesis and the promised chemical synthesis of a self-replicating virus – many biologists believe that physics and chemistry are adequate to the full explanation of living things, or, in other words, that the complex phenomena of living things – including their *apparent* purposiveness, and including awareness, memory, desire, and thought – are in principle fully explicable in terms of molecular structures and biochemical reactions. Let me, for now, appeal only to the phenomena and simply assert that they are all mistaken.[2] In the study of organisms, teleological notions are so far from being undermined by or reducible to mechanical and materialist notions as to be the indispensable principles both of formulating a mechanistic inquiry and of rendering its findings intelligible. As we shall see, it is *history* and *not* mechanism that presents the real challenge to teleology.

III. THE EXPLICIT TEACHING OF *THE ORIGIN OF SPECIES*: NATURE HAS NO END

The more challenging and problematic question about plants and animals is whether their teleological character reveals anything about the natural backdrop before which they purposively come and work and play. Is nature as a whole purposive? Are plants and animals, in some sense, also *purposes* of nature? Why are living things purposive? How did they and how do they come to be that way?

Before Darwin, there were several grand alternatives competing as an answer. Some held that the cause was naturally and eternally resident in living things, whereas others held that it had been implanted in them by an intelligent being or designer. Some looked to a cause largely within the material of a living thing, while others turned to the organic form of the thing itself. Some views credited combinations of matter and form, and of nature and design. Perhaps the two grandest and most influential alternatives are these: The Biblical view of a teleological and created world with its various forms specially created after God's plan, and the Aristotelian view of a teleological but eternal nature with its various forms keep in being, generation after generation, by the immanent workings of eternal species (*eidē*).[3]

Neither the teleology of special creation and design nor the teleology of the eternal *eidē* could survive unchanged once Darwin's theory of evolution joined with them in the struggle for existence. However, and this I want to emphasize again, the view of living beings as purposive, as self-producing, self-maintaining and self-reproducing, is in no way undermined by Darwin's theory. In fact, in certain respects, the teleological emphasis is accentuated.

Darwin himself, in addition to seeking an account of the origin of species, was mainly interested in explaining their teleological character, *which he took for granted.* This is the task he set for himself as recorded in his Introduction to *The Origin*:

In considering the Origin of Species, it is quite conceivable that a naturalist, reflecting on the mutual affinities of organic beings, on their embryological relations, their geographical distribution, geological succession, and other such facts, might come to the conclusion that each species had not been independently created, but had descended, like varieties, from other species. Nevertheless, such a conclusion, even if well founded, would be unsatisfactory, until it could be shown *how* the innumerable species inhabiting this world have been modified, so as to *acquire that perfection of structure and co-adaptation which most justly excites our admiration* ([5], p. 66; emphasis added).

Darwin was interested in accounting not only for the internal purposiveness of plants and animals, but also and especially for the perfection of their structure and for the perfection of their usefulness to each other – i.e., for their relative purposiveness or *externally* teleological character. He wanted to account for why everything was so perfectly ordered, for why everything *appeared* to be designed.

What is the Darwinian account? The core of Darwin's theory, as presented in *The Origin of Species*, has two major theses: the Transformation of Species and Natural Selection. The first holds that species are neither eternal nor separately created, but have come into being by descent with modification from pre-existing species. The second holds that natural selection has been the major *means* of modification. The argument of *The Origin* proceeds in the reverse order. In the first part of the book, Darwin puts forth his theory of natural selection; the argument draws much of its rhetorical force from the ingenious opening chapter in which human artfulness in animal breeding is used as the mystery-solving analogy for the "craftsmanship" of Nature. In the latter part of the book, Darwin convincingly shows that a theory of descent with modification – which he now assumes, but no longer argues, is due to natural selection – is superior to a theory of separate creation in accounting for numerous mysteries in the geological record, in geographical distribution, in the classification and morphology of organisms, and in embryology.

The theory of natural selection, Darwin's most novel proposal, comprises four elements: variation, inheritance, the struggle for existence, and natural selection proper. Though offspring by and large resemble their parents, variations occur frequently, and in all parts of the organism. These variations can themselves be transmitted to the progeny of the variant, i.e., they can be inherited. Some of these variations may be "beneficial" – i.e., they may enhance the individual organism's chance for survival under the conditions in which it finds itself; others may be "injurious." Since all species tend to reproduce themselves in geometrical proportions, more offspring are produced than can be supported by the available resources; death of large numbers of each species is a necessary occurrence. Variations which give to a plant or animal some advantage in this struggle to survive – say, an increased ability to escape the notice of predators, or a greater resistance to disease, or a greater efficiency in securing food – will give those individuals a greater chance of surviving, and those surviving individuals will transmit the advantageous traits to their offspring. This preservation of favorable heritable variations, and the rejection of injurious variations, Darwin named *natural selection.*

In concluding this brief summary of Darwin's theory, let me emphasize that there are two aspects to change through natural selection: the *origin* of novelty and the *preservation* of novelty. Variation (or mutation, as we now generally call it) is responsible for the *appearance* of novelty (both the "beneficial" and the more common "injurious" sorts), while natural selection proper – i.e., the selective elimination of the less fit – is responsible for *preservation.* The second step is dependent on the first; as Darwin put it,

Unless profitable variations do occur, natural selection can do nothing ([5], p. 132).

This fact led Samuel Butler to complain that Darwin overrated natural selection, that "The 'Origin of Variation,' whatever it is, is the only true 'Origin of Species' " ([2], p. 263). Of this, more later.

What are the implications of Darwin's theory for teleology? To a first approximation, with the theory of natural selection, Darwin succeeds in providing a non-teleological account of the origin of and basis for the teleological character of organisms. Purposive and co-adapted beings and races rise and change and fall by processes that are themselves *not* purposive, out of a nature that is seemingly purposeless and aimless.

The following two important qualifications should, however, be noted: (1) In Darwin's hands, the teleological character of organisms is reduced to the single dimension of self-preservation, with the consequence that one looks for and only at the *survival* value of all aspects of structure and behavior. This over-zealous effort at a retrospective "teleological" explanation later bothered Darwin himself, as he records in this passage from *The Descent of Man:*

I now admit . . . that in the earlier editions of my 'Origin of Species' I perhaps attributed too much to the action of natural selection or the survival of the fittest . . . . I did not formerly consider sufficiently the existence of structures, which, as far as we can at present judge, are neither beneficial nor injurious; and this I believe to be one of the greatest oversights as yet detected in my work. I may be permitted to say, as some excuse, that I had two distinct objects in view; firstly, to shew that species had not been separately created, and secondly, that natural selection had been the chief agent of change . . . . . I was not, however, able to annul the influence of my former belief, then almost universal, *that each species had been purposely created; and this led to my tacit assumption that every detail of structure, excepting rudiments, was of some special, though unrecognized, service.* Any one with this assumption in his mind would naturally extend too far the action of natural selection . . . ([3], pp. 441–442; emphasis added).

(2) Darwin's non-teleological explanation – variation, inheritance, struggle for existence – not only assumes but even depends upon the teleological character of organisms. The desire or tendency of living things to stay alive and their endeavor to increase their numbers – which are among the minimal conditions of the theory – are taken for granted and are unexplained. It would only be part of an explanation to say that those beings with no tendency to maintain and reproduce themselves have died out. Why are the other ones – the self-maintaining and reproducing beings – here at all? Why should a non-teleological nature generate and sustain teleological beings?

Yet, despite these qualifications, Darwin's theory of natural selection is not a theory of nature acting to an end, but of nature acting of necessity and through chance coincidences of its necessitated workings and effects.

Let us then make explicit the implications of Darwin's theory for the grand alternatives of teleology due to design and teleology due to self-actualization of immanent and eternal *eidē*. The doctrine of special creation and design was the view Darwin explicitly tried to overthrow, and to a large extent he succeeded. Yet the argument from design has not and need not become extinct, but survives by retreat and modification: the retreat is to the first beginnings, with design postulated only at the origin of the world; the modification is to consider specially created only the special "laws of nature" impressed upon matter, which both necessarily and designedly thereafter gave rise to the world as we know it, with its profusion of living forms. Darwin himself was sometimes at least partially disposed to this opinion:

> Another source of conviction in the existence of God, connected with the reason and not with the feelings, impresses me as having much more weight. This follows from the extreme difficulty or rather impossibility of conceiving this immense and wonderful universe, including man with his capacity of looking far backwards and far into futurity, as the result of blind chance or necessity. When thus reflecting, I feel compelled to look to a First Cause having an intelligent mind in some degree analogous to that of man; and I deserve to be called a Theist. This conclusion was strong in my mind about the time, as far as I can remember, when I wrote the *Origin of Species* . . . ([4], Vol. I, pp. 312–313).[4]

And this is certainly what Thomas Huxley had in mind when he praised Darwin, as did Asa Gray, for his reconciliation of Teleology and Morphology:

> But perhaps the most remarkable service to the Philosophy of Biology rendered by Mr. Darwin is the reconciliation of Teleology and Morphology, and the explanation of the facts of both, which his views offer. The teleology which supposes that the eye, such as we see it in man, or one of the higher vertebrata, was made with the precise structure it exhibits, for the purpose of enabling the animal which possesses it to see, has undoubtedly received its death-blow. Nevertheless, it is necessary to remember that there is a wider

teleology which is not touched by the doctrine of Evolution, but is actually based upon the fundamental proposition of Evolution. This proposition is that the whole world, living and not living, is the result of the mutual interaction, according to definite laws, of the forces possessed by the molecules of which the primitive nebulosity of the universe was composed . . .

. . . . The teleological and the mechanical views of nature are not, necessarily, mutually exclusive. On the contrary, the more purely a mechanist the speculator is, the more firmly does he assume a primordial molecular arrangement of which all the phenomena of the universe are the consequences, and the more completely is he thereby at the mercy of the teleologist, who can always defy him to disprove that this primordial molecular arrangement was not intended to evolve the phenomena of the universe ([4], vol. II, pp. 201–202).

This theory of "design on the installment plan," as John Dewey ridiculed it, is perhaps only a vestigial rudiment of the full blown account, but it remains undefeatable.

As for Aristotelian teleology, at least insofar as it rests on the eternity of the species, it has suffered perhaps even more, although if one considers the world of our experience – and the world since the beginnings of civilization – and ignores the immense lapse of time that preceded, the *stability* and *permanence* of the *eidē* in generation is by far the most impressive phenomenon, much more so than the appearance of significant novelty. Nevertheless, if Darwin is right about descent with modification – and, no doubt, he is right about that – then the *eidē* are not eternal. And if he is right about natural selection, the *eidē* may be mere outcomes of a non-teleological nature, even though they still can be the principles (*archai*) or the ends (*telē*) of change that rule over the generation, growth, and reproduction of individual organisms. Species come and go – albeit very, very, very slowly. In the beginning, there was only one or perhaps a few species – out of these have come millions of species. And if Darwin is right about natural selection, the process is not yet finished; evolution continues and will continue. It is a process without a clear directedness, with no definite term or end or goal, and *a fortiori*, not a process for the sake of its completion.

Let me emphasize that each of the two parts of Darwin's theory has significant yet distinctive implications for teleology. Although any showing of the transformation of species would be fatal to notions both of the eternity of species and their special creation as fixed species, [5] it is Darwin's special theory, natural selection, which, at least on first, and also on second glance, seems to rule out even an ordered and goal-directed unfolding of created nature. Lamarck and others had put forth theories of evolution, of unfolding, but the changes that appeared in Lamarck's account, even those in response to change in the environment, were the results of nature working from *within* organisms. For Lamarck, organic change was purposive, directed either for the sake of adaptation to changed conditions, or toward the goal of achieving greater complexity. For Darwin, descent with modification was not an evolution – an folding of a prearranged plan – and the term "evolution" does not occur in *The Origin of Species.*[6] The changes that occur, or more precisely, the changes that are preserved, are not due to causes *within* organisms, but rather are the result of competition *between* and *among* organisms. There is no necessary, let alone goal-directed, connection between the variations which nature sportively tosses up and the circumstances which deem some of them "advantageous" and which lead to their preservation.

Thus, to sum up, though plants and animals and man are purposive beings, they are not nature's purposes, nor anyone else's. Moreover, they are part of a process that began by accident, proceeded by necessity, and has no end. Nature is blind and dumb, aimlessly but persistently going nowhere.

This, at least, is what the theory of natural selection tells us. But is it true? Are all aspects of evolution really without a direction and really open-ended? We should have another look at *The Origin of Species* and also at the evidence.

IV. *THE ORIGIN OF SPECIES* RECONSIDERED: THE DIRECTIONS OF EVOLUTION

The Origin of Species is replete with teleological terms and passages, both explicit and implicit, not only about the functioning of individual animals but also about the overall course of evolution. Terms such as "useful," "important," "purpose," "adapted," "fit," "the good of each being," "profitable," "harmful," "beneficial," "injurious," "advantageous," "good," "tendency," "success," "welfare," "improvement," "perfection," "low" and "high" in the "scale of nature," and "absolute perfection" occur frequently, almost on every page. Most of these terms are treated as self-evident and apt; Darwin apparently felt no need to define or discuss most of them nor to defend their use. One might say they flowed naturally into his account.

Consider some selected passages that illustrate the range of these teleological references. (1) On *adaptation* and its *excellence*: "How have all those exquisite adaptations of one part of the organisation to another part, and to the conditions of life, and of one distinct organic being to another being, been perfected?" ([5], p. 114). (2) On *directedness*: "Every organic being is constantly endeavoring to increase in numbers," ([5], p. 217). (3) On *absolute perfection*: "Natural selection will not necessarily produce absolute perfection; nor, as far as we can judge by our limited faculties can absolute perfection be everywhere found" ([5], p. 233); and again: "The wonder indeed is, on the theory of natural selection, that more cases of the want of absolute perfection have not been observed;" ([5], p. 445). (4) On *purpose* and its *alterability*: "The swimbladder in fishes. . . . shows us clearly the highly important fact that an organ originally constructed for one purpose, namely flotation, may be converted into one for a wholly different purpose, namely respiration;" ([5], p. 220). (5) On the *tendencies* of natural selection: "There will be a constant tendency in natural selection to preserve the most divergent offspring of any one species" ([5], p. 444); and again, "Hence we may look with some confidence to a secure future of equally inappreciable length. And as natural selection works solely by and for the good of each being, all corporeal and mental endowments will tend to progress towards perfection" ([5], p. 459).

Even the sub-title of the book invites teleological consideration: "The preservation of favoured races in the struggle for life." Did Darwin really mean to say the "preservation of *favoured* races," rather than "the preservation of presered races" or simply, "the preservation of races?" What is favored about the "favoured races"?

Many of these teleological terms and passages are difficult to understand, especially in the light of the theory Darwin is propounding. Most of Darwin's followers have purged these notions from the theory. The current neo-Darwinian theory, which defines natural selection as differential reproduction (i.e., leaving more offspring) rather than differential survival, and which focuses on gene pools and gene frequencies rather than on plants and animals, makes these notions seem even more problematic. Nevertheless, the terms and passages appear in Darwin's book in abundance. We must, like good scientists, ask why. Did Darwin intentionally include these teleological references? Were they deliberate attempts to coat the bitterness of his teaching, either for himself or for his readers? Were they rhetorical devices to win adherents? Was Darwin simply sloppy of speech, confused, or undecided? Was he merely the victim of a flowery nineteenth century English prose? Or was Darwin, the devoted naturalist and lover of nature, simply recording his impressions of the appearances, of a nature whose beings indeed shine forth "that perfection of structure and co-adaptation which most justly excites our admiration"?

We can not readily answer this question, at least not in general. I suspect that in some cases, Darwin was indeed merely whistling in the dark, e.g., in the surprising last sentence of the chapter on the "Struggle for Existence":

> When we reflect on this struggle, we may console ourselves with the full belief, that the war of nature is not incessant, that no fear is felt, that death is generally prompt, and that the vigorous, the healthy, and the happy survive and multiply ([5], p. 129).

In other cases, it is clear that Darwin is undecided, while in yet others, he seems to use certain terms against his own strictures (e.g., a memorandum in his copy of *The Vestiges of Creation* says "Never use the word 'higher' or 'lower' ").

But it may be more interesting to consider the meaning of these teleological passages than to continue to speculate about how they came to be in Darwin's book; knowing their origins can not teach us whether or not they are true. It may turn out that we should be grateful to Darwin for recording and preserving his intuitions, his sentiments, his uncertain musings, or even only his fond hopes, which a more scrupulous devotion to his theory might have caused him to delete. I propose to examine two such passages, one in which I will argue Darwin was teleological without adequate justification, a second in which he was not teleological enough. The first passage deals with the small workings of natural selection at the level of individual animals; the second deals with the full-sweep of evolution.

Darwin says in many places that "natural selection can act only through and for the good of each being" ([5], p. 133).

> As man can produce and certainly has produced a great result by his methodical and unconscious means of selection, what may not nature effect? Man can act only on external and visible characters: nature cares nothing for appearances, except in so far as they may be useful to any being. She can act on every internal organ, on every shade of constitutional difference, on the whole machinery of life. Man selects only for his own good; *Nature only for that of the being which she tends* ([5], p. 132; emphasis added).

Let us leave aside what Darwin means by Nature, and also the vexed question of whether Natural Selection is indeed an *agent* of change, i.e., whether Natural Selection is the name of a cause or of a result, and let us consider the meaning of the phrase "the *good* of *each being*." There are two problems: what is *good*? and *whose* good is it?

The chief good that appears to rule over natural selection is *survival*. But this is a good which natural selection can but very imperfectly give to any living being, since every living being dies. Moreover, those variations which give it advantage over its competitors arise not by natural selection but by mutation; *if* these variations are indeed beneficial, and *if* they are inherited, "natural selection" means that they will be preserved and conferred as "benefits" *on the offspring*. Thus, each individual may be said to be the beneficiary of *previous* natural selection, but no *living* being is tended or served by the *current* action of natural selection.

Then perhaps one should say that natural selection acts for the good of each *kind* of being, i.e., for the good of the *species*. Temporarily, even for long periods of time, this may be so. But for this improvement to *remain* an improvement depends upon the stability of the environment: introduce climatic changes or new plants or animals, and yesterday's gift of natural selection may today turn out to be a kiss of death. Moreover, in the long run, the species itself is not a permanent or stable entity; this is the whole point of the story. Species, like individuals, come and go. Enough natural selection will produce not an improved but a different species. "Life" continues, but beings *and* kinds equally perish. If survival is the good, it cannot be *their* good. Is there possibly some other "being" whose good it could be?

Our difficulty in identifying *whose* good is survival must lead us to reconsider whether in fact mere survival can be the good, or the only good, of living things? If permanence and stability are the end, why should life have appeared and persisted at all? I quote from Whitehead:

In fact life itself is comparatively deficient in survival value. The art of persistence is to be dead. Only inorganic things persist for great lengths of time. A rock survives for eight hundred million years; whereas the limit for a tree is about a thousand years, for a man or an elephant about fifty or one hundred years, for a dog about twelve years, for an insect about one year. The problem set by the doctrine of evolution is to explain how complex organisms with such deficient survival power ever evolved ([9], pp. 4–5).

The same point is picked up and enlarged by Hans Jonas. Life, he says,

is essentially precarious and corruptible being, an adventure in mortality, and in no possible form as assured of enduring as an inorganic body can be. Not duration as such, but 'duration of what?' is the question. This is to say that such 'means' of survival as perception and emotion are never to be judged as means merely, but also as qualities of the life to be preserved and therefore as aspects of the end. It is one of the paradoxes of life that it employs means which modify the end and themselves become part of it. The feeling animal strives to preserve itself as a feeling, not just a metabolizing entity, i.e., it strives to continue the very activity of feeling: the perceiving animal strives to preserve itself as a perceiving entity – and so on. Without these faculties there would be much less to preserve, and this *less* of what is to be preserved is the same as the *less* wherewith it is preserved ([8], p. 106).

If natural selection aims at some good, if evolution has direction, it must go beyond mere survival.

Perhaps we can understand better how evolution works in the small, if we can see where it is headed in the large. Though Darwin speaks little of the overall directions of evolution, and occasionally cautions us against trying to generalize, we can extract from Darwin's own words that there are at least three prominent tendencies in the evolution of life, which I shall name *Diversity, Plenitude,* and *Ascent: Diversity,* the tendency to greater and greater variety ("Nature is prodigal in variety, though niggard in innovation",

[5], p.445); *Plenitude*, the tendency to more life; *Ascent*, for now, a mystery. Diversity and Plenitude are related. Natural selection tends to produce what Darwin called "the divergence of character" – more and more widely differing species; and divergence of character leads toward plenitude, i.e., more organisms:

The advantage of diversification in the inhabitants of the same region is, in fact, the same as that of the physiological division of labour in the organs of the same individual . . . In the general economy of any land, the more widely and perfectly the animals and plants are diversified for different habits of life, so will a greater number of individuals be capable of there supporting themselves ([5], pp. 158–159).

More animals, more species, tending to complete and fill the democratic polity of nature.

But Darwin speaks not only of the range and breadth of animal profusion. Without proper justification from his theory, and not without some reticence, Darwin also records for us the notion that some living beings are *higher* than others:

Thus, from the war of nature, from famine and death, the most exalted object which we are capable of conceiving, *namely the production of the higher animals, directly follows* ([5], p. 459; emphasis added).

What does Darwin mean by "higher"?

It is hard to be sure what he means, generally and in this passage, by "the higher animals." In some passages, the term seems to imply a ranking of structure and activity. For example:

The embryo in the course of development generally rises in organisation: I use this expression, though I am aware that it is hardly possible to define clearly what is meant by the organisation being higher or lower. But no one probably will dispute that the butterfly is higher than the caterpillar ([5], pp. 420–421).

and again,

But in some genera [of cirripedes] the larvae become developed either into hermaphrodites having the ordinary structure, or into what I have called complemental males: and in the latter, the development has assuredly been retrograde; for the male is a mere sack, which lives for a short time, and is destitute of mouth, stomach, or other organ of importance, excepting for reproduction ([5], p. 421).

But in other cases, "higher" seems merely to mean "the latest and the surviving":

There has been much discussion whether recent forms are more highly developed than ancient. I will not here enter on this subject, for naturalists have not yet defined to each other's satisfaction what is meant by high and low forms. But in one particular sense the more recent forms must, on my theory, be higher than the more ancient; for each new species is formed by having had some advantage in the struggle for life over other and preceding forms. If under a nearly similar climate, the eocene inhabitants of one quarter of the world were put into competition with the existing inhabitants of the same or some other quarter, the eocene fauna or flora would certainly be beaten and exterminated; as would a secondary fauna by an eocene, and a palaeozoic fauna by a secondary fauna ([5], pp. 336–337).

Yet despite Darwin's hesitancy when he explicitly considers the meaning of the term "higher," and despite the fact that his theory only clearly permits the application of terms "higher" and "lower" on the basis of success in the struggle for life, Darwin himself appears to be sensitive to, and, at the very least, unwittingly points *us* toward a higher view of "higher." We are led to remember our common sense view that some survivors are higher – i.e., more fully developed and more fully alive – than others, that there is a hierarchy among living beings, that moss, barnacles, grasshoppers, and chimpanzees, though all equally prospering and successful, are nevertheless unequal in some decisive respects. Moreover, if one looks at the appearance of organic beings over the span of evolution, it is clear that there has been, in general, an upward trend, and this despite the fact that the life of the higher forms is often more precarious than the lower. To be sure, "low" forms persist and flourish, but higher and higher types have successively appeared.

Higher types in what sense? From here on, I shall present, in a crude and sketchy fashion, some free speculations, which, though not foreclosed by Darwin's theory, go beyond it in some important respects. To repeat, "higher types in what sense?" *Higher in terms of soul.* Somehow, as thousands upon thousands of species came and saw and went under, the processes of evolution not only produced more and more organisms, new species and new forms – or, in other words, produced both more soul and more forms for the expression of soul – but also *higher grades* of soul.[7]

Though the first beginning of life lies in darkness, we may surmise that the first organisms could only maintain and reproduce themselves; theirs was the life of nutrition and growth. After this, there were only two or perhaps three more *big* "surprises" in the hierarchical history of soul: (1) the emergence of sensitivity and awareness, i.e., of the sensitive soul, (2) the emergence of locomotion, and with it, desire, i.e., of the full-fledged animal soul, and finally, (3) the emergence of speech and intellect, of the rational soul.

It is not enough to call these new and distinct powers merely more complex or more organized forms of life: they represent and make possible new and essentially different ways of life. Sensing is not just a more complicated form of nourishing; thinking is not just a complicated form of sensing or moving. Though these higher kinds of soul may have emerged, and probably did emerge, gradually by cumulative differences of degree, the difference of degree eventually became a difference in kind. A certain critical threshold in development and organization was reached and crossed, and a new activity appeared. A ready analogy for such a gradual emergence of new levels of soul is available in the development of a human being from a single cell – the fertilized egg – which, as far as we know, neither senses nor desires nor thinks.

The ascent of soul has meant the possibility both of an ever-greater awareness of and openness *to* the world, and of an ever-greater freedom *in* the world. The growth of soul has produced a hierarchy among living things in their powers both to be affected by and to affect the world. The hierarchy of soul is a hierarchy of openness and purposiveness. Touching, smelling, tasting, hearing, seeing, moving toward, exploring, desiring, remembering, judging, speaking, naming, thinking – all these forms of *openness* to things, to others, and to ourselves are the pride of the human soul. The freedom from immediate preoccupation with necessity provided by human reason and its arts, and the inner awareness of this freedom and of our purposiveness, permit us the opportunity to set our own *purposes*, to have goals, plans, dreams, decisions, projects, and activities that soar far beyond mere survival. Yet to claim these things for man is not to deny some or many of them to animals or even plants. Some openness and purposiveness is seen all along the rise, even at the bottom.

V. THE END OF THE CLIMB: BEYOND CHANCE AND NECESSITY?

If we have correctly identified soul as at least one "subject" or "being" of evolution, and if we have correctly identified, historically, an emergence of ever higher forms of soul in evolution, we can now ask some teleological questions about this overall process. Has evolution itself an end, a completion? Is this end a cause, or an effect of some other cause, or a mere outcome of some accidental process? Are openness and purposiveness in some sense purposes of nature? (These questions, though prompted by Darwin, are foreign to his thinking though not to Lamarck's.)

First, is there a natural term to evolution at least in its tendency of Ascent? The answer would appear to be man; at least it is hard to see how it could be anything else. Man is the peak, both in possessing the *highest*, and also in possessing the *complete range*, of faculties of soul. Even looking to the future, what could be higher than man? To be sure, many more species will come and go. Man himself will no doubt undergo many changes. Certain changes in the human constitution have been occurring fairly rapidly over the past century, albeit more due to human interventions than to natural selection in a strict sense: the average height has risen, the average age of menarche has decreased. The frequency of genes for myopia and diabetes in the human population continues to rise, as eye glasses and insulin have compensated for the disadvantage of these genetic defects. In the future, modern science may find a way to graft chlorophyll-containing tissue onto our foreheads so that we may make our own food, or, alternatively, to endow us with paper-digesting enzymes so that we may more fully digest the great books we so much enjoy. But even these somewhat farfetched possibilities would represent mere wrinkles of shape and form, compared to the great powers of soul. Like air-conditioning and power-steering, they would add nothing essential to the Model-T. Granted, we could be, and might in time become, more intelligent, more alert, less forgetful, more energetic, etc., but can we imagine for ourselves or for soul anything really new? What great innovations can we imagine?[8]

That we, or at least I, can't imagine any startling new faculties of soul may indeed point to a deficiency in our imaginations that the next great emergence will rectify. And one should never be comfortable predicting the end of change or history – though I can be bold with impunity, knowing that I will not be proved wrong in my lifetime. But in fact, there *are* a few things we can imagine, but some are not *decisively* novel, and others strike me as out of reach. To the former class belong new faculties of sensation, responsive to X-rays, radio waves, television, gravity, and magnetism, and to whatever it is that those people with extra-sensory perception now perceive, if they do indeed perceive. To the latter class belong bodily immortality and the ability to be in two places at the same time. And while both of these fantastic possibilities would radically change the nature of things, they might not greatly add to the kinds of activities we might pursue forever or in two places at once. For these reasons, I suggest that with nutrition, reproduction, sensation, appetite, imagination, emotion, and intellect, the story of the *Ascent* of soul may already be complete, though improvements can surely be made, and though evolution and its tendencies to Diversity and Plenitude may continue without bound.

What is responsible for this upward trend in evolution? This remains an important and fascinating question, even if I am roundly mistaken in suggesting that the ascent has reached the summit. Let us recall the distinction between the origin and the preservation of novelty, between the first appearance of a 'higher' form of life – say the first sensitive being, the first animal with the rudiments of touch – and the preservation of that higher form. Let us provisionally say that the origin of sensitivity was due to chance. But is it due to chance that sensitivity was preserved or that it began to flourish? Could it not be said that, if it is a sensible world, sensing beings would be at home in it, or to put it in Darwinian terms, sensing beings would – other things being equal – have an advantage over their insensitive forebearers? Ought we be surprised, should we regard it as an accident, that, in a visible, odorous, and sounding world, the powers of sight, or smell, or hearing once they appeared should have been preserved, magnified, perfected? Likewise with intellect. However accidentally intellect first appeared, is it surprising that it should have been preserved in a world of cause and effect, past and future, means and ends, all of which can be brought to consciousness and used to advantage in a being endowed with memory, a sense of time, self-awareness, and the ability to order means to ends in securing the future? If it is an intelligible world, is it surprising that an intelligent being, once one appears, will be at home in it, or to put it in the less complete terms of survival, will be likely to survive and flourish? Is it not only not surprising, is it perhaps even necessary that such a world would be hospitable for the maturation of soul, for evolutionary ascent?

Having gone this far, let me try out a most speculative speculation: Let me suggest that the kinds and levels of soul complement and answer to the kind of a world this is, and that by evolving to complement the things that are, soul *completes* the things that are. Are not the looks and beauty of a flower incomplete until there exists a seeing being open to and aware of this beauty? Are the laws of nature in their intelligibility fully intelligible, and hence *fully themselves*, before they are discovered by an intelligent being? And is the unfolding and emergence of soul complete until there exists a being who discovers the emergence of soul, and what is more important, understands soul, independent of whether and how it emerged?

Now one may ask, "What have the sense of beauty or knowledge of nature, or for that matter, philosophy or music or mathematics or morals, – what have all these things to do with natural selection?" Surely, these activities of human beings, which we admire and which are dependent on human rationality, go well beyond the uses of reason that may have enabled intelligent beings to survive. Even if we grant that reason was not accidentally preserved – that intelligence, in the form of arts and cunning and planning, naturally helped its possessors to survive – is it not an *accident* that these powers of intelligence were turned to other uses? Or is it possible, or even likely, that faculties which are first preserved for the sake of living already have or soon acquire, so to speak, a life of their own, and continue to flourish for the sake of living well?

Let us turn, in some closing speculations, from preservation and flourishing to the first appearance of life and of each level in the ascent of soul. Is it true, as we provisionally asserted, that the beginnings were due to chance? What *is* due to chance is, perhaps, the time and circumstance in which life first arose, and also the beings in which the first steps of further ascent were taken. But what emerged on these occasions were possibilities already present as possibilities in the lower forms of life, and ultimately, in matter itself. Everything present to the higher animals, including man, must have been "present" *potentially* in pre-animate matter. Matter, if not actually alive, was potentially alive; given the right circumstances, it came alive *on its own.* The accident of circumstance merely *released* the *inherent capacities* of matter to become organic, to form structures which produce and maintain and reproduce themselves. Analogies may be found in the releasing of developmental capacities by the chance union of egg and sperm, or in the bursting into flame of wood brought to the kindling temperature, which flame maintains itself as long as further material is available for combustion, or in the releasing of the potentiality to form water caused by the chance joining of hydrogen and oxygen gases. Just how this releasing took place, just what circumstances made possible the emergence of life, we may never know. But at least formally, some account like this one must be correct. Matter must not be thought of simply as inert and passive, as resistance to acceleration, as little billiard balls, as waves or particles. Matter has character; matter has possibility; matter is prefiguratively alive. For all his errors on this subject, Aristotle was perhaps closer to the truth even about matter than are the modern physicists:

> Animals and their parts exist by nature, and so do plants and *the simple bodies*, for example, earth, fire, air, and water; for we say that these and other such exist by nature: . . . All things existing by nature have *in themselves* a principle of motion and of rest . . . ([1], p. 25; emphasis added and translation amended).

Whether or not Aristotle correctly identified these principles of motion and rest, and whether or not he correctly identified the material elements, the insight that the elements no less than the plants and animals were self-moving – each in its own way – strikes me as fundamentally sound, and useful in thinking about the connectedness of nature lifeless and nature alive.

To be sure, pre-animate matter probably had numerous potentialities including many that were never realized – although if my earlier account is correct, most of the most interesting possibilities may have been perfected. One probably cannot infer, strictly speaking, that the emergence of soul and of the human soul was *necessary* simply from the fact that it happened. One probably cannot even say that life or intelligent life would come to be, if nothing obstructed. One comes closer to the truth, perhaps, if one says that life, then sentient life, then intelligent life came to be and would most likely come to be if circumstances were not unfavorable to their release. Given enough time, given a durable deck of cards, given a tireless dealer and shuffler, sooner or later the cards will be dealt out in the right order, Ace-to-King, spades, hearts, diamonds, and clubs – if not always, at least in most such protracted dealings. In and from the beginning, the ascent of soul was, as they say, in the cards.

The University of Chicago,
Chicago, Illinois

NOTES

[1] Originally a Formal Lecture delivered on October 11, 1974, at St. John's College, Annapolis, Maryland. The support of the National Endowment for the Humanities (grant AI-9919-73-428) is gratefully acknowledged.

[2] A more complete defense of this assertion would examine the following questions, among others: (a) *On explanation:* What do scientists and philosophers of science mean by "explanation"? Is "to know" or "to understand" identical with "to explain"? If explanation is to be understood in opposition to "description," is it true that description is no part of understanding? (b) *On being and becoming:* What or how much can we learn about *what* something *is* from learning about its genesis? Can one ever deduce "the what" from "the out-of-what" or "the from-what"? For example, knowing only that Hector is "son of Priam," can we know or predict that he is also "breaker of horses"? Can one, in principle, know what a chicken is from knowing the full base sequence of chicken DNA? Or does the *what* of a chicken – its looks and its power – require an account of its own? (c) *How to put the questions about aliveness and wholeness:* Is not the mechanist–vitalist quarrel misconceived and inappropriate to the phenomena of life? Vitalism can only be understood in relation to mechanism, to which the vitalists themselves adhere in dealing with non-living things. But here both sides err, for both believe that material and efficient causes are sufficient to fully understand, say, machines. Therefore, the vitalists are obliged to try to account for living things in terms of some mysterious, thing-like, addition to a machine (see, e.g., the speculative writings of Hans Driesch). But is not the question about the purposiveness of living things (and machines) better formulated in terms of part and whole, or material and form?

[3] It is difficult to translate the Aristotelian notion of *eidos* (plural, *eidē*). The usual translations, "form" or "idea," are bound to be misleading, given the long history of

these English words as technical philosophical terms. Though clumsy in English, "looks" would be a more appropriate translation. It would preserve Aristotle's appreciation that the distinctive look of an animal reveals what it is, that the "inside" shines forth and makes itself known in its presentation on and to the "outside" – in its appearance and in its visible activities. *Eidos* refers both to the *what* of the particular animal, as distinguished from its *hyle*, or material, *out of which* it is composed, and thus also to the *kind* or *species* of animal it is. Each dog is what it is because it is dog. See J. Klein, "Aristotle, An Introduction", in J. Cropsey (ed.): 1964, *Ancients and Moderns*, Basic Books, New York, pp. 50–69. For an excellent modern account of animal "looks" and their relation to animal being, see A. Portmann: 1967, *Animal Forms and Patterns* (transl. by H. Czech), Schocken, New York.

4 Consider also the following similar passage, from Darwin's correspondences with Gray: " . . . I cannot see as plainly as others do, and as I should wish to do, evidence of design and beneficence on all sides of us. There seems to me too much misery in the world. I cannot persuade myself that a beneficent and omnipotent God would have designedly created the Ichneumonidae with the express intention of their feeding within the living bodies of Caterpillars, or that a cat should play with mice. Not believing this, I see no necessity in the belief that the eye was expressly designed. On the other hand, I cannot anyhow be contented to view this wonderful universe, and especially the nature of man, and to conclude that everything is the result of brute force. I am inclined to look at everything as resulting from designed laws, with the details, whether good or bad, left to the working out of what we may call chance. Not that this notion *at all* satisfies me. I feel most deeply that the whole subject is too profound for the human intellect. A dog might as well speculate on the mind of Newton. Let each man hope and believe what he can" ([4], Vol. II, p. 312).

5 Would not Aristotle have been greatly perplexed by the very title of Darwin's work, *peri eidōn geneseōs*, the genesis of the *eidē*? (Though *genesis* is not a proper translation of "origin," Darwin's book is really about "origin" understood as *process* – coming-to-be – rather than as *source*. *The Genesis of Species* would have been a more precise title.)

6 The word "evolved" occurs once; it is the very last word in the book: "There is grandeur in this view of life, with its several powers, having been originally breathed into a few forms or into one; and that, whilst this planet has gone cycling on according to the fixed law of gravity, from so simple a beginning endless forms most beautiful and most wonderful have been, and are being, evolved" ([5], pp. 459–460). One can only suspect that Darwin's theological waverings were here leaning toward belief; in later editions he inserted "by the Creator" after "originally breathed." This may account for the novel and unprepared for use of the term "evolved."

7 The reader may well be puzzled by the sudden appearance – not to say resurrection – of the notion of soul, a notion long deemed either meaningless or useless in biology (though preserved, albeit with shrunken meaning – consciousness or mind – in psychiatry and common discourse, as *psyche*). By soul I mean nothing mystical or religious, not a disembodied spirit or person, not one of Homer's shades, not a ghost in a machine, not, indeed, a "thing" at all. Rather, following Aristotle's reflections in *Parts of Animals, Book I*, and in *On the Soul, Book II*, I understand by soul the integrated vital powers of a naturally organic body, always resident in such a body while it is alive ("animated", "ensouled"), even when such powers are not actively at work (e.g., in sleep, before

maturation). The subsequent paragraphs in the text will, I hope, make my meaning more clear. I should emphasize that, thus understood, soul is not a possession of man alone, as it seems to have been for Descartes. All living things have soul. It is precisely Darwin's theory, which regards the life of man as continuous with the rest of living (and non-living) nature, that invites one to reconsider whether a notion of soul might be necessary in order to understand the aliveness of all living things.

[8] This is not to deny the great influence of changing cultural conditions on the way man lives. Nor do I mean to rule out great future changes in social organization, including improved social and technical means for harnessing native human capacities.

BIBLIOGRAPHY

1. Aristotle: 1969, *Physics* (transl. by H. G. Apostle), Indiana University Press, Bloomington, Indiana.
2. Butler, S.: 1878, *Life and Habit*, A. C. Fifield, London.
3. Darwin, C.: 1936, *The Descent of Man*, Modern Library Giant edition, Random House, New York.
4. Darwin, C.: 1888, *Life and Letters*, F. Darwin (ed.), 3 vols., John Murray, London.
5. Darwin, C.: 1968, *The Origin of Species* (first edition), Pelican Classics edition, J. W. Barrow (ed.), Penguin Books, Harmondsworth, Middlesex, England.
6. Dewey, J.: 1910, *The Influence of Darwin on Philosophy and Other Essays in Contemporary Thought*, Henry Holt, New York.
7. Gray, A.: 1874, 'Charles Darwin: A Sketch', *Nature*, June 4, 1874; reprinted in A. Gray: 1963, *Darwiniana*, Harvard Univ. Press, Cambridge.
8. Jonas, H.: 1968, *The Phenomenon of Life*, Dell Publishing Co., New York.
9. Whitehead, A. N.: 1962, *The Function of Reason*, Beacon Press, Boston.

MARJORIE GRENE

INDIVIDUALS AND THEIR KINDS: ARISTOTELIAN FOUNDATIONS OF BIOLOGY[1]

The Aristotelian universe is eternal, not developing. Man begets man, for ever and ever. Professor Balme would adjure us to remember Libya, where individuals of differing species hybridize [1, 12]; still, this remains a small anomaly in a generally stable nature, in which species imitate the imperishable spheres by going round and round in their own imperfect way. Insofar as he is not an evolutionist, moreover, Aristotle's biological thought remains irreconcilably at odds with ours. Yet at the same time, while evolutionary thought, that is, in effect, Darwinism, remains the comprehensive framework for advancing biological knowledge, there is something in the philosophical foundations of modern thought that is disturbingly anti-biological and something in Aristotelian ontology that somehow seems better suited, if not to the practice of biochemistry, bioenergetics and the like, still to the pursuit of the more traditional branches of biology, which do still exist and on which in a sense their more sophisticated sister disciplines depend. This disparity makes me uncomfortable; I have tried a number of times to approach the question, what Aristotle has to teach the modern metabiologist, but the answer, perhaps the correct formulation of the question, still eludes me. Without much hope of succeeding better than before, let me try once again, relying chiefly on Aristotle's programmatic statements in P.A. I, together with some of his remarks about the τὶ ἦν εἶναι and about form and matter in *Met Z–θ*.

Richard Owen in his lecture "On the Nature of Limbs" declared:

> The recognition of an ideal Exemplar for the Vertebrated Animals proves that the knowledge of such a being as Man must have existed before Man appeared. For the Divine mind which planned the Archetype also foreknew all the modifications. The Archetypal idea was manifested in the flesh, under diverse modifications, long prior to the existence of those animal species that actually [that is, now] exemplify it. To what natural or secondary causes the orderly succession and progression of such organic phenomena may have been committed, we are as yet ignorant. But if, without derogation to the Divine Power, we may conceive the existence of such ministers and personify them by the term *Nature*, we learn from the past history of our globe, that she has advanced with slow and stately steps, guided by the archetypal light amidst the wreck of worlds, from the first embodiment of the vertebrate idea, under its old ichthyic vestment, until it became arranged in the glorious garb of the human form ([10], p. 212).

Ever since then, I suspect, biologists, and especially evolutionary biologists,

S. F. Spicker (ed.), Organism, Medicine, and Metaphysics, 121–136.

have been terrified of falling into Owen's typological trap. To Darwin in particular it was to be a bitter disappointment that Owen – whose description of the homology of the vertebrate limb seems to provide one of the pillars of evolution – should stand so firmly and even angrily opposed to his theory. And the fear of sounding like Owen has thus come to haunt biological thinkers, so that any one who takes any quasi-"essentialist" stand is declared to be typological in this vicious sense. Biologists have therefore been inclined to go along with a style of thinking about science that in fact invalidates the foundations in taxonomic and experimental practice of their own kind of knowledge. For biological procedures depend in the last analysis on the biologist's confidence that nature itself is divided in some way congruent with the way he cuts it. *Some* sort of belief in natural kinds seems to be fundamental to the procedures of biology. And if we look for support for such a belief in the history of philosophy, we do seem to find it first and foremost in Aristotle or in thinkers derivative from Aristotle.

There are two obvious strengths in Aristotle's position as a metabiologist. First, he provides, if with his usual terseness, a conceptual apparatus for acknowledging the existence of individuals of the same kind, and second, he recognizes that what is by nature is "always or for the most part," that some individuals in fact fall short of the perfect fulfillment of the characteristic life-style of their kind, and that therefore biological cases need to be assessed, normatively, with respect to a standard of what is right for the kind in question. Yet both these strengths seem to carry with them, if not weaknesses, at least obscurities. First, if it is Coriscus and Socrates who exist, yet each of them *is* primarily man – his *atomon eidos* – and if this form is the same in both of them, one wonders what sort of "same" this is. One needs to ask, in other words, what really is involved, both ontologically and methodologically, in recognizing a specimen of a species. And second, given that it is its *eidos* that makes the individual knowable as the individual it is, and even makes it what it is, one wonders how it is that one can distinguish scientifically between better and worse specimens of the same *eidos*. For if it is the *eidos* that both is and is knowable, and if the *eidos* is identical over its instances, how can individuals of the same *eidos* be said to differ in ways that are cognitively important or even recognizable? It almost looks as if a satisfactory account of strength one would defeat a satisfactory account of strength two and *vice versa*.

What I want to do here is to reflect a bit further, in a rather rambling fashion, about these Aristotelian claims (or implicit claims) and the relation between them: that is, the claim that each natural entity *is* in some fundamental way its τῆε and the claim of what one may call, following Canguilhem's usage, biological normativity [2]. In conclusion I shall have to append a few remarks about yet another difficulty: the relation of Aristotle's concept of *eidos* and the τῆε to some of his statements about perception.

First, with respect to the τῆε and the cluster of concepts that go with it (form, formal cause, actuality and *telos*), I want to approach Aristotle's method, and its underlying ontology, by contrasting with it the kind of basically Humean methodology that seems to be presupposed by much if not most modern thinking about the procedures of the natural sciences and *a fortiori* of the biological sciences, the latter especially in view of the terror of typology to which I have already alluded.

What I want to contrast is, on the one hand, a concept of scientific method (and hence also biological method) as starting with particulars and moving *via* their resemblance to general laws about them and, on the other, a concept of biological method that begins with organized individuals and recognizes each of them as belonging to one and the same kind. In the one case we have particulars resembling one another in respect to some property or other, in the other case individuals individuated as the very individuals they are through belonging to the same kind. (Let me admit in passing that matter is what enables us to distinguish individuals of an identical kind yet remind you at the same time that the individual is what he is, the very being he actually is, primarily in virtue of *eidos* rather than *hyle*.)

It is the first of the two models I have just mentioned that has been canonical for modern thought. As compared with the second, Aristotelian methodology, I want to argue, it inevitably fails to provide an adequate conceptual basis for the knowledge of living things. Let me look briefly at its operation, first in Hume, then in Kant, and then, even more sketchily, in recent thought about biology.

Every simple idea is derived from a simple impression of which it is a faithful, if fainter, copy. This correlation, together with the constant conjunction (i.e., contiguity) of some impressions and ideas, accounts, compulsively, for the catalogue of beliefs we come to hold. Simple impressions, the ultimate units of experience, are bare particulars, the *minima sensibilia* Berkeley had so confidently sought for. Some of these resemble one another, and when the resemblances go in pairs we come to expect a resembling idea (or impression) of one sort to follow one of the resembling impressions (or ideas) of another sort. We have had a_1, b_1, . . . a_2, b_2. . . . a_n, b_n, so given a_{n+1}, we expect b_{n+1}. But what sort of sorts are these? How do resembling simples group themselves together? Presumably they resemble one another with respect to some one identical property. If we have only simple impressions and ideas of which to compose experience, however, how can we recognize an identical property? Simple impressions and ideas, to begin with, are declared wholly to resemble one another: in respect to what? Suppose Imp_1 is a least bit of yellow. (If I were really a Humean I could not even say that, indeed I could not say anything, but that is beside the point.) Then id_1 is a least bit (but fainter) of the same yellow. What sort of idea (impressions?) is "same"? The treatment of identity in *Treatise*, Book One, Part Two and under philosophical relations in Part Three wholly evades this question. An identical impression or idea is just a persistent one, not the same one at a different time, let alone a different one of the same kind. The only attempt at a reply to the question, it seems to me, is the treatment of the distinction of reason in Part One: the white (or black) globe (or cube) of marble. But that is a very unsatisfactory treatment. If we can really recognize the same or different shape or color apart from the sequence of aggregates of simples, then there is something more to experience than Hume's description allows. And if we cannot, then we are already committed, as we soon will be in any case in Part Three, to blind custom as the guide of human life. We have no rationale for justifying our belief in an external world, or other persons, let alone plants and animals, whose self-ordering, as distinct from our associative ordering of them, we might confidently recognize. The familiar moves of modern philosophy have begun: the problems of counterfactual conditionals, or confirmation theory, of grue or bleen or brown ravens: it is all contained in the effort to make resembling particulars shuffle themselves into a pattern of experience. And the resulting pattern, see Part Three, Section 14, is our habit, not nature's. As Leszek Kolakowski puts it:

> The picture of reality sketched by everyday perception and scientific thinking is a kind of human creation (not imitation) since both the linguistic and the scientific division of the world into particular objects arise from man's practical needs. In this sense the world's products must be considered artificial. In this world the sun and stars exist because man is able to make them *his* objects, differentiated in material and conceived as 'corporeal individuals'. In abstract, nothing prevents us from dissecting surrounding material into fragments constructed in a manner completely different from what we are used to. Thus, speaking more simply, we could build a world where there would be no such objects as 'horse', 'leaf', 'star', and others allegedly devised by nature. Instead, there might be, for example, such objects as 'half a horse and a piece of river', 'my ear and the moon', and other similar products of a surrealist imagination ([5], pp. 57–8; [14], p. 358).

This seems a poor grounding for any science, let alone biology.

Even if we could introduce the concept of *one* identical property ranging over particular simple ideas, moreover – like this very shade of yellow, or this very shape or size of least spatial bit (whatever that may mean) – such simple-property identities would not suffice for the recognition of living things. As Aristotle emphasizes in his account of division in P.A. I, no one differentia will select out the kind of *atomon eidos* one needs to be aware of in order to single out an organism.[2] It is a certain life-style, a way of functioning in its proper place (its ecological niche?) that distinguishes a member of a given species, and it takes a very complex account to pin this down – if indeed an exhaustive account is possible.

Kant's reply to Hume, further, equally fails to do justice to the foundation of biological practice, or to the character of biological explanation. True, Kant tried, belatedly, in the second part of the *Critique of Judgment*, to expound the presuppositions of biological knowledge. But the definition of an organism as an entity in which everything is reciprocally means and ends has somehow a non-biological ring. What Aristotle makes plain in P.A. I (640 a 33–5) is that we must *first* understand what it is for a man to be – or a placental dogfish or an octopus or what you will – and *then* show that because his (or its) *ergon* is such and such, these other properties or processes are necessary to his (or its) existence. The τῆε is, admittedly, the *telos* of the living thing; but it is in its character of formal cause that it is primary, and the end, which is the actualization of the form, in turn governs the whole development of the organism. This is not reciprocity of means–end relations, but priority of form, which is end, to means. There are indeed subordinate means–end relations within the organism and insofar as these subsystems interact with one another there is a kind of reciprocity. Yet here, too, it seems to me, in a given internal context the relation is asymmetrical. The lung is designed for the exchange of gases; the exchange of gases facilitates delivery of O_2 to all working organs, including, of course, the lung itself. So the lung is both the means for O_2 transport and one of the organs to be supplied by its own activity. Or, for example, its innervation serves it, but it also serves the nervous system. Yet to study the function of a given organ or organ system is to study the mechanisms subordinate to *its* operation, not, primarily, to show how other organs at the same level in the hierarchical ordering of the whole function equally for its sake. It is the subordination of means to ends, matter to form, physico-chemical processes to organizing principles, that most clearly characterizes the distinctive arrangement of every kind of living thing.

Where Kant does introduce an asymmetry between mechanism and teleology, moreover, he has it backwards. For him (as also admittedly for many modern biologists and biochemists, as well as philosophers) it is the mechanical (when-then casual) account that is scientific, while the teleology is what Waddington called "an appearance of end" imposed by the biologist as a merely regulative maxim ([12], p. 190). Yet what is characteristic of organized systems is precisely (1) that they are scientifically explicable on more than one level and (2) that although the existence of the lower level is a necessary condition for the existence of the higher (there are no non-enmattered living things), it is the *ergon*, the way the system works, that makes it the kind of living thing it is. This relation is one of what I have elsewhere called ordinal complementarity. It is the higher level that is both epistemologically and ontologically (though not chronologically)[3] first. Kant did not see that at all.

Indeed, it is already clear from the central argument of the first Critique that Kant's interest leaves the most characteristic biological phenomena – the *explicanda* of the sciences of life – out of account. The chief question of the Analytic is the question: what makes possible experience of an object, and/or what makes possible the objectivity of experience. But this is objectivity neutrally conceived: as ranging over any possible sector of space and time or, via the Analogies, over all experience. There is no possibility here for differentiation, which is the most obviously perceptible characteristic of biological *explicanda*. Biological explanation, therefore, clearly holds little interest for him.

Moreover, secondly, Kant, though refuting Hume's scepticism, does so by strengthening still further the thesis that any order given to nature is *our* order. I wish the Transcendental Object were strictly coordinate with the Transcendental Unity of Apperception, but I must admit it is not so. Kantian *a prioris* are wholly the products of the human mind. Yet however deeply our concepts shape our theories (and of course no one would deny that they do), the way living things and parts or behaviors of living things present themselves outruns those conceptions. It forces itself on us, not as data of inner and/or outer sense, but as what, in any given case it is. The squirrel sitting to attention in my garden or the jay that used to eat out of my hand: no synthetic unity of the manifold captures their presence.

The trouble is, thirdly, that Kantian nature, though rationally organized, is no more alive than Humean experience is. To live is to live bodily, but for Kant the body remains simply a part of the external world.

Modern metabiological thinking, finally, either reverts to Hume or, by turning synthetic into analytic *a prioris*, attenuates Kant. Darwin was delighted that Mill approved his method. But Mill's methods are Hume without sceptical leavening. And it seems that modern biologists, in some numbers, at least, are in an even sadder case. They still want to start from particulars, but finding Hume's associationism purely "psychological," they resort to a Popperian *modus tollens* and the purely logical property of falsifiability. The slogan "Sir Karl Popper has shown us what science is" is heard far and wide in more than one land. Yet as a methodologist Popper, whether $_1$, $_2$, $_3$, or $_n$, undercuts precisely the kind of "aesthetic recognition" characteristic of field biology at least as strikingly as do Hume and Kant.[4] Let me mention just one instance of what happens when Popper's account of science is presented as canonical for biology. Francisco Ayala, of the Davis genetics department, has written a concluding "philosophical" chapter for a cooperative text in evolutionary biology.[5] He begins with the usual obesiance to Popper, endorsing the falsifiability thesis, and then goes on to give a perfectly reasonable and convincing account of the growth of genetic and evolutionary knowledge in the past century that is entirely unpopperian. It tells us how much more is known about the history of life and its basic mechanisms than was known in Darwin's time, and yet how Darwin's great theory underlies all this advancing knowledge – a story that simply cannot be told in terms of the *modus tollens* model. There must be something that can be done in trying to describe and explain the understanding of living nature in terms less incoherent than this.

And that should bring me at last, explicitly, to Aristotle. But let me insert, parenthetically, one more point about the particular-resemblance model of scientific thought and its influence in biology. If the psychological least units of Hume have gone the way of all sense data, and if logical manipulation has replaced resemblance as the alleged mechanism of scientific reasoning, the reliance on least particulars nevertheless persists: evolutionary biologists, though admitting that evolution works through phenotypes, find their "untimate" explanations in changing relative gene frequencies, and genes are very satisfying, discrete, isolable bits.

In Aristotelian biology, on the contrary, it is not *particulars* but *individuals* that we must recognize as primary, and it is their *atomon eidos*, which is identical over any number of individuals, that makes them the individuals they are. What is fundamental is the existence of an organized being, a this-*such*, informed in such and such a way. It is this kind of being that exists, and for its sake the processes leading to its maturation take place. ". . . the matter is potentially since it might go into the form, but when it is actually, then it is in the form." (*Met* 1050 a 15–16) Thus even in nature, actuality must be prior to potency, since only the control of what might be by what is can give to process the orderliness that permits a given form of thing, a given life-style, to come into being a single time, let alone again and again. Max Delbrück has suggested that Aristotle be posthumously awarded the Nobel prize for having discovered DNA [3]. A whimsical notion, to say the least! It would be difficult to be more mistaken about reproduction and heredity than Aristotle was; but what he did recognize was the fundamental role of information in the existence of a living thing. To be a ζῶον is to be organized in such and such a way, and for a ζῶον to develop is for it to follow instructions well enough to express its specific form: for its matter to have been activated by its form and to have come to be *in* its form – form, of course, not primarily as shape, but as organizing principle or function. This does look like a much more promising beginning for biology than do Humean impressions or even Kantian objects. And identity of form looks like a firmer foundation for explanation than mere similarity – which has somewhere to rest on identity if it is not to remain simply a primitive – and unintelligible – idea.

But now of course we get ourselves into what may be an equally serious difficulty, certainly a traditional difficulty in the interpretation of Aristotle. Coriscus and Socrates are identical, as are Socrates the child and Socrates the man. So are Socrates the child or man and the philosopher who drank hemlock, or the person who was of all the men of his time the best and wisest and most just. Those identities may have troubled pre-Kripkean logicians and may still puzzle some. But the identity of form of Socrates *and* Coriscus – and Xantippe and Plato and each of us – that seems to an ordinary and unlogical mind a much more difficult concept to deal with. What *is* the *atomon eidos*? Apart from the individuals whose *eidos* it is it appears to have only conceptual existence, yet it is the very same *eidos* in each case that makes the individual the individual he (or less excellently she) is.

The resolution of this tension between *eidos* as individual existence and *eidos* as known in definition (or at the terminal point of a correct division), and hence known as universal: the resolution of this tension appears, for Aristotle, to reside in the being of man as knower: as able to comprehend what it is for an octopus or a duck or a man to be, even though what there is is this octopus, this duck, this man, not octopus or duck or man in general. Given the stability of Aristotelian nature, this may work – though even for Aristotle there seems to be a residual problem. Consider the mind of a given biologist, say George Williams. Now if Williams' mind becomes the forms it knows, and it knows the universal, then this individual mind, though the actuality of this biologist, is at the same time the actuality of all the species he knows (in this case, I think, especially fish), and of these as universal, rather than as individual. So these forms, universalized, must actually be as the mind of this very biologist. Thus in the case of an animal that knows other living things, the individual form is not only individual but universal, and expresses not only its own *eidos*, but many others. "The mind becomes all things." Yet, at the same time, surely, what *is* is individual and *one atomon eidos*, not universal and many forms at once.

Aristotle managed, it seems, to keep these problems under control, if only in a delicate and elusive equilibrium. In any case, I do not want to press them further here. Rather, before going on to the problem of normality, I would like to ask briefly how *we* might treat, in our terms, the conjunction of statements (a) that organized individual things belonging to natural kinds are what there is in nature and (b) that such natural kinds are known by us although they have no existence apart from the individuals that happen to "express" them. I am not here asking a logical question, such as David Wiggins deals with, for example, in his paper on "Essentialism, Continuity, and Identity,"[6] but rather a metabiological question: how can we defend statements (a) and (b) in the light of our evolutionary concept of how the species now inhabiting this planet came into being and will eventually cease to be? The answer to (a) is relatively easy: by natural selection acting on members of already existent populations in such a way as to bring about certain relative gene frequencies rather than others. That is a crude statement, but for our present purpose let it do. Then in terms of the same evolutionary principles we can see that had we not achieved some recognition of the ways in which our surroundings were sorting themselves out we would not have known our predators from our prey and would not have lived to tell the tale. If, like Painted Jaguar, we had consistently confused Stickly-Prickly with Slow-and-Solid, we would not have survived. Evolutionary epistemology is a slippery slide, but this little bit of it: a defense of a minimal belief in natural kinds, I think we can and must allow. It can keep us at least in the same kind of precarious equilibrium in which the eternity of species and the essentialist bent of human intellection kept Aristotle.

The second major advantage of Aristotelian over modern metabiology lies in Aristotle's recognition of the normative character, not only of the biologist's judgments, but of his subject-matter itself. The *eidos* is the norm from which some individuals may deviate to some extent or other. When the matter is actual it is in its form, yet the form masters the matter to greater or lesser degrees. A little less (in Aristotle's account), you have a female; still less, a more conspicuously monstrous example of the kind; less still, a being of the genus only, where the form scarcely shows. Aristotle's teratology in its details, like his embryology, sounds at best quaint, yet the principle he is applying is fundamental. Nature has somehow got itself sorted into kinds, these kinds turn up in their individual instances, with fair but not total regularity, and the exceptions deviate somehow from the standard embodied in the "best" case.

Modern discussions of "normality" have often tried to eliminate this last, "normative," component, but with less than conspicuous success. Let me look briefly at the way some of this controversy runs.

First, "normal" often means "healthy." "Health" and its contrary "disease" are on the face of it normative concepts. However, efforts have been made recurrently to remove that allegedly "unscientific" taint from the meaning of these terms. Indeed, if disease can be "causally" accounted for (in a fashion that Aristotle would have recognized as roughly Empedoclean or Democritean; I am here speaking in terms of a modern, when-then notion of causality) and if health is defined as the absence of disease, this task of purification would appear in fact fairly easy to accomplish. There are, however, at least two difficulties with this apparent solution. First, if disease can be reduced to cause and effect relations, so can health. Other things being equal, air pollution causes difficulties in breathing; other things being equal, clean air facilitates "better" breathing. How do we tell the difference between such cause and effect sequences unless we first have some notion of what health amounts to as some kind of "well-being"? Second, there are abnormalities that are not in any reasonable sense unhealthy. To take the respiratory system again, for example, the occurrence of anomalous lobes is just an oddity, a phenomenon without clinical consequences. So "normal" cannot just mean "healthy."

The most "objective" alternative is a statistical one, and that seems to be what Aristotle has in mind too when he calls the natural what is "always or for the most part." Normal pulse, normal height, normal blood pressure is what one finds in most cases. Thus it is tempting to fall back on statistics and interpret "normal" as meaning in some sense most frequent or "average," whether the mean, the median or the mode. This seems to be, and has seemed to many to be, the obvious way to banish any normative component from the concept of normality and so make it scientifically "respectable." And of course statistical considerations do play an important role in the effort to express with precision the relation of the normal to its contrary, especially in the context of health and disease. A clinical sign, which is often a measurement, is *a fortiori* understood in terms of a convention based on statistical findings. But does "normal" *mean* "average" or "most frequent"? Can its apparently normative component be eliminated in favor of a norm-free statistical concept? It seems not. Here again there are obvious counterexamples. In conditions of an epidemic or of widespread malnutrition, although the diseased or unhealthy are more common than the healthy, we do not want to call them "normal." Even though "normal" is not synonymous with "healthy," there does seem to be some kind of connection between the two concepts. Thus, once more in the context of respiration, we could ask, is black lung normal in mining communities? It is by now abundantly clear, I take it, on statistical evidence, that respiratory disease occurs more frequently among smokers than among non-smokers. Is it therefore normal for smokers to suffer from respiratory disease? In a way it is, but in a way it surely is not. In other words, to use "normal" *only* to mean "what usually happens" would get us into some rather strange corners. Statistical assessments of frequency, like appraisals of "health," have a bearing on the meaning of normality, yet cannot be identified with it.

A text I have recently been reporting on, *The Normal Lung*, by J. F. Murray, reflects this peculiar tension [6]. On the one hand one can find (p. 92, Fig. 4–10) a figure representing "volume–pressure curve of a normal lung" or reference to "the normal dog" (p. 88), and so on and so on. But when the term is used only statistically, or at least sometimes when it is used only statistically, it acquires quotation marks. To take one example, Dr. Murray is explaining why it is difficult to examine the responsiveness of the central chemoreceptors. CO_2 response curves have been used for this purpose, he reports, but they have certain limitations; among these: "(3) there is a wide range of 'normality' " (p. 240). Why the quotation marks? Because when one has only a scatter of data and no standard to measure them by, one does not have a norm, and does not know what the normal really is. Of course the proponent of normality as purely statistical would answer, in other cases the statistics do show a narrower range of common occurrences of certain values and so we can say, purely statistically, what is normal and what is not. It is only the fact that in this case we cannot get a nice curve that turns normality into "normality." Such a reply would raise the very large issue, which I certainly cannot deal with, of the nature of statistical inference. On this occasion I can do no more than express my hunch that statistics can only be used in some context where one already has some notion of what one wants to measure and why, and that in particular the trained clinician's assessment of the normal or of the healthy serves as informal measuring rod for the way he uses precise measurements of the variables he has chosen and their statistical formulation in order to make more tractable the distinction between normal (and/or healthy) and abnormal (and/or diseased) that lies at the foundation of his craft. When, as in CO_2 response, there is a wide range of measurements, they do not let themselves be captured by the net of clinical judgment, and so normality gets away – one can only speak of "normality" instead.

The normal, then, is not identical either with the healthy or with the most frequent, although it has some relation to both. What other meaning is there? Some remarks by C. F. A. Pantin in his *Relations Between the Sciences* [9] may perhaps help us out here. He refers to a distinction Whewell made between two kinds of "directors" for classification. One may classify in accordance with a strict definition, or in reference to what Whewell called "types," and Pantin, rather alarmingly, says he agrees with Whewell's general description. "But," he continues, "we must guard against confusion about what is the nature of the type. The type we use in recognition as our director is not a material object. *It is the machinery of our trained perception*" ([9], p. 89, my italics). How can one translate "type" into the scientist's trained perception? The type, if there is such a frightening object, must be somehow "in" nature: it is *Hydra viridis* or *Parasilurus aristotelis* or *Spirea vanhootiens* or the normal dog or the normal lung. The perception, however, was Pantin's or Whewell's. We seem to be caught here between a vicious "essentialism" on the one hand and "subjectivism" on the other. Perception, however, should be understood here in a realistic (as distinct from phenomenalist) interpretation. What Whewell and Pantin between them were saying is that biologists operate through learning, in a highly organized real world, to assess more precisely how that world is organized, and they do this not only by more exact saying, but also by more exact seeing, real, embodied seeing in interaction with really existing kinds of things. And in this sort of live, embodied, exploring situation, the situation constitutive for science, the concept of the normal remains, I fear, incurably normative. One cannot escape, nor should one try to do so, the context of skilled assessment that is always inherent in the practice of science, and in particular of biological science.

To acknowledge the normative ingredient in the concept of the "normal," however, is to approach to some degree at least the concept of the *τῆϵ*, what it is (was? was understood?) for a given thing to be. And it is this that is by nature, that is always or for the most part. True, "normal" is sometimes synonymous with "healthy"; true, again, the "normal" sometimes means just the usual in one or another statistical sense – and, as Aristotle's phrase "for the most part" suggests, statistics are almost always useful in establishing what the normal is. But there is also a concept of "normal" that is biologically most fundamental, and yet very difficult to get biologists or philosophers to admit, except surreptitiously, to their conceptual repertoire. And that is the normal as the standard for a certain kind of thing, or process, or organ, or tissue – the right way for it to be the sort of living system, or part of a living system, that it is. But again, I must insist, this "normative" meaning of "normal" does not have to be "typological" or in some obscurantist way "essentialist," let alone "utopian," as one pair of writers have called it [7]. And this for at least two reasons. First, the non-eternity of natural kinds does not entail (as even Darwin sometimes thought it did) their non-existence. Nor, secondly, does the concept of kind as norm for its instances entail essentialism in the sense that all the characters of a kind necessarily belong to it in an all-or-none, bean-bag way. The characters used to "define" a species, for example, form a set of alternative sufficient conditions, such that any one may be lacking, or any one may be present in a lesser or greater degree. To take just one case, in the European oyster *Crassostrea angulata* some individuals have an umbonal cavity in the left valve and some do not. One taxonomist has argued that "the presence or absence of a cavity" is not, therefore, "a specific character," but another argues that variability on a continuous scale down to zero need not exclude eligibility as a specific character [11]. That seems fair enough: species need not be characterized by the museum taxonomist's key. The field naturalist may be guided by the kind of trained perception Pantin was talking about. Nevertheless, the whole set of conditions that, to one degree or another, characterize a given species, do form a class – or family, in a familiar philosophic (not taxonomic) meaning of that term – which can be referred to in order to describe the general nature to which each member of the species is said to approximate when we make the judgment that the individual is a member of this, and not some other, kind. And it is class membership of this sort that can be said to be exemplified more or less exactly, more or less, well, whether for species, organs or billogic processes. A person with an anomalous lobe is certainly a member of the human species, but an abnormal one, just as a blind person, of course, is human, but abnormally so, not only because blindness is rare, but because the human eye is the sense organ of one of our five senses, indeed, of our dominant sense. One needs a normative concept of natural kinds, whether of a species, or of a given structure or function characteristic of a given group, in order to understand such abnormalities, and to assess them as deviations, to one or another degree, from the type they represent.

When I speak here of "types," however, I raise – for myself, at least – not only the ontological problem correlative to Aristotle's schema of individuals identical in kind, but also two further, epistemological difficulties inherent in the normative interpretation of *eidos* and the τῆε. First, there is the methodological problem that I mentioned at the outset. If it is the form that is known, how can we know the deviations from it? How can we know the failures of form to master matter? Of course even the definition of a natural substance has to include a reference to its matter, but only a general reference, to the right *kind* of matter. To saw, a saw has to be made of metal, not wood. But a saw with specially blunt or fragile teeth can be known, as a saw, only for its power to saw, not for its deficiencies. How can there be any scientific knowledge, in Aristotle's sense, of deficiencies, which, as deficiencies, are failures of the form to perform as it ought? There was a controversy about Spinoza's *Ethics*, in which it was debated whether the *Ethics* itself exemplified the third, the second, or only the first kind of knowledge. Similarly, I want to know how, in the light of Aristotle's thesis that only *eidos*, existing individually but identical in all individuals *of* an *eidos*, is knowable, deviations from *eidos* assessed in the judgment of abnormality or normality can be scientifically known at all. In what sense does the treatise on the *Generation of Animals*, for example, belong to the body of natural science? Strictly speaking, only the *De Anima* should be science, because only it deals directly with form as such.

That brings me to my last, and worst, quandary. If deviations from *eidos* are not scientifically knowable, how do we come to be aware of them at all? They have a foundation somehow in our awareness of the world around us, presumably if not in intellection or scientific inference, then in perception. But again, there seems to be a problem here. Either, as is suggested in *Post Anal* II, 19 (110 a 17–b) it is the *eidos* (and worse yet, the universal) that is the object of perception, not indeed man but still Callias as man; and so, again, it would not be the deviations from the *eidos* that are the target of perception, but the *eidos* itself. Or else we abandon the *Post Anal* statement and are thrown back on the doctrine of aisthēsis in the *De Anima*, which seems to me, in this context, even more difficult to interpret. Perception appears to come about through the interactions between the real sense organs of real animals and real perceptibles, in the appropriate media. But what is *certain* in perception, according to Aristotle, is this white color, not Callias, or this shape, not Callias, or this sound, not Callias speaking (let alone, in any of these cases, Callias as man). And so it seems to be even for Aristotle single qualities, grosser Humean impressions, rather than really existent individuals with which each sense brings us into contact. Nor are the common sensibles powerful enough to undo the damage done by the exposition of sensory certainty. Note: I am not suggesting that a more realistic, or transactional, theory of perception would itself be compatible with a claim to sensory certainty. I do not believe it would. But what troubles me is that Aristotle does want to claim certainty for the information conveyed by special sensibles, and that he can do so only by withdrawing the claim that the object of perception is an *atomon eidos* like man or octopus. On the contrary, in accordance with the *De Anima* the object of perception must be at best an *eidos* like yellow or one cubit long, and I do not think that will help if we want to suggest that deviations from the best or normal in living things are apprehended by perception, even though not by *nous* or *epistēmē*. Granted, if Callias is bilious, his complexion is yellow, not ruddy, or if he suffers from polydactyly, his fingers are six to a hand instead of five, and we can perceive the difference in shape. But the assessment of normality and abnormality, what is appraised via the "machinery of the biologist's trained perception," is not always or even usually specifiable in terms of such segregated and itemizable sensory units. To understand the *eidos* that makes the individual this kind of an individual is already a subtle and only partly specifiable task. To grasp deviations from the *eidos* may sometimes be to perceive differences in color, texture, shape or the like, but it may also entail a kind of global diagnostic perception of a general failure of form to master matter, an assessment not readily reducible to the discrete data of the special sensibles as the *De Anima* describes them.

In short, I still find, intuitively, Aristotle's metabiology right in stressing the priority of form to matter and in allowing a normative ingredient in biological reality – but when it comes down to it, I do not know how to bring together these two insights into a coherent methodology. The first if defensible, in a lame kind of way, but the second appears to be incompatible both with the first and with the only foundation Aristotle seems to provide for it in his own theory of perception. It may be that only a radical return to a more ontological, and less epistemological, consideration of the aims and achievements of the sciences will permit us to do justice to Aristotle's metabiological insights.

University of California,
Davis, California

NOTES

[1] This paper is dedicated to Hans Jonas on the occasion of his 75th birthday. It was first presented at a conference on Aristotelian biology held at Princeton University on

Dec. 3 and 4, 1976. Its aporistic character surprised the writer; I hope that with further thought I may find some resolution of the difficulties I have presented. Meantime, I am thankful to a number of people in a number of places with whom I have discussed its central problem(s), but hesitate to involve them in my own aporia by naming them. Let their anonymity represent my gratitude.

[2] See David Balme's discussion of Chs. 2–3 in [1], pp. 101–119.

[3] This of course counter to Aristotle.

[4] The term "aesthetic recognition" comes from C. F. A. Pantin [8].

[5] Read in MS; see ([4], Ch. 16).

[6] See [5].

BIBLIOGRAPHY

1. Balme, D.: 1972, *Aristotle's De Partibus Animalium I and De Generatione Animalium I*, Claredon Press, Oxford.
2. Canguilhem, G.: (in press), 'La question de la normalité dans l'historie de la penseé biologique,' in *Boston Studies in the Philosophy of Science*, 24. D. Reidel, Dordrecht, Holland.
3. Delbrück, M.: 1971, 'Aristle-totle-totle', in J. Monod and E. Borek (eds.), *Of Microbes and Life*, Columbia University Press, New York, pp. 50–55.
4. Dobzhansky, T. H.; Ayala, F. J.; Stebbins, G. L. and Valentine, J. W.: 1977, *Evolution*, W. H. Freeman & Co., San Francisco.
5. Kolakowaski, L.: 1968, *Towards a Marxist Humanism*, Grove Press, New York.
6. Murray, J. F.: 1976, *The Normal Lung*, Saunders, Philadelphia.
7. Offer, D. and Sabshin, M.: 1966, *Normality: Theoretical and Clinical Concepts of Mental Health*, Basic Books, New York.
8. Pantin, C. F. A.: 1954, 'The Recognition of Species', *Science Progress* **42**, 587–598.
9. Pantin, C. F. A.: 1968, *The Relations Between the Sciences*, Cambridge University Press, Cambridge.
10. Rudwick, M.: 1972, *The Meaning of Fossils*, American Elsevier, New York.
11. Stenzel, H. B.: 1963, 'A Generic Character, Can It be Lacking in Individuals of a Given Genus?', *Systematic Zoology* **12**, 118–121.
12. Torrey, H. B. and Fellin, F.: 1937, 'Was Aristotle an Evolutionist?', *Quart. Rev. Biol.* **12**, 1–18.
13. Waddington, C. H.: 1957, *The Strategy of the Genes*, George Allen and Unwin, London.
14. Wiggins, D.: 1974, 'Essentialism, Continuity, and Identity', *Synthese* **28**, 351–359.

CHARLES HARTSHORNE

THE ORGANISM ACCORDING TO PROCESS PHILOSOPHY

Hans Jonas is a very interesting – I am tempted to say fascinating – as well as learned philosopher. He has written perceptively about biological problems and also about the type of philosophy in which I have most confidence. I have decided, however, not to comment in much detail upon his views. This is partly because I am not sure that he is at his best in evaluating process philosophy or in analyzing "the phenomenon of life." I like him best either as an historian of ideas or as a moralist and philosopher of religion. He seems to me to stand for much that is most valuable in our spiritual heritage and also for much that is soundest in the "modern temper." In both respects I think he is closer than he fully realizes to process philosophy.

I. SENSE DATA AS PHYSIOLOGICAL

I begin with some epistemological points. Each of us who tries to understand organisms is, or has, an organism, a human body. This body is that part of the material world that each of us knows in the most intimate way in which such a part can be known. One can see, hear, and touch one's body, as one can other bodies; but in addition one can feel, enjoy, suffer it by physical pleasures, pains, kinaesthetic, and other bodily sensations in a manner not open to us with the rest of the material world. More than that, since the most immediate material conditions of all our sensations are inner bodily processes, one can argue that these processes are the most direct data, the most sheerly given realities, in all our sensory awareness. This thesis is misunderstood if taken to mean that what we initially *know* in sensory experiences are inner bodily events, from which knowledge we then infer or surmise possible truths about the environment. Experiencing, sensing, is one thing, knowing (if the word is given a pregnant meaning) is another. Infants do the first. How much can they be said to do of the second? Nor is the point that we can set up infallible "protocol sentences," from which to infer statements about the rest of nature. The having of data in the sense I intend is something that occurs in animals lacking the human capacity to make statements. This capacity is no absolute power with unlimited access to data. What we can in detail tell ourselves or others that we experience is not identical with what we do experience. The power consciously to analyze experience and "put it into words"

S. F. Spicker (ed.), Organism, Medicine, and Metaphysics, 137–154.

is one that infants and subhuman animals lack and none of us possesses in ideally perfect form. But all of us, and all the higher animals at least, do have experiences, and these have data, in the senses I am trying to give these words. Now I submit that A. N. Whitehead is right in holding that there is no reasonable way to distinguish between what is sheerly given in sense experiences (as such only more or less accessible to definite conscious detection and description) and the physiological conditions of these experiences. I shall set out some reasons for this identification.

(1) It seems clear that the idea of a "given" is vacuous unless it means something that must be or must occur if the experience is to be or occur. But this requirement is met by the physiological conditions. Even the sensory content of dreams must have such conditions. We dream of being cold when we are cold, of hearing a sound when there is such a sound. But in other cases, artificial or internal stimulation of nerves can produce sensations without the normal external correlates. The neural correlates are the absolute requirements for sensation, not environmental conditions. If it should be objected that the idea of a given implies not only that the thing given must be there if the experience occurs, but that it must be there *for* the experience in the sense of being consciously detectable in or by it, then I repeat what was said in the previous paragraph: The power to know what we experience is no absolute power. Like Leibniz, Freud, Peirce, Bergson, Whitehead, and many others, I think that we often flatter ourselves in this respect. Infants cannot say what they experience; we are better off than infants in this respect. But how much better off? If introspection were the absolute power that many philosophers, usually without quite saying so, take it to be, there would be less controversy about the use of introspection in psychology. It is indeed an implication of "the given" that adult human beings have some capacity to detect it in their experience. Data must be *more or less* accessible to such detection. The position I am defending holds that while inability to detect an item is not, in general, proof that it is not given, the bare possibility that the experience might occur and the thing said to be given not occur is proof that it is not a datum. How detectable items are depends on their prominence, the degree of analytic power of the subject, direction of interest and attention, and the like. But by definition they must be necessary for the experience. This can be shown only for inner bodily processes.

(2) In some types of case we are indeed well aware that the data of sensory experiences are physiological, as when we tell a doctor how and where it hurts. The doctor is interested in knowing this because it is evidence about our physical condition. (Referred pains are not counter-instances, for the

"where" is initially in a single sensory field and location in public, intersensory space is a learned and fallible procedure.) To deny that pain is a potentially cognitive function seems remarkably arbitrary. Physical pleasure is equally cognitive. In principle there is no demonstrable difference between any sensation and any other so far as its immediate evidence is concerned. Always this evidence is of something happening inside the body. Absolute immediacy for the distance receptors as such cannot be shown. Nor need it be assumed to explain our knowledge. Given natural selection and evolutionary adaptation in a universe with a considerable degree of causal order, it is no mystery that normal animals do a fairly good job of judging correctly about (responding correctly to) their environments, using the not quite direct evidence, the sensory content, available to them. Of course they do this.

(3) There is a third requirement if something is to be accepted as datum of experience: The thing must be independent, for its existence or occurrence, from the experience. Without this assumption the escape from solipsism can only be a *tour de force.* It is a dismal way to begin an epistemological inquiry to allow even a logical possibility of an experience which is aware only of itself or its own creations. I agree with Thomas Reid, as against Hume (and Russell), that an experience whose data are merely its own ideas or impressions is nonsensical. Not even dreams are correctly so described. There is a creative aspect of experiencing, but what is thus produced is precisely not the data but the awareness of the data. Granted *X*, there may then be produced the awareness of *X*, and this will be something additional, a new reality of which *X* has become a constituent. With Whitehead, I interpret this as an argument, not the only one, for the view that the experience–experienced relation is one of temporal succession: first the reality to be sensed or intuited, then the intuiting of it. No proven fact contradicts this.

(4) Here we break, and to our advantage, with all traditional two-aspect theories of mind and body, according to which the two aspects occur simultaneously. Rather, we take the subject–object relation to be one of later–earlier. What is sensed or felt is what has just happened in the body, not what is happening precisely now. Thus the bodily condition is cause in the normal temporal sense. And so we escape from the supposed need to assume an absolutely peculiar form of relatedness in the mind–body case.

Note, however: If the given is to be antecedent cause and yet independent of its effect (the human experience), then strict determinism must not apply. The necessary condition must not be in the strict sense "sufficient." For "necessary and sufficient" is the same as sufficient and necessary, and thus it implies complete symmetry of dependence between cause and effect. The phrase, therefore, contradicts our natural intuition of one-way dependence of effects on causes, as well as our intuition that experiences depend on things experienced, not conversely. In process philosophy strict determinism is rejected for all cases.

II DATA AS PAST WHEN GIVEN

If the reader is shocked by the idea that sense awareness gives us the past rather than the exact present, I ask him to consider not only the fact that events outside the body are demonstrably prior to our perceptions of them, but also the basic fact of experience that in memory we seem to have intuitive awareness of what has already happened, especially of what has just happened. (The notorious "mistakes of memory" go back far enough in time to be explicable by functions other than memory pure and simple.) The argument that the past "no longer exists" and so cannot be intuited proves either nothing or too much. For if the past is non-existent, then there are no causes of present happenings. If the past exists sufficiently to have effects now, it exists sufficiently to form data of present experiences. If experience-of-*X* includes *X*, so does effect-of-*X* include *X*. The denial of reality to past events is ill-considered. Bergson was one of the first to do justice to this point. But Peirce too was aware of it.

III. FROM PHYSIOLOGICAL DATA TO ENVIRONMENTAL KNOWLEDGE

The view of experience we have reached is as follows: Every experience has its data, and these are the same as its independently real, temporally prior and necessary, but not strictly sufficient conditions. (They suffice for its possibility, not for its actuality. Even given the conditions, its occurrence, precisely as it is, was contingent and might not have been.) Apart from memory, the data are inner-bodily, presumably largely neural. From sensing, feeling, or intuiting just occurrent neural processes, we judge, believe, surmise (in good part correctly, having learned to do so in infancy and childhood with bodily equipment evolved so as to make this normally probable) that certain processes are going on around us in the world. It is by a special sophisticated form of this procedure that we learn about neural processes as what we are sensing. In all this a reasonable but not an absolute degree of causal order is assumed. The status of this assumption I shall not inquire into in this essay. I have dealt with it elsewhere [3].

Since all knowledge of the extra-bodily world is indirect, it follows that our ideas about other animals (also about plants and minerals) must be derived by analogy from direct awareness – not, as subjective idealists argue, from awareness of that very awareness or of mere ideas, but from awareness of certain physical processes inside our skins as directly sensed or felt. This might seem to mean that our knowledge of "other mind" is in no case direct. However, this does not quite follow, as we shall see.

The world as given in sense awareness is through and through a matter of what are termed "secondary qualities" – red, sweet, shrill, painful, ticklish, etc. Every one of these qualities is set aside in the account of the world given in textbooks of physics and chemistry, save so far as the qualities are employed as signs, in principle more or less dispensable, of temporo-spatial structures which alone are recognized by science as making up the world apart from our kind of animal. As actually given, the structures in question are merely outlines (or relations) of secondary (or tertiary) qualities. Thus the highly abstract, colorless, tasteless, value-free world of science is conceived by a radically truncated, though in some ways much extended, analogy with the world as given.

I do not mean that science has no good reason for its abstractive procedure. The procedure becomes arbitrary only if it is accompanied by the assertion that in truth nature, apart from animals, simply is such an abstract affair, with temporo-spatial structures and their causal inter-relations, but devoid of qualities in the sense in which secondary qualities are such. On this matter Russell agrees with process philosophers; he admits that nature cannot be as abstract as science depicts it as being. Structures cannot exist without qualities ([4], pp. 168, 246 ff.). But he holds that we are incurably ignorant concerning the qualities belonging to natural kinds apart from our kind of animals. I do not see that we must admit quite so absolute an incapacity to know, at least in principle, what is to be meant by quality, apart from the mere relationships, mathematically expressible, given by the equations of science, and apart from the sensed qualities in our experiences. I think we can know in principle that the two questions, "How does it feel to be *X*?" (or "What experiences does *X* have?") and "What qualities does *X* have?" are one and the same question, save that the first formulation is the more illuminating. These are not two mysteries but only one. Here as so often Peirce and Whitehead agree. I shall sketch the reasoning.

IV. THE PERVASIVENESS OF THE PSYCHICAL

That some parts of nature involve no form of experience or feeling, however primitive or other than the human form, is an absolute negation which no possible observation could ever support. Those who think otherwise are invited to tell us their criteria for the total absence somewhere of anything like experiencing or feeling. For the absence of elephants there are criteria or for the absence of protein molecules (very high temperature, for one); but for the total absence of anything psychical such as memory or enjoyment there are no criteria that do not merely beg the question. I have been saying this for years and have yet to receive a clear answer. The notion of mere matter is vacuous, representing no possible knowledge. We do not know that it would be like for there to be such a thing.

Please note: "Mere matter" is not equivalent to "insentient things," things that do not sense or experience. Of course, there are such things, e.g., tables and chairs. What philosopher has ever held that these entities experience or feel? But the total absence of sentience in tables would mean not only that they do not sense or feel, but that their invisibly small constituents, if they have them, do not. On what ground would the latter statement be made? Tables are too inert and too lacking in dynamic unity to be regarded as sentient. But molecules are not inert or lacking in dynamic unity. Leibniz was so much clearer about these questions than some philosophers today! The author of the article "Panpsychism" in the *Encyclopedia of Philosophy* pays no attention to the distinction just made [1]. He also pays no serious attention to the most important recent representatives of the psychicalist doctrine, particularly Peirce, Bergson, and Whitehead.

The argument can summed up also as follows. The less directly given must be explained by principles illustrated in or somehow derived from the more directly given; nothing like mere matter, devoid of qualities of feeling, is directly given. The body is the most directly given material thing, and, as given, it is composed not of mere matter but of psychical life: pulses of feeling, pains, pleasures; feelings of sweet, sour, tickle; and the like. I wrote a book on this subject over forty years ago which is still in print [2]. Nor, I think, has its main thesis been refuted. Broadly similar views about the given may be found in Croce, Berkeley, Bradley, Bosanquet, Whitehead – writers sufficiently different from one another – and in some psychologists, e.g., Spearman.

Our contention is not that only the body is directly given. One's own past experiences, at least those immediately previous to the present experience, are so. We call this experience of past personal experience 'memory.' Perception could be termed 'impersonal memory' since, while it also gives the past, it gives other portions of the past than one's own previous experiences.

Memory in this generalized sense, common to personal and impersonal memory, is the whole of our direct acquaintance with reality. 'Introspection,' as both Ryle and Whitehead (agreeing for once) maintain, is not an additional avenue to reality but is rather one way of utilizing personal memory, especially in very short run or most immediate cases.[1]

All the forms of direct experience have in common that their data are past and are irreducible to mere matter. As given, they have psychic life in the general form for which the handiest label is feeling, as including sensing, emotion, desire, some form of valuing. This is a phenomenological assertion in support of which many authors (including Peirce and Bergson, though not, alas, Husserl) could be cited. I came to it before my study of the history of philosophy and when I was not thinking of any philosopher who asserted it. Simply experience seemed to me like that. It still does.

The data of memory are previous personal experiences. The data of perception are also previous experiences, but not personal ones. Consider a physical pain or pleasure. It is a datum of one's awareness. Set aside the notion of mental pains or pleasures; for example, a woman weary of carrying a child might feel relief at the onset of labor pains. Such mental satisfactions or dissatisfactions are not what I intend by physical pleasures or pains. Sexual pleasure may be of both kinds. Some thinkers seem curiously unable to distinguish between physical feelings and the mental ones that may, in special cases (some of them more or less abnormal, as in masochism), accompany them. So, once more consider a physical pain or pleasure. It is a datum of awareness if there are any such. It may be sharply localized in phenomenal space. It is, in Whitehead's phrase, an objective feeling form, not a subjective one. True, it is a feeling that the human subject has come to feel, but so that, in Whitehead's admirable language, the entire experience is one of feeling *of* feeling. There is a duality and, according to the hypothesis that I adopt from Whitehead, the doubleness is quite real and definite since one feeling follows the other in time. The human feeling is subsequent; the subhuman feeling that it feels is antecedent. We suffer because that sort of feeling, on a more primitive level, has already been felt, not by us but by the microconstituents of our nervous system.

Of course, it is not by direct intuition that we know the meaning of "microconstituents of our nervous systems." Intuitively we have only the duality of subjective and objective feeling, quality of our awareness, and quality of what we are aware of. Our bodily constituents, the details of which we know only through science, are given in emotional terms and no others, direct intuition being too indistinct to disclose them.

V. EXPERIENCE AS IMMEDIATE SYMPATHY

In other language, what we have been saying is that the subject-to-immediate-object relation is one of sympathy in a very literal sense, participation by one subject in the feelings of other, and temporally prior, subjects. Whitehead's "physical prehensions" are his technical label for what in ordinary language are acts of immediate sympathy. I deeply believe that there is no other key whatsoever to the ultimate relation of mind to its objects or to its body. When Wordsworth spoke of "seeing into the life of things" he knew only vaguely what he was talking about, but he was vaguely right and faithful to experience in so speaking. To be alive is, in principle, to share in life not simply one's own. The cell theory might have taught us this long ago. But philosophers often pride themselves on not learning, even by way of suggestion, from science. Wittgenstein is one source of this lamentable way of thinking. The theory that life is essentially participatory is as old as Buddhism, but it is made easier to conceptualize by modern science, which has disclosed the elements of illusion in the idea of mere bits of stuff hurtling about in space, each being what it is simply in itself, with no necessary reference to anything else. This old atomistic myth has lingered on in the "logical atomism" of Hume and Russell.

The formula, then, for an animal mind is that of life participating in the life of at least some of the bodily parts. Allowing for difficult borderline cases, this formula can be taken to cover macroscopic or multicellular animals in general. There are enormous differences in complexity. Our human "symbolic power," centered in language in the human sense, is so unique in degree that it matters little whether or not we call it a difference of kind. Thus a bird can learn a short human melody but nothing like a symphony. And even the impressive work with chimpanzees learning a sign language will probably put them in the class of fairly young children, not of human adults. The other rivals of humanity, the whales and porpoises, may be similar but are more difficult to study.

Macroscopic plants present a basically different problem, and one that has been misconceived by most philosophers. Aristotle stumbled close to the truth when he said, "A tree is like a sleeping man who never wakes up." And indeed an animal in (dreamless) sleep is an almost vegetable creature, exchanging materials (air) with the environment, circulating fluids, and growing or repairing parts, but as a whole only potentially an experiencing subject. We now know that the basis of the animal's possibility for waking experience, the central nervous system, is a subsystem of cells able so to coordinate its parts that they present a suitable array of lesser lives for the participation of the whole-life. In dreamless sleep the nervous system does not quite achieve this coordination, though it makes some contributions to the vegetative functions. The logical conclusion is neither the Aristotelian one that there is a special non-experiencing "vegetative soul" (what could that be?) nor the dualistic or materialistic one that there is in plants no psychical life at all, but rather that the subjectivity which is the active agency in this case is not only on a lower level than with macroanimals but that it is wholly restricted to the microconstituents, presumably the single cells. In growth it is not the tree that does things; simply the cells produce additional cells or repair or enrich their own parts. To talk of the tree doing things is shorthand. As Whitehead picturesquely puts it, a tree is a democracy, a pure democracy without even a temporary ruler or representative. Leibniz, with creditable hesitation, supposed the contrary, that a plant has a "dominant monad," but a botanist of his time (or was it in Fechner's time?) made the right criticism that a plant has less unity than this implies. Fechner, without hesitation, and I think less creditably, assumed that all the experience in the tree belongs to the tree as a single subject.

Some would argue that since our experiences depend upon the nervous system, the logical conclusion is that without such a system there is no experiencing, whether in the whole or in the parts. If this reasoning is sound, why not the following? Since in animals digestion is achieved by subsystems of cells called stomach and intestines, therefore there can be nothing analogous to digestion in single-celled animals. The same is the case with the possession or lack of possession of lungs and the utilization of oxygen. Rather, the proper conclusion is only that what many-celled animals achieve by subsystems of cells, single cells may achieve in a more primitive way by arrangements of subcellular parts. The test of subjectivity is dynamic unity, and this seems as clear in cells or protozoa as in a person but by no means as clear in a tree as in either a cell or a person.

Coming at last to "inanimate" nature and following the same general pattern of thinking, we see that if there is in a tree no whole-life capable of sympathetic participation with life, meaning here experiencing, in the parts, there is *a fortiori* no such whole-life in a rock, a house, or, probably, a star or planet (in spite of Fechner's eloquent attempt to persuade us of the contrary). The doings of such things seem mere statistical outlines of the doings of their constituents. As J. Clerk Maxwell said with regard to a gas, it is as with a swarm of bees: at a distance it may appear to be a single thing that scarcely moves, but seen close to each bee proves to be highly active and to move in an individual way. Thus the apparent inertness and lack of organized unity of animate things that suggest their lack of psychic life is compatible with their consisting of active, organized, unitary, and sentient parts that are unseen. Accordingly, traditional dualisms and materialisms are groundless. They commit a mixture of the fallacy of division and the fallacy of confusing "we fail to observe the presence of such and such" with "we observe the absence of such and such." Let it be ever remembered that Leibniz was the man who first avoided both of these fallacies. Alas, he fell into others.

There is a final problem. If there are least parts with no subparts, then these partless units cannot be organisms in quite the same sense as ordinary units, those with parts, may be. If such a partless unit is a subject, it must, it seems, be an "unembodied one," the real instantiation at last of this old idea. Not quite, however. A body is the set of those other subjects which a given subject most intimately experiences. Electrons may experience neighboring protons or neutrons, say. In other words, here the role of body is taken over by neighbors. Thus conceptual continuity is preserved even in this extreme case. In a general way I am here agreeing with thoughts of Peirce, Bergson, and Whitehead. Believe it or not, there does exist a new type of metaphysics, perhaps as truly unique to the twentieth or late nineteenth century as relativity or quantum physics. That our culture has not been made aware of this is a signal example of the difficulty of the philosophical task.

VI. MERE MATTER: A CATEGORY MISTAKE

A basic conviction common to process philosophers and all who, like Leibniz, hold that the true pluralism is of subjects, not mere objects, is the contention that when we compare an animal, a plant, and a rock as each a single thing, differing however radically in quality, we are making a "category mistake," a mistake as to logical types. An animal is a single agent in a sense in which a plant, a rock, or a gas is not. There are qualitative differences here also, but they cannot be clearly understood until the category difference is taken into account. To keep within one logical type one must compare an animal with a cell of a tree, not with a tree, and with molecules or atoms, not with rocks or gases. Also, when it comes to particles, there is an additional type difference in that the self-identity of individuals through change is questionable on their level. There are particulate events rather than particulate things; however, in process philosophy there is no absolute requirement that all events must form careers of identifiable individuals. Here the break with Leibniz is sharp, but the Buddhists took a similar path long ago without help from quantum physics. And they were psychicalists too in a vague sense and process philosophers of sorts. Thus the new philosophy is in some respects very old.

Jonas is inclined to think that process philosophy underestimates the difference between even the lowest biological organisms and the constituents of inanimate nature. I suggest that he underestimates the resources of this philosophy for conceiving almost infinite, in a sense truly infinite, differences in spite of the "monistic" ontology. What Whitehead calls "societies," corresponding to the individuals or things of common sense, can vary more widely than our imaginations can readily follow, both in degree and kind of "social order," that is, their degree and kind of identity through change as well as in the levels of feeling or experiencing of their momentary members or "actual entities."

Granting, for the sake of argument, that, as Leibniz held, we can understand singular realities only to the extent that we can see their analogy to ourselves as singulars, it is to be expected that such understanding will be most difficult to achieve as we consider cases most unlike ourselves. With other persons we know roughly what they are, for they are roughly what we are; with nonhuman animals we must make more allowance for differences from ourselves; with single cells still more, with molecules much more still; with particles not only extreme qualitative differences but, as noted above, probably a logical-type difference as well in the partial absence of identifiable socially ordered sequences, that is, individuals enduring through change. At the opposite extreme, in the attempt to understand deity, we have again the difficulty of combining analogy with utmost difference. The radically superhuman and the extremely subhuman are not likely to be the easiest topics for human thought.

The difficulty of interpreting the inanimate is not mitigated by a materialistic or dualistic language. If the stuff of nature at large is not any form of experiencing, this can only mean that, whether we call it "matter," "energy," or (as Democritus did) bits of "being" in the void of "nonbeing," we know of it only that it is something or other with the abstract relational patterns that physics defines through its formulae. In the only directly given instances of being we have access to, the relational patterns are filled in by the stuff of feeling and the like. If we discard this in our view of nature at large, there is nothing to put in its place.

It is not only Whitehead who attributes some minimum of "historicity," in Jonas' word, to inanimate nature. Bergson also argues that without something like memory (or, I add, perception as impersonal memory) there is no way to relate the moments of time to preceding or succeeding moments. But both writers have a vivid conception of the vast differences in the kinds and degrees of identity through change which this involves.

VII. QUANTA OF BECOMING AND THEORY OF RELATIONS

Professor Jonas is troubled by Whitehead's epochal theory of becoming. The reason for this theory is a version of the old Zeno argument against change assumed to be continuous. Henrik von Wright has, apparently independently, argued for a similar view as the only way to avoid contradiction in describing becoming, the only way to have definite terms with definite natures for relations of succession. Otherwise, in any time, however short, a process will be both *p* and *not p*, and nothing will conform to the idea of entities with definite and consistent properties. Whitehead's argument was that (assuming continuity) before a process taking up a second can occur, a process must occur taking up half a second, and before that one taking up a smaller amount of time, and so on, implying a nonfinite series which, though beginningless, must have been run through before anything identifiable can happen. The difficulty is not to get to the end of the series but to get into it. With the admission of momentary actualities, which become rather than change and which correspond to successive finite segments of physical time, the specified difficulties vanish.

There is here a difference between Bergson and Whitehead. For the latter it is (in spite of some careless Whiteheadian sentences) simply false to say that successive moments of process "interpenetrate." And Bergson's own view of open future and closed past, or of present memory as embracing the past but not *vice versa*, his theory of becoming as cumulative creation, is contradicted by the use of symmetrical expressions such as *inter*penetrate. The earlier penetrates the later but surely not the later the earlier. Over and over Bergson rightly rejected the doctrine of *mutually external* moments of process and treated the contrary extreme of *mutually internal* moments as thereby justified, as though mutual internality were the only escape from mutual externality! This is no more logical than it would be to hold that, since a formal logic of propositions, all of which are mutually independent of one another, would be absurd, we must adopt one of propositions all mutually implicative or equivalent. Instead, logicians operate with the assumption that some pairs of propositions are mutually independent, some mutually implicative, and some such that, while *p* entails, depends upon, the truth of *q*, the converse dependence does not obtain.

Alas, most controversies about internal and external relations have fallen into this amazing fallacy of treating contrary symmetrical extremes as exhausting the possibilities. On the externalist side we have Occam, Hume, and Russell; on the internalist, Spinoza, Bradley, Royce, Blanshard, and in Ancient China the Buddhist Fa Tsang. I see no logical justification for the neglect, common to both parties, of the asymmetrical case. In the last hundred years only Peirce and Whitehead are even relatively clear on this point, and even they are sometimes forgetful or confused about it.

VIII. A CAUTIOUS INTERACTIONISM

Among the advantages of the new metaphysics is the way it enables us to escape the harshness or crudity of the old issues between epiphenomenalism, two-aspect views, interactionism, and materialistic identity. In these controversies several values are at stake. There is the value of keeping open the path of scientific inquiry. This value is threatened by crude declarations that causal explanations are incompatible with teleological ones, or that a caused action is not really an action but merely a happening. In process philosophy there is no absolute dichotomy between happenings and actions. All happenings are also actions, so far as the singular dynamic agents are concerned, though in most happenings nothing remotely approaching the human level of conscious purpose is involved. Also causes of happenings or of actions are simply necessary preconditions that are only approximately or probabilistically "sufficient" to determine the outcome. There is no absolute contrast between "mechanical" and nonmechanical, for nothing is, in the sharp sense, merely mechanical. Always there is in present happening an element, however slight, of here and now determination not fully and uniquely implied by the preceding situation and causal laws. Always, however, the conditions do make a difference, whether or not they tie down results to precisely what happens, and it is worth knowing what the conditions are and what difference they make. Hence, science has plenty to do searching out necessary and probabilistically sufficient conditions without worrying about the idol of absolute determinism.

As for interactionism, the scientific import of this is easily exaggerated. Our theory of participation makes it clear that neural processes powerfully influence experiences, for they furnish their direct data (except so far as these are our own remembered past experiences, and they exercise selective agency over which memories assume prominence in consciousness). Thus the processes are conditions of the experiences. Is there a reverse influence? Process philosophers affirm that there is; but how strong is this influence? If participation is the way the world hangs together, then the only possibility of explaining an influence of human experiences upon cellular processes is in the supposition that cellular processes are themselves experiences participating in our (antecedent) experiences. However, there seems reason to consider this reverse participation or influence as incomparably weaker, in one sense, than the other. Taking any one cell, an experience of ours has far greater influence on it than it has on our experience, and the experience influences a multitude of cells directly, while each cell influences directly only a few neighbors. But taking the nervous system and the body as a whole, the comparison is very different. It is the higher forms of life that can adequately participate in the lower forms, not *vice versa*. (Were it otherwise it would be the lower forms that had science, not we at the pinnacle of terrestrial life.) From this it seems to follow, so far as I can make out, that "epiphenomenalism," while not strictly true (how many propositions are that?), is less false than some critics of it suppose. Our cells can do a lot without us, as one can see from the fact of dreamless sleep when our experiences are in abeyance, but we can do nothing without them. We must participate in their feelings, they can dispense rather largely with ours. There is, therefore, a stronger case for our bodies influencing our minds than for our minds influencing our bodies. Combining this consideration with the vast and, for our intelligence, virtually infinite complexity of the nervous system, it seems to follow that too much is made of the need to assert the absolute independence of neuromuscular processes from influence by our experiencing in order to safeguard or promote scientific investigation.

As the well-known geneticist Sewall Wright pointed out to me long ago, an influence of mind on body need not, to be significant, involve more than extremely small en..gy differences. It is like pushing a tiny button, or not pushing it, to launch, or withhold, a mighty missile. Still earlier Maxwell hinted at a similar idea. As Jonas also has pointed out, an almost infinitesimal change in angle can produce as great a change as you please in outcome.

IX. ROBOTS AND THE PROBLEM OF QUALITY

Another issue that takes on a different character when viewed in process terms is that of thinking machines and man-like robots. The issue is not whether we are mere machines, for there are no mere machines. At least on the microlevel there is always a lowly form of life or mind, i.e., of self-activity, which Plato and Aristotle identified with mind or soul, but which they did not realize is pervasive in nature. The issue is only whether in robots there is any whole-life, or whether instead the life in them is solely in the imperceptibly small parts. On the participatory theory there can be such whole-life only if life in the parts is sufficiently coordinated. Moreover, in animals the coordination is on a supramolecular level, a harmony of cellular activity. In robots there is nothing like this. (Or, if there should be, we can consider crossing that bridge when it seems a probability for the reasonably near future.) There is a kind of coordination, but it is not such as to make possible a simulation of life, experience as at least emotional and sensory; rather, the aim is at simulating intelligence as derivation of conclusions from premises or at simulating the functions of the brain as feedback mechanism to guide bodily movements. But experience is no mere device for getting conclusions from premises or for controlling movements; it is something in itself, indeed it is the principle of reality. And it is feeling, enjoyment, basically, with intelligence a complication and contingent development. Experience is the end of ends.

We are told that if a robot acted in all ways like a human being, we should have to accept it as equivalent in awareness. Since I take it as an *a priori* truth that absolute duplication of human behavior by a machine radically different in internal structure is a contradiction in terms (every difference makes a difference), I wonder how I am to react to this proposal. And I think that the evidence of a robot's experiences, or lack of them, must include the facts of its internal construction. Behavior is not the end to which the nervous system is mere means; behavior is, in part, a means to preserving and stimulating the nervous system and its accompanying experiences and, in part, a means to preserving and stimulating the nervous systems and experiences of other animals. If the cellular structure of animals is the thing that makes their whole-experience possible, as I believe it is, then I want the robot to show analogies to that structure if it is to pass as a fellow conscious being.

The value of structure is found, not in structure alone, but in quality of experience. In most of the talk about thinking machines nothing but structures are mentioned. But on that level of abstractness not even the ideas of causality, time, or space are intelligible. This is the real message of Hume's analysis of causality. What has present happening to do with past or future happening? In experience it is memory, perception, and goal seeking that tie events together; in materialism we have merely the assertion that something or other does this.

The neglect of secondary or tertiary qualities by science leads some to suppose that these qualities characterize only the subjective experiences, or these and the bodies, of animals. The specific qualities yes, since we can have no direct evidence that they qualify things outside our bodies. However, we do (as I have argued above) have direct evidence that they qualify bodily constituents. Moreover, the fact that science builds its world picture out of mere temporo-spatial–causal relations shows only that our human power to deal with qualities is more limited than our power to deal with quantities and geometrico–causal relations. And for a good reason! The so-called primary qualities are more abstract, and it is the concrete that is difficult to handle conceptually, not the abstract. (For this reason mathematicians mature early.) We can conceptually run through possible geometries, possible relational patterns, and there is no clear limit to our power to explore possibilities of this kind. But what can we do to explore possible qualities? A man born blind can do rather little to imagine color experiences though, as Helen Keller guessed, he can do more than some suppose. But to conceive mathematically definitely a certain type of geometrical pattern, we need only sufficient logical power. And we have learned that the geometrical structures of the world hang together in certain lawful ways that can be more or less sharply defined. How do the qualities of feeling in the world, outside our type of organism, hang together? That is a mystery by far less accessible to intellectual plus observational inquiry. If we knew *that* we should indeed know everything. For quality includes relational pattern, as every artist knows. Quality could not be other than more or less mysterious. It is not we who could know how all things feel. We are not God "to whom all hearts are open." We might, however, know *that* all things – except loosely organized composites – do feel.

X. DEATH, OBJECTIVE IMMORTALITY, AND GOD

Professor Jonas suggests that Whitehead, by his "immortality of the past" doctrine, has too easily done away with the tragedy of death. But here he is hardly making vivid to himself what for Whitehead is the real situation. Apart from God, the immortality of the past is cold comfort indeed, being but a generalization of the standard notion of "social immortality" which any philosopher can preach if he wants to. Granted, the human species is itself not known to be immortal, and indeed it clearly could come to an end, whereas for Whitehead the creative process as such could not do so, and therefore in a generalized sense all actuality is for him strictly everlasting. However, with the generalization goes a diluting. My vivid experiences are palely, inadequately, and more and more palely and inadequately as time goes on, prehended by later experiences, some (until I die) belonging to my personally ordered society, the rest not. More and more the prehensions are "negative," relegating aspects to the irrelevant background of experience. Only God has adequate prehensions of the past, only his prehensions are entirely positive. Thus, unless one can accept not only Whitehead's categories but also his idea of deity, the tragedy of death remains nearly as it is in other philosophies that reject personal immortality in the usual sense.

Even with God, there remain deeply tragic aspects. "Always there is the dream of youth and the harvest of tragedy." That we are mortal is something with which I for one have no emotional quarrel; but that the time and manner of our dying is subject to chance intersections of creative actions that no one, not even God, can know or determine in advance, is harder to face with equanimity. Premature death, ugly ways of dying, or lingering half alive, these are evils, even though mortality as such is not. Lovers all face the question: Which of us dies first? If it is *A*, *B* suffers bereavement, and *A* perhaps fails to carry out some of his or her major plans in life. For both selfish and unselfish reasons one wants and does not want to survive the other. In any case it takes a high level of self-transcendence to find sufficient the preservation of one's experiences, not ultimately in one's own experiences, or even in those of other human persons, but in those of deity. True, this self-transcendence is no more than should all along have been read into the imperative, love God with all your being. Nevertheless, some insist that only if one can count on endless additions to one's own experiencing, or endless further experiences of one's friends, has death lost its sting. So I cannot feel that this philosophy trivializes the meaning of mortality.

The faith of Jonas that we can contribute to the divine life, and in that way achieve permanent significance, seems formally close to Whitehead's, apart from being more vaguely and tentatively expressed. At least the two ways of thinking seem to be within talking distance of one another, which is more than one can say of some philosophical confrontations.

University of Texas at Austin,
Austin, Texas

NOTES

[1] For Ryle's view on introspection as retrospection, see his *The Concept of Mind*, 1949, Barnes and Noble, p. 166.

BIBLIOGRAPHY

1. Edwards, P. (ed.): 1967, 'Panpsychism', *The Encyclopedia of Philosophy*, Collier-Macmillan.
2. Hartshorne, C.: 1934, *The Philosophy and Psychology of Sensation*, The University of Chicago; reprinted in 1968, Kennikat, Port Washington, N.Y.
3. Hartshorne, C.: 1973, 'Creativity and the Deductive Logic of Causality', *Review of Metaphysics* **27**, 62–74.
4. Russell, B.: 1948, *Human Knowledge: Its Scope and Limits,* George Allen and Unwin, London.

STRACHAN DONNELLEY

WHITEHEAD AND JONAS: ON BIOLOGICAL ORGANISMS AND REAL INDIVIDUALS

For students of Whitehead's speculative philosophy (esp. [3] *Process and Reality*), thanks to recurrent critical discussion over the past half century, the fundamental Whiteheadian notion of 'actual entity' has become reasonably intelligible.[1] The notion of actual entity, which refers both to God and to finite, episodic 'actual occasions,' is Whitehead's reformation of the traditional notion of substance. Actual entities are the "final, really real things" of Whitehead's cosmos. They are its concrete real individuals, world-experiencing, self-creative, self-functioning, working together in constituting the creative advance which is the evolving cosmos, realizing among themselves that emergent order and those novel, concrete values which are the cosmos' ongoing significance. So much is reasonably clear. But does this doctrine make sense – real, ontological sense? Can there really be real individuals such as conceived by Whitehead? This is a question which demands further critical discussion. It is a question not about the coherence of Whitehead's system, but about its adequacy to our experience, to our good ontological, if not common, sense.

This question has a specific importance relative to Whitehead's central aims. Above all else, Whitehead wishes rationally to understand our status as living and valuing creatures, to understand our true significance and that of the natural universe, to understand the meaning of civilization and to promote certain future directions of civilized effort. No doubt, these are noble and important aims. And in fulfilling these aims, the philosophic notion of actual entity holds a privileged position. The actual entity (or actual occasion) is the grand home of reconciliation – that concrete entity in which the dominant persuasions of modern science concerning nature (e.g., compulsion, lawfulness, necessity) and the dominant persuasions of civilized man concerning life (e.g., freedom, effective teleological activity, aims at the realizations of novel values) meet and are reconciled. Nature and life are seen in their mutual requirements, as factors of a concrete individual. The phenomena of life and value, the activities of living and valuing, are saved, spared from the omnipotent grimace of an ironclad, natural necessity impervious to free, self-determined functionings and realizations of value. We need not fear being written out of the universe, explained away. Our lively selves, with all the

S. F. Spicker (ed.), Organism, Medicine, and Metaphysics, 155–175.

significance and meaning which attend life's necessary freedom, can, and do, peacefully coexist with the natural universe – ontologically, if not practically. Yet, despite the nobility of Whitehead's saving aim, can we accept this home of reconciliation, the Whiteheadian doctrine of actual entity?

I wish to examine a specific critical question. Can there be real individuality – in the sense of a real concretum involving self-constituting creative activity, aiming at and realizing some particular value for itself and for others (Whiteheadian criteria) – apart from biological organisms (at least some, if not all, organisms)? If not, Whitehead's speculative philosophy, in all its phases, is undermined (on ontological grounds). His pre-speculative assertions are in doubt; his conceptual scheme is thrown in jeopardy, and his interpretation of life and our value experiences are suspect. We are no longer philosophically saved – at least by Whitehead's system. However, the door may be opened upon a new, more secure philosophic salvation.

To explore this critical question of the relation between real individuality and organism, I wish to examine certain reflections of Hans Jonas, a philosopher who, in *The Phenomenon of Life* [1] and *Philosophical Essays* [2], is also centrally exercised by the problem of real individuality and whose findings are at odds with White head.

I. JONAS ON INDIVIDUALITY

Jonas holds that individuality implies something more than an entity being stubbornly there, a mere identical "it" firmly ensconced in a spatio-temporal scheme. He wishes to distinguish between real individuality and particular existence as such, which is possessed by all concrete entities. To warrant the title of a real individual, the concrete entity must also be internally related to itself. Individuality must have a hold in the entity itself, be something internally and essentially owned by it ([2], p. 186).

Jonas endeavors to take a stand "within": to understand what it is, or means, to be an individual from the inner perspective of the individual itself. From this exercise of rational imagination (a phenomenological enterprise), he concludes that individuality, as opposed to some mere identity (i.e., individuality in any ontologically significant sense), is the prerogative of a specific mode of existence, a peculiarity clinging to entities which take up that mode of existing. These entities are biological organisms ([2], p. 186–7). Their mode of existence is organic existence, which essentially and fundamentally involves metabolic activity. As we shall see, this activity *qua* an existential necessity is decisive *re* the ultimate ground of individuality.

To support his conclusions, Jonas proposes initial criteria of individuality.

Only entities whose being is their own doing, their own task, are individuals. Such entities are "delivered up to their being for their being, so that their being is committed to them" ([2], p. 187). So far, Whitehead and Jonas agree, and Whitehead could (from Jonas's perspective) legitimately posit actual entities *qua* real individuals for whatever philosophic purpose he wishes – e.g., to account for the ultimate constituents of the world, or for ordinary physical objects and their spatiotemporal interrelations. But then comes the decisive break, a basic ontological divergence. For Jonas, individuals are committed to keep up their being by ever renewed acts of it. Otherwise, they are not (truly) individuals. For Whitehead, a real individual has no such opportunity nor onus. An actual occasion, *qua* a *subject*, arises and perishes with its one, epochal shot at self-existence. Ontologically, this means that a real individual cannot be a biological organism, considered in its full, ongoing integrity, and that an organism cannot be a real individual, with the particular inner integrity enjoyed by an individual.

In this divergence of philosophic opinion, the whole inner landscape of an individual changes. For Jonas, the inner being of an individual is irrevocably exposed to the alternative of not-being. It continually constitutes itself or achieves its being in answer to this constant, potential imminence ([2], p. 187). This means that individuals are entities "that are temporal in their innermost nature, that have being only by everbecoming, with each moment posing a new issue in their history" ([2], p. 187).

With this shift in ontological perspective (which initially moves on the level of "pre-systematic" discourse), essential characteristics of a real individual, as recognized or posited by Whitehead, must be newly understood and newly interpreted. For example, the particular, concrete identity of an individual cannot merely issue from a novel form of definiteness (an 'eternal object'), epochally aimed at and achieved, thereafter undergirding the stubborn particularity of the individual's objective immortality, its existence *qua* experienced (as a given datum in the future creative advance). Rather, the identical individual remains (through time) an acting agent – primarily, for itself, an experiencer and an actor, not an object experienced ([2], p. 190). The individual's identity undergoes a sea-change, *qua* a philosophic problem. Identity is a protracted effort, the self-creation of a particular integrity through continuous performance ([2], p. 187). If the definiteness of this identity be unique, it is not unique once and for all.

Further, uniqueness of character is not the fundamental issue (as it is for Whitehead). Vital integrity, self-identity through time, is what is at stake. Particular individuality, difference from all other reality, is a peculiar existential necessity, rather than a metaphysical obligation (perhaps serving some cosmic purpose). "*Difference* from the *other*, from the rest of things, is not adventitious and indifferent to them [individuals], but a dynamic attribute of their being, in that the tension of this difference is the very medium of each one's maintaining itself in its selfhood by standing off the other and communing with it at the same time" ([2], p. 187). In sum, individuality is an existential achievement, won in the face of a world to which the individual is necessarily committed. Whitehead says as much. But his individuals, though free to be themselves, are metaphysically guaranteed (e.g., by the Category of the Ultimate – creativity, the many, and the one – and by the cosmic functionings of God, most specifically in setting up an individual's initial subjective aim). There is no existential risk involved, no possibility of *their* non-being before the term of their original performance. In that integrity is not the challenge of *continuing* integrity, it is not such a nerve-wracking, existential issue. This is one of Jonas' explicit complaints against the adequacy of Whitehead's philosophy ([1], p. 95).

Jonas philosophically interprets the nature and origin of characteristics exhibited by individuals – the ascending range of biological organisms, up to, and including, man. We will follow his investigation in as far as it pertains to our critical questioning of Whitehead.

However, there remains the initial, outstanding bone of contention. A natural universe filled to the brim with momentary real individuals stands against a natural universe which harbors relatively few, temporally enduring individuals, all else being the seemingly "dead" matter of inorganic nature, involving passive, extrinsic, or external identities – identities which merely happen or come about due to concatenations of blind forces. How are we to choose between these rival philosophic declarations? How are we to recognize real individuals and to distinguish them from pretenders to the throne of individuality?

The issue is complex. For Jonas, we are natively equipped with an epistemic arsenal which, if not conclusively deciding the outcome, nevertheless decisively controls or sets limits to philosophic speculation. We are real individuals, living creatures, ourselves. We know at first hand the joys and horrors of internal identity, the struggle to maintain bodily (and psychic) integrity. We are our own clues to other individuals. Certain external signs speak for internal identity; others do not. Life recognizes life and its opposite, non-life ([1], p. 81). What these signs are, we leave to later.

For Whitehead, real individuals, at the very least beyond the momentary me, cannot be recognized at all. We do not explicitly or consciously experience single actual entities (except perhaps God), but nexūs of actualities. Actual entities, as conceived by Whitehead, are speculatively *posited* so to interpret our creaturely, "macro" selves and all else that we consciously experience (all our pre-speculative accounts). As we shall see, Jonas feels that Whitehead speculateth too much – and wrongly.

I.1 Organisms and Internal Identity

According to Jonas, organic identity, which *can* be observed by external signs, is a necessary and sufficient clue to internal identity (which *cannot* directly be observed) and thus to individuality. Anything which exhibits organic identity is an individual. Anything concrete which does not, is not an individual, but a mere particular. This is Jonas's contention. As such, he places on himself the philosophic burden of explaining organic identity and all that it implies.

By the fact of metabolic regeneration, organic identity must be different from physical identity – an identity of its own kind. During the course of its life, an organism is continually changing its material constituents, physical identities, while remaining more or less the same ([2], p. 190). As opposed to Whitehead, Jonas accepts in some form the (more) classical conception of physical identity – some concrete, identical it (energy or "matter"), continuously adventuring (blindly) through space-time, recognized only by the (greater or lesser) continuity of its spatio-temporal determinations ([2], p. 188). (Whether all nature is "alive" or, rather, whether we may with epistemic legitimacy speculate about the concrete reality of inorganic nature is perhaps the final, real bone of contention between Whitehead and Jonas, from which all else follows – Jonas finding it more judicious to assume an agnostic stance towards "that which we cannot know.") Whitehead, philosophically determined by the discontinuity demanded by his atomic theory of actualities (and perhaps by the quantum theory of modern physics), explains physical identity (over space-time) in terms of organic identity ([2], p. 189). What remains identical is an ongoing *nexus* of episodic actualities physically inheriting from one another, repeating or reaffirming certain forms of definiteness.

Jonas will have none of this inner dimension of experiential inheritance as the foundation of physical identity. He wishes to save such a dimension for organic creatures ([1], p. 81). Seemingly, the physical record of the identical it does not change. We have no warrant to attribute organic identity unless the concretum's identity is somehow *threatened* – unless there is a threat to its existence, felt and overcome by the entity, and recognized by us lively observers ([2], p. 189). Molecules and stones, or their ultimate constituents, do not evidence such internal threats.

Beyond the demand of epistemic agnosticism, Jonas wishes to reserve the rights of a "dead" nature, so that organisms (and their organic identities) remain the natural curiosities and the profound ontological puzzle they are ([1], p. 95). He further wishes to preserve the fundamental role of death, of the threatening imminence of not-being, in the organism's self-constituting activities – something he feels Whitehead inadequately accounts for ([1], p. 95). He thereby discharges a philosophic debt to Heidegger. This is not to say that, by so defending the uniqueness of the organic realm, Jonas does not bring philosophic difficulties down on his own shoulders. How can there be interaction between organic and inorganic nature (e.g., in metabolism, vegetable and even animal)? How can life arise at all? These remain genuine, if unanswerable philosophic puzzles for Jonas, which Whitehead's doctrine of actual entities and their prehensive relatedness (their "experiential" bonds) were meant to resolve.

Notwithstanding these philosophic perplexities, organic identity does not rest on physical identity, though the organism essentially requires its passing material constituents ([2], p. 191). There is no physical record of *its* continuing identity. Organic identity is the identity of a *form* (through time), not of matter ([1], p. 81). Further, it is not the continuing identity of a mere form, but of a self-constituting agent who realizes this living form of itself and continually does so ([2], p. 194). Such a living form is ontologically "the whole structural and dynamical order of a manifold" ([2], p. 192). This living form is what remains the same, thanks to the activities of an immanent agent continually replacing the materials of the structural and dynamic manifold.

The temporal continuity of this living form, this half-way house for physical identities, only transiently resident, is the telling sign, observable from some external perspective, of the internal identity of a real individual, a true "ontological one" (versus a merely identical it) ([1], pp. 80–1). We recognize such individuals because we share something in common – metabolic activity, with all this entails, of which we have an intimate, inner acquaintance (because we are continually living through it).

"With all this entails" is the crucial foothold for Jonas, a provincial Archimedean point from which to weave an ontology of individuality and of organic life – a prelude (perhaps) to a general ontology, an alternative to Whiteheadian cosmology. Metabolic activity signifies, and originates, the ontological state of needful freedom ([1], p. 80). Organic identity has to do with the relation between form and matter, in which the living form takes a certain priority over matter ([1], p. 79). It has an independence or freedom relative to any one passing collection of its material. This is the organism's fundamental freedom vis a vis the inorganic universe ([1], p. 81). But to realize itself and to continue itself, the living form needs material constituents and their constant refurbishment. Thus this origin of freedom is balanced by the origin of need ([2], p. 191). A certain independence from the material universe speaks likewise for a certain dependence on that universe. The ontological individual, by virtue of its very self-constituting activity, stands in the relation of needful freedom to its surrounding environment ([2], p. 193).

According to Jonas, this basic freedom of organism, this certain independence of form from its own matter, as realized in metabolic activity, is an ontological revolution in the history of matter ([1], p. 81). Thereafter, the development and enhancement of this freedom or independence is "the principle of progress in the evolution of life, which in its course produces new [ontological] revolutions – each an additional step in the initial direction, that is, the opening of a new horizon of freedom" ([3], p. 81). But each new freedom is a conditioned, or a relative freedom. It is balanced by new necessities, new needs, by new kinds of dependency on a world from which the organism never escapes. Freedom is always needful freedom. The very variety and scope of an organism's needs, generated by its particular mode of existing, is the mark of its excellence ([2], p. 193).

It must be emphasized that organic identity implies a necessary, integral, reciprocal relation to the ontological one, to the individual who is, and constitutes, the living form through its various and necessary material manifestations. This augurs Jonas's decisive break from Whitehead's philosophy. The identity of the organism must be located in something which exercises and enjoys the freedom of *continuing* living form ([2], p. 194). The individual, as the originator and re-originator of the form, requires "*internal* identity as the subject of its existing *in actu*" ([2], p. 195). The living, dynamic form, passingly enmattered, itself must be understood as in service to the self-related identity ([2], p. 195). Metabolism is the function of the lively, crisis-ridden organism, not organism the function of metabolism ([1], p. 78). The organism's existence is its own performance. It accords its own identity, thanks to an immanent teleology, which marks the residing and presiding individual ([2], p. 193).

This fundamentally means that a real individual, according to Jonas, must be a more substantial, lasting creature than Whitehead's epochal actualities. Further, thanks to its dialectical relations with its own material substratum, *it* (in a sense different and more radical than for Whitehead) engenders the characteristics of its individual self. "This ontological individual, its very existence at any moment, its duration and its identity in duration is, then, essentially its own function, its own concern, its own continuous achievement" ([1], p. 80). In sum, according to Jonas's lights (in opposition to Whitehead), an individual's unity, potentialities of unique character, and potential freedom do not arise from God, *qua* the guarantor of individuality. They arise primordially in metabolic activity, in the ontologically novel, dialectical relation of form to matter. Here is Hegel really submerged in the mud. Here is a functional identity, but an identity which keeps functioning.

Jonas emphatically underscores the essential finitude of organic nature, the necessary dynamic relation of an organism to its environment. Need, indigence and insufficiency are no less unique distinctions of life than its powers ([2], p. 195). This is the ontological message of metabolism. Whitehead might agree – for *his* biological organisms, but not for his real individuals. Their initial deficiency is not so radical. They are originarily equipped with measures to deal with presented contingencies, more or less happily, and are metaphysically spared from mid-stream annihilation. Jonas explicitly brings to task Leibniz – and less assuredly Whitehead – for failing to weave life's insufficiency more deeply into the fabric of their philosophies ([2], p. 195). If Whitehead looked to bodily feelings as the clues to his theory of real individuals, he should have looked deeper – to the basic crisis of metabolic activity.

I.2 Jonas and the Characteristics of a Real Individual

For Jonas, the ontological clues of metabolism, the signs of life's insufficiencies and powers, are philosophically decisive. It is with metabolic activity that we gain primordial philosophic explanation of all essential characteristics realized by individuals ([1], p. 82). The characteristics are explained *re* their origins and their necessary mutual requirements. Such an investigation transforms the philosophic enterprise itself, as conceived by Whitehead and others – transforms the landscape of a real individual: both the nature of its fundamental characteristics and their explanation. At the center of this transformed landscape is the demand for temporal, enduring, worldly integrity (identity). We further summarize Jonas's explanations, so far as they pertain to our investigation.

Teleology and potentiality. As we have seen, metabolism signifies the twin, correlative virtues of life: liberty and existential necessity. An organism's being is suspended in possibility, the possibility to be or not to be ([2], p. 195). This is the origin of potentiality in its most emphatic, ontological sense, as it is enjoyed by real individuals. To actualize the possibility of being, metabolism necessarily requires the use of the external environment. Thus the "world" originates, as a realm of real possibilities, as a counterpoise to existentially necessary activities. The world supplants or overlays the prior existence of a realm of external necessities, a more or less orderly whirlwind of blind forces. Possibility (in its meaningful, human sense), subjective and objective, comes on the scene with organic existence.

The origination of a world as a realm of possibility not only speaks for the advent of an individual in need. It necessarily marks the advent of teleological activity, the ontological answer to the individual's innermost possibility, which itself is a reflection of need. In short, need is the ultimate ground of possibility – entities *qua* possibilities for subjects, including the subjects themselves. Teleology necessarily implies possibility (in this restricted, important sense) – and vice versa. Thus this means that need and teleological activity originate, or come upon the cosmic scene, together. "*Teleology* comes in where the continuous identity of being is not assured by mere inertial persistence of a substance, but is continually executed by something *done*, and by something which *has* to be done in order to stay on at all" ([2], p. 197). This demand for constant regenerative activity further augurs the ontological birth of the phenomenon of concern ([2], p. 197). And finally, the basic concern and *telos* of an (organic) individual, the fundamental teleology of an individual, which is an answer to its wanting, is of a particular nature. It is "the acting out of the very tension of the polarities that constitute its being and thus the *process* of its existence as such" ([2], p. 197). Here is teleological activity where it really counts – in service of continuing existential performance, the organism necessarily pulling itself up by its own bootstraps, or, rather, by using its world.

It is to be particularly noted that, in Whitehead's scheme, teleological activity and concern exist prior (though perhaps minimally) to the organic realm and, correlatively, require God for their ultimate philosophic explanation. Have teleology and concern been ripped out of their proper natural context and let loose upon the universe, to find another explanatory home – a home where they ill-fit and from which they long to escape? Seemingly, there are other unwilling aliens as well, other expelled souls longing for their native soil.

The self and *the world*. For Jonas, metabolic activity enacts within itself the cardinal polarities which characterize a real individual – not only the polarity of being and not-being, of freedom and necessity, but also the polarity of self and world ([2], p. 196). He wishes to emphasize that "selfness," which is always self-constituting activity, and "worldliness" *originate* with metabolic activity. This is the cosmological and ontological origination of "experience": both the experiencer and the experienced (*qua* experienced). Here is the original transformation of the mere universe into a world – such transformation always being effected from the side of the organic subjects. The world only exists for subjects. This basic Whiteheadian insight is echoed by Jonas. But there is a mere, "dead" universe, as well as a world lit up by organisms. Further, the basic opposition of an autonomous self *formaliter* and an external world *objectivé* is not metaphysically secure. Transcendence and autonomy must be (ever) won in the face of the world. They are an intrinsic aim of basic existential activity in a sense far more radical than Whitehead would have it. Success is never guaranteed.

Again, we can see Jonas's basic divergence and his introduction of a note of existential drama – a drama enjoyed (and suffered) only by organisms.

> The introduction of the term "self," unavoidable in any description of the most elementary instance of life, indicates the emergence, with life as such, of internal identity – and hence, as one with that emergence, its self-isolation as well from all the rest of reality. Profound singleness and heterogeneity within a universe of homogeneously interrelated existence mark the selfhood of organism. An identity which from moment to moment reasserts itself, achieves itself, and defies the equalizing forces of physical sameness all around, is truly pitted against the rest of things. In the hazardous polarization thus ventured upon by emerging life, that which is not "itself" and borders on the realm of internal identity from without assumes at once for the living individual the character of absolute otherness. The challenge of "selfhood" qualifies everything beyond the boundaries of the organism as foreign and somehow opposite: as "world," within which, by which, and against which it is committed to maintain itself. Without this universal counterpart of "other," there would be no "self" ([2], p. 196).

Whether all organisms – and organisms at all times – feel such existential anxieties is another question. Unfortunate are those organisms which are visited frequently by such Hobbesian (or Gnostic) moods. There are sunny days when the world, and its resident creatures, invite pleasant commerce.

Nevertheless, Jonas's philosophic message is clear. Whitehead, being a good natural organism, may have held out for too many summer days thus adversely infecting his philosophy. In the service of life he has (it is charged) misread ontological structures of his basic evidence. Selfhood is foremost a challenge, not a metaphysically guaranteed opportunity – the challenge of winning the bodily self in the face of, and by means of, the external other.

As such, the actuality potentiality dynamics of an individual are read differently by Whitehead and Jonas. "The becoming of the one self" has different meanings. For Whitehead, inner activity has to do with the novel realization of the one self, the creation of its epochal wholeness. With Jonas, wholeness is coterminous with activity. The organism acts to regenerate its (previously achieved) wholeness – a wholeness which is essentially and internally threatened, irrespective of any external threats. The impending loss of wholeness primordially animates activity ([2], p. 191). This is the potentiality that basally directs activity and which transforms the external environment into relevant possibilities for self-integrating activities.

For Jonas, this is a paradox of organic individuality which finds no direct echo in Whitehead's theory of real individuality nor (perhaps) in his philosophy as a whole. The organic individual exists at any one time in a one simultaneous composition of stuff – its necessary physical, or material, basis. Yet it is not identical with this basis, for, while coinciding with it, it is already in the act of passing beyond it ([2], p. 191). The organism lives as a whole through its entire life-span, which means living as a whole in the various material compositions through which it passes. It is no classical substance, but a changing subject of change, which fights to keep its integrity through change.

Further, in that an organism is essentially a "moving beyond the given condition," an openness or a horizon is intrinsic to its very existence ([2], p. 197). "Concerned with its being, engaged in the business of it, it must *for the sake* of this being let go of it as it is now so as to lay hold of it as it will be" ([2], p. 197). Such a basic, forward-looking concern necessarily means, most pointedly with the emergence of animal existence, the advent of temporal and spatial horizons – the inner horizon of temporality and the outer horizon of spatiality, both yoked together by the fundamental, active concern. In short, Jonas can give an account – in a way that Whitehead cannot – *why* there is world-experience. He need not resort to a category of the ultimate and other supporting metaphysical principles. There are cosmologically provincial, non-metaphysical reasons – i.e., creatures living under the onus of metabolic activity.

Unity and the problem of wholeness. As expressed above, Jonas feels that, with living organisms, nature has sprung an ontological surprise. Here are systems of matter which are real unities of a manifold, inexplicable by the mere concurrence of the forces which bind their parts together ([1], p. 79) – i.e., inexplicable on purely mechanistic, "billiard ball" terms. Wholeness is no mere accident, no mere phenomenal or extrinsic by-product of subterranean machinations. Organisms are real, composite unities

> in virtue of themselves, for the sake of themselves, and continually sustained by themselves. Here *wholeness* is self-integrating in active performance, and form for once is the cause rather than the result of the material collections in which it successively subsists.

> *Unity* here is self-unifying by means of changing multiplicity. ***Sameness***, while it lasts, . . . is perpetual self-renewal through process, borne on the shift of otherness ([1], p. 79; emphasis mine).

Such active self-integration is the essence of a real individual and founding for its ontological conception.

In short, 'self-unification' and 'the drive for integrity' are given rival metaphysical interpretations by Jonas and Whitehead. Whereas Whitehead stresses epochal concrescence to concrete unity (the movement, within the one subject, from relative disjunction to determinate conjunction), Jonas emphasizes self-initiated activity in service of continuing a unity, of continuing an integrity. Unity comes out of unity, integrity out of integrity. What remains to be seen is how Whitehead accounts for the particular unity and integrity of an organism, and to what extent this account is convincing. Thus, the question of whether Whitehead adequately accounts for the wholeness of an organism must be deferred until the next section. We must leave open the question as to how Whitehead accounts for the "individuality" of a biological organism, and whether or not it might be considered a concrete "really real" entity. Jonas's central concern with the problem of 'wholes and parts' will be revisited in that context.

Nevertheless, we might briefly consider relevant advantages of Jonas's philosophic stance. Unity and individuality are active, existential achievements, with no metaphysical guarantees. As such, there are, and can be, greater and lesser degrees of unity and individuality, depending on the grade of the organism involved and *on its particular performances.* "To be an individual means not-to-be-integrated with the world, and the less so, the more it is an individual. Individuality implies discontinuity" ([2], p. 204). This fundamental message of organic existence, already presaged on the metabolic level, lends intelligibility to certain organic developments – and to the world which opens forth on the basis of these developments.

These organic developments basically concern the centralization of organic individuality, as evidenced in animal existence. This augurs the introduction of new needful freedoms, new ways of serving basic metabolic functions – for example, the concomitant development of powers of motility, perception and emotion ([2], p. 198). Such developments necessarily imply the centeredness of the one self, with its various powers. The existential functions of the (complexly diverse) sensing self, the emotional self, and the motile self must be pooled in, or emanate from, a single center if they are to issue, as coordinated, in informed, controlled action, serving metabolic (and other) needs ([2], pp. 190–200).

This centeredness necessarily implies greater individuality, a more fixed internal identity, a more pronounced unity ([2], pp. 201–203). An organism enjoying the advantage of a developed centeredness is less dependent upon, and thus less integrated with, its immediate environment. A far-seeing animal with far-reaching, sustained emotions may roam far in pursuit of food – and in avoidance of enemies. That it must do so, once committed to such a mode of self-nurturing, goes without saying. And there can be perceptual, emotional and motor breakdowns. The animal may find itself longing for the more domestic, less adventurous life of a plant. Nevertheless, during its season of success, the animal enjoys a self-constituted identity and integrity beyond that enjoyed by a plant. The animal may pick the time and place of its foragings, reverse its field, and take time-outs to rest comfortably with itself ([2], p. 201). Thanks to its concentrated centeredness, evidenced by the development of a central nervous system ([2], p. 200), the animal may run around the world – perceptually, emotionally and physically – without losing its hard-won sameness ([2], p. 203). It is the same animal which whets its appetite by gazing upon prey from a perch on the rock, and finds itself in hot pursuit of its dinner. A plant could not tolerate such uprootedness, without losing a hold on itself.

Jonas holds that individuality and freedom in organic existence apply fully only on the level of "centered" animal existence ([2], p. 201). The centeredness of animal organization underscores the " 'indivisibility' which lies in the literal meaning of 'individual' " ([2], p. 204). Yet it is the organism as a whole which is centered, which is the central agent controlling its own parts and functions ([2], p. 200). Centeredness does not mean the loss of the organism's full integrity *qua* an ontological individual. Individuality does not escape into some ethereal center, to preside over a material country alien to itself. The living form, and thus the individual, extends to the frontier of its body – barring some catastrophe, in which case it must pull in its boundaries.

Individuality is always biological individuality. This necessarily means the tension of selfhood and world ([2], p. 205). The accentuation of this tension is the accentuation of life itself. What results is a more pronounced (organic) self standing over and against a more pronounced world ([2], p. 205). The two necessarily go hand in hand, for both result from the self-constituting activities of the one organic individual.

II. WHITEHEAD'S BIOLOGICAL ORGANISMS: THE PROBLEM OF ORGANIC WHOLENESS AND INTEGRITY

We wish to take our interest in the relation of biological organism and real individuality – and our critical glance at Whitehead – a step further. Jonas's claim that only biological organisms (most obviously, animal organisms) are real individuals carries a correlative assertion. We can only concretely understand organisms in virtue of their being real individuals. Their self-constituting, internally-related activity is what engenders their concrete wholeness and integrity – that which manifests and expresses their real, dynamic individuality. Such an assertion or conceptual perspective can be turned back on Whitehead. In short, how adequately does he philosophically interpret biological organisms, with their particular kind of wholeness and integrity? If his system of interpretation should fail this test of adequacy, this would be another blow to his theory of actual entities (as the final, concrete individuals of the universe in terms of which all else is to be understood).

Again, Whitehead's systematic interpretation of biological organisms (including man) has been rendered sufficiently clear. Biological organisms are enduring nexūs, internally related groups, of constituent, episodic actual occasions, more or less complex modes of togetherness of Whitehead's fundamental, experiencing entities. More specifically, internally complex organisms (such as animals and man) are 'living structured societies', which harbor within their complexly patterned selves sub-nexūs or sub-societies which account for organisms' central control of themselves and for the dominant soul of the human organism. The resident actual occasions constitute and sustain the enduring organic society in virtue of their episodic, creative activity, in virtue of imposing dominant complex characters on the experiencings of one another. The society is sustained and nurtured by this genetic inheritance imposed by its members on themselves (along with, less proximally, requisite, more or less stable impositions from the extra-societal environment). Novel functionings of resident members effectively nurture the whole society in novel ways, but only within the confines of the dominant social order which must be sustained.

It is within this general interpretative context that Whitehead accounts for the diversity of biological organisms and for the diversity of their modes of functioning, more or less complex. All depends on the modes and subtlety of organization (togetherness) and on the opportunities for experiential activities which thereby arise. Correlatively, it is within this general interpretative context that Whitehead must account for organic wholeness, including the integrity of the human organism. The adequacy of this interpretation is our specific concern.

It is not difficult to recognize that an individual actual entity is an organic whole. The one self-constituting subject is involved in each constituent prehension (the subject's active "taking account" of items of the universe other than itself), mutually adjusting them, integrating them into its final, integral, composite self. The one subject stands over and above its various constituents; it is more than the mere summation of its parts. It is a real (ontological) whole. But can this be said of Whitehead's biological (in particular, animal) organisms? They are complexly composite entities, but seemingly there is no one subject presiding over the composition – a single subject which is more than its organic, "material" parts and which survives their metabolic repair.

Rather there are millions of centers of life, millions of epochal subjects, coordinating among themselves ([3], p. 128). A Whiteheadian biological organism is a natural nation with individual citizens, passingly performing requisite, coordinate functions, forming particular sub-societies – a division of labor essentially serving the national purpose, the integrity of the organic nation in internatural relations (and warfare). But *who* is the national purpose for? Seemingly, for no singular who. It is for all the democratic individuals, though a fortunate aristocracy (with animal organisms) seems to enjoy the better of the bargain. Even though they have their own crucial social duties, they get the best (and perhaps the worst) of the national experience. Yet the nation is not theirs, save experientially. No one aristocrat owns the organisms, *qua* an active becoming subject. The organism as a whole is always "other" for any one experient – beyond its own immediacy. Further, the social coordination as a whole importantly depends on the persuasive agency of God (via the initial subjective aims of the constituent actual occasions) – one nation, under God.

If in political philosophy it is dangerous or pernicious to conceive a real nation on the model of a natural organism, so in philosophical biology it may be unfortunate to conceive a real organism as analogous to a national society. There may be certain advantages – e.g., explaining the so-called dissociation of personality and the seemingly autonomous functionings of organs ripped out of their organic contexts ([3], p. 128–9). Yet, as is becoming increasingly evident, such organs *in situ* do not function in complete autonomy. A yogi can take striking command over his bodily functionings. Does this speak for a one organic self, of which conscious activities are but a component tip, and which usually effects its own divisions of labor? Or does it speak for many subjects inheriting from one another?

Here, of course, we are at the critical, thorny, perhaps irresolvable problem of the integral self – how far it dips down into the body, how enduring it is in time. Is it true that the enduring human self is a society of actualities, epochal real selves, inheriting past members' memorial experience? Do I not *have* a personal past, rather than inherit it? 'Inheritance' seemingly connotes gaining possession of something that I did not previously own. But do not I, the same me, own or possess my past – a past which frequently leaps up in my face? Memory seems different from inheritance, something more personally mine, lurking in the bowels of my self, constituting much of myself. Am I more of an integral "person" than Whitehead's interpretative scheme would allow?

In a similar fashion, do I really inherit from my (subjectively other) body? I have a body; I *am* my present bodily self. The bodily me perhaps inherits from the outer world, but seemingly not from itself (myself). I seem too close to myself for that. It is not that hand, stuck on that arm, which recoils from the hot radiator and informs me that it has done so. *I*, my whole organism, recoils. In short, this is a point which Jonas wishes to stress. The self-centered individual which is the organism exists to the boundary of its body ([1], p. 78). Things are either "inside" (me) or "outside" (the others), with the skin patrolling the frontiers – whether or not these be the cavernous avenues meandering through my body or the more exposed flanks.

With regard to the problem of wholes and parts, Jonas finds the modern analytic tradition of science and philosophy wanting ([1], pp. 200ff). We must judge if Whitehead has successfully overcome this deficiency. In the modern tradition, wholes are understood as mere summations of parts, mere aggregates, which are not the (final) cause of the parts' organization. "The aristocracy of form is replaced by the democracy of matter," the higher explained in terms of the lower ([1], p. 201). This is decidedly not the case with a biological organism, where the living form is the reason for the organization of the parts and for their metabolic turnover ([1], p. 78).

In good Whiteheadian fashion, Jonas digs deep into the seventeenth century and uncovers a neglected side of Spinoza's philosophy: his theory of complex bodies (mind-bodies). Finite individuals are not substances – Spinoza's early reaction to Cartesian dualism – but are modes of the one substance. The individuality, the distinct, continuous identity, of a finite mode is marked by its continuing form of determinateness, not by its substantiality – a form realized through relation (causal communication) with the modal others, a continuance evidencing the mode's basic conatus ([2], p. 212). "Form, continuity, and relation are integral to the concept of individual and provide a clue to the meaning of its identity" ([2], p. 212).

Such modes, in Spinoza's theory, may be highly complex – i.e., be composed of other individual modes in determinate forms of interaction or communication. This *form* of mutual determination, as long as it continues, speaks for the individuality, the individual conatus, of the complex mode. Constituent individuals may change as long as the complex form survives. Thus, Spinoza accounts for the possibility (which may be a necessity) of the metabolic activity of the *one* individual ([2], p. 212). The whole, the determinate pattern, is more than its parts. The pattern lasts; the parts pass. And it is the business of the mode to continue its formal pattern and replace its passing parts.

Further, in that all modal determinations are won on the basis of interactions or causal, communicative relations with the wider, world environment, there can be various levels of "acting and sufferings," depending on the kind of body involved ([2], p. 212). "Since the individual is a *form of union*, there are qualitative grades of individuality, depending on the degree of differentiated order" ([2], p. 215). With highly complex biological organisms, this means that a highly articulated and differentiated communicative interaction with the world is enjoyed by the enduring individual, thanks to the complexly ordered individuals "ridden" by itself. Sophisticated abilities of suffering (sensitive reaction) and of activity strictly parallel one another, and are correlative to the uniquely appropriate body.

> In such a stratification, the variability of being which compositeness enjoys as such, is communicated upward cumulatively, and with each supervening level is raised to a higher power, so that the uppermost level representing the totality in question is the beneficiary of all its subordinate members ([2], p. 216).

The decisive point is that such complex organisms are themselves modal wholes. As real individuals, *they* continue "their conative life" in virtue of metabolic activity, of interchange with the larger environment, productive as much as produced, as much "natura naturans" as "natura naturata" ([2], p. 216).

The question remains, do Whitehead's biological, animal organisms pass such tests of organic wholeness? As we know, he claims no total, enduring, embodied subject for such organisms. Their organic wholeness cannot be accounted for by (a single) subjectivity. Within his conceptual scheme, a form of definiteness (a complex character), as *severally* enjoyed (experientially) by each constituent actuality, and as imposed on one another, must foot the bill. Seemingly, this must mean that there is a feeling *for*, a feeling *of*, the bodily integrity (the society) in each constituent occasion, more or less dominant in its experience.[2] Bodily integrity would be a realization of a form of definiteness in the subject's objective datum, provoking subjective response. The whole is in the part, and the part in the whole. This is a necessary part of the occasion's world experience. Other (intra-bodily) social functions are conformed to, and perhaps are determined by, this form of definiteness, realized by the past members of the society and as a project for new members.

Yet is this satisfactory? There (always) is *only* an experience of the whole (*objectivé*) – either as settled in the past or as a project for the future. And this is only from a limited perspective within the body. Further, contemporary bodily occasions are atomic. Though tied to a common past and a more or less common future, they are (according to Whitehead) discontinuous relative to one another. Among themselves, they do not even form a society (though they belong to one). This seems a rather loose confederation. If a nation is an ideal entity, existing only in the minds of its citizens, so seems an organism, considered as a lively, immediate whole. Perhaps by transmutation (an actuality's activity of taking a nexus of former actualities as one, in virtue of a common character they severally exhibit), the organic entity is taken as one. But, though a "real entity," this is still not a real individual, with *its own*, self-constituted integrity. The organism never really is, *qua* a subject *formaliter.*

Whitehead has interpreted biological organisms on the basis of his doctrine of actual occasions, his final, concrete experiential entities, experiencing the experience of former experiencers. Experiencing implies mental functioning, no matter how minimal or "unconscious." With such a scheme of interpretation, has Whitehead really, successfully overcome the subjective bias of modern philosophy? Has he truly escaped Hegel and the idealist tradition? Can experience be all that there really is (all final, "really real" existents being experiential entities, experiencing and experienced, no matter how objectively and stubbornly "there")? Seemingly, we experience because we are bodily organisms, because it is necessary to our mode of material existence (of which experiencing is a partial function), to our maintaining individual integrity by metabolic regeneration. Likely, we are not organisms because the creative advance, spurred by God, wishes to stage particular intensities of experience for its own (or his) enjoyment (Whitehead's final judgment). The only applause we can safely bank on (and that is not assured) is from other living organisms – because they are lives like ourselves, swimming in the soup of organic existence, needful of one another. Primarily, we experience to live, not live to experience. What we want is our liveliness touched, enhanced, and secured, not the mere, unending accumulation of more and more experiences.

Whitehead wavers on this point. Practically, he keenly appreciates the practical and erotic demands of organic life. Theoretically, he occasionally seems to lose his practical sense. For example, metabolism, that phenomenon which so seriously concerns the philosophic reflections of Jonas, gains only a passing glance from Whitehead – and as not such a serious affair. The reason that there are metabolic entities is that the universe has staged concatenations of actual occasions enjoying the unconscious conceptual activity of reversion (conceptual entertainment of novel forms of definiteness relevant to the subject's worldly situation), which disturbs the equilibrium of their immediate environment ([3], p. 126). In the name of continuing social order (of the organism), they must repair the damage. They are (in the last analysis) spurred to reversion by the primordial nature of God, by his subjective aim after intensity of experience. Thereby, the organism's robbery of the external environment, including its feasting upon other organisms, is cosmically, if not ethically justified ([3], p. 125) – all in the name of self-justifying experience (ultimately to be enjoyed by God). Little is said about the organism being resigned to robbery by its very mode of existence, willing experiencer or no. The chill is taken off organic existence, the necessity of metabolic activity.

In sum, a philosophic understanding of the organic wholeness, integrity and individuality of an organism – the adequate understanding of our organic natures and existence – may be only one sacrifice at the altar of a philosophy of organism which is too much of a "philosophy of experience" – of a philosophy which commits itself (in pre-speculative assertions) to the fundamental assumption that there is nothing ultimately real or actual but experience and experiencers.

III. CONCLUSIONS

If our critical discussion and recurrence to Jonas carry persuasive force, certain conclusions seem inevitably to follow. Whitehead's interpretative scheme fails the test of adequacy on two counts. Whitehead fails to account adequately for real individuals since he fails to see that real individuality is necessarily tied to organic (biological) existence. He fails adequately to interpret biological (at the very least, animal) organisms because he fails to see that they *are* real individuals, that they are more radically one, *qua* self-functioning subjects, than he allows. Biological existence and real individuality (in their correlative, infinite shades of complexity and intensity) go hand in hand. If this mutual implication is correct, this is telling evidence against Whitehead's doctrine of actual entities and the scheme of interpretation built on its basis.

Yet whether our critical remarks have telling force, and whether Whitehead or Jonas has given the more philosophically adequate explanation of real individuality and biological organisms, are not the fundamental issues. They are but a part of a larger, far more important issue. This concerns the nobility or dignity of Whitehead's philosophic aims: to understand our status as living and valuing creatures; to understand the meaning and significance of ourselves and our universe; to understand our ultimate worth and duties.

Beyond all questions of final adequacy, Whitehead, amidst much darkness and doubt, endeavored to take ourselves *qua* living, natural organisms and our cosmic setting seriously, endeavored to recognize their true contribution and value in the lives we realize. And he forged an impressive conceptual scheme to do so. If we can possibly view individuality and organisms more adequately than Whitehead, this is only because he has helped to clear the way and we can stand on his shoulders. If we (and Jonas) are critical, it is only to take his project further – to take the mutual implication of life, organism, and real individuality even more seriously. Herein may be the fulfillment of his aims: to understand the human organism as a knowing, valuing, acting creature; to understand knowledge, valuation, and action as the natural performance of that real individual which is the integral human organism; and finally, to understand the human organism, in its lively integrity, as a religious, ethical, and aesthetic animal, with organic life coming fully to itself in its religious, ethical, and aesthetic dimensions. But this fulfillment awaits a philosophic imagination equal to, or even greater than Whitehead's.

New York City,
New York

NOTES

[1] See such culminating critical studies as William Christian's *An Interpretation of Whitehead's Metaphysics* (Yale University Press, New Haven, 1967) and Ivor Leclerc's *Whitehead's Metaphysics* (George Allen and Unwin Ltd., London, 1965).

[2] Leclerc, *op. cit.*, p. 219.

BIBLIOGRAPHY

1. Jonas, H.: 1968, *The Phenomenon of Life: Toward A Philosophical Biology*, Dell Publishing Co., New York.
2. Jonas, H.: 1974, *Philosophical Essays: From Ancient Creed to Technological Man*, Prentice-Hall, New Jersey.
3. Whitehead, A. N.: 1967, *Process and Reality*, The Free Press, Macmillan, New York.

BETHIA S. CURRIE

THE REDEFINITION OF DEATH

In response to the report of the "Ad Hoc Committee of the Harvard Medical School to Examine the Definition of Brain Death," Jonas included in his essay "Philosophical Reflections on Experimenting with Human Subjects," a section dealing with the redefinition of death as "irreversible coma" ([5], pp. 105–131.) In this section, Jonas states what he later repeats in "Against the Stream: Comments on the Definition and Redefinition of Death," that he sees nothing wrong with allowing a patient in a coma which has been determined to be truly irreversible to die with dignity; that is, that he does not object to disconnecting the life-support systems that are maintaining the marginal life of such a patient. What he does oppose is pronouncing the patient "brain dead" and leaving him connected to the life-support systems so that the patient may become available for vivisection under a different designation.

In this first response to the redefiners, Jonas' opposition is based on his argument that we do not know the exact borderline between life and death, and that this borderline may be – and probably is – unknowable. If all that is in question is allowing the patient in a truly irreversible coma to die with dignity, he says, then it does not matter whether we know this borderline or not. Nature will cross it if we simply abandon technological supports and say "Let nature take its course".

On the other hand, if we are going to pronounce the patient "dead" and leave the life-support systems connected, then we absolutely *must* know the exact borderline between life and death, because until we do we cannot be sure that we are not arresting natural processes and keeping the patient "this side of what may in truth be the final line."

In "Against the Stream," ([5], pp. 132–140) Jonas begins by addressing himself to three charges brought against him by some members of the medical profession in private exchanges:

(1) that his reasoning regarding "cadaver donors" counteracts sincere life-saving efforts of physicians;

(2) that he is countering precise scientific facts with vague philosophical considerations;

(3) that he is overlooking the difference between death of "the organism

S. F. Spicker (ed.), Organism, Medicine, and Metaphysics, 177–197.

as a whole" and death of "the whole organism", which contains a related distinction he is also said to have overlooked: the difference between spontaneous and externally induced respiratory and other movements.

Jonas answers the charges. He points out that the first begs the question. The so-called "cadaver donors" are only "cadavers" if the redefinition of death as "irreversible coma" is granted before the argument begins, and Jonas, of course, opposes the redefinition. He rejects the second charge on the ground that the vagueness is not *his* vagueness, but the vagueness of the subject matter. He is arguing *about* a vague *condition.* He adds:

> Giving intrinsic vagueness its due is not being vague. Aristotle observed that it is the mark of a well-educated man not to insist on greater precision in knowledge than the subject admits, e.g., the same in politics as in mathematics. Reality of certain kinds – of which the life–death spectrum is perhaps one – may be imprecise in itself, or the knowledge obtainable of it may be. To acknowledge such a state of affairs is more adequate to it than a precise definition, which does violence to it. I am challenging the undue precision of a definition and of its practical application to an imprecise field ([5], p.134).

The third question, with its subsection, is the focus of Jonas' essay, and for those whose interest lies in a response to this charge, the best advice is "Read Jonas."

We will concentrate on the "vagueness of a condition," specifically, the condition of coma. We will make a particular point of answering the claim made by these members of the medical profession that they can offer "precise scientific facts." Jonas has pointed the way with his statement that we do not know the precise borderline between life and death, and that it is probably impossible *ever* to know that borderline, and when he remarks that the comatose patient may have remaining sensitivity situated elsewhere than in the brain, and vulnerable to suffering. This is another way of saying that our knowledge of the brain and central nervous system and their relation to the rest of the body, is not sufficient to assure us that there is no possibility of remaining suffering in the supposedly "dead" patient.

We will begin by offering evidence from a noted Russian neurologist, S. A. Sarkisov ([24], pp. 138–9; 148; 185–6; 220; 241–2) with the intent of making a quite explicit counter-charge that our ignorance of basic processes and functions of the brain belies the claim to possession of "precise scientific facts" in an area in which knowledge is absolutely essential if the redefinition of death as "irreversible coma" is to be accepted. We will also raise some questions about the limitations of the principal technological aid in determining brain function in findings of "brain death", the electroencephalograph (hereafter referred to as the EEG).

We will review very briefly the guidelines suggested for positive finding of "brain death", and will then discuss some cases in which these guidelines were met, but the patients were *not* "brain dead", and the apparently "irreversible comas" were reversible, and the patients recovered, thus indicating that it is possible to exhibit all of the signs of "brain death" without being brain dead.

To stress the fact that it is not possible in many cases to predict with accuracy whether or not a particular coma will prove to be irreversible, we will discuss some cases in which attending physicians anticipated no reversal and admit that subsequent reversal was "inexplicable".

We will point to some of the dangers in over-reliance on the EEG as a definitive determinant in positive findings of "brain death", and finally, we will discuss some cases in which totally unresponsive and apparently unconscious patients were conscious and aware even while being pronounced clinically dead.

The purpose is not to attack the medical profession, but to stress that it is a disservice to the profession for some members to claim that "precise scientific facts" are available where they are not. It is also the intention to stress the dangers to patient and physician in leaving "brain dead" patients connected to support systems for an indefinite time, as suggested by Dr. Willard Gaylin in his article "Harvesting the Dead" ([3], p. 23). Gaylin's article is not chosen at random. Gaylin cites and even quotes Jonas in this article, mentioning specifically two points made by Jonas:

(1) a defence of the sacrosanctity of the human body;

(2) the unknowability of the precise borderline between life and death.

Gaylin is inaccurate on the first point. Jonas does *not* defend the sacrosanctity of the human *body,* but of the human *person.* Gaylin, unlike Jonas, regards personhood as residing in the functioning brain, so that when the brain "dies" the "person" dies. Jonas asserts that the human being is an organism, and that personhood belongs to the whole organism, body as well as mind, even if higher functions of personhood may be said to be seated in the brain. The comatose patient is still a person, therefore, although a partial person, and he retains a residue of the sacrosanctity accorded to the human being. He must not be reduced to thinghood and used as a mere means.

II. OUR KNOWLEDGE OF THE BRAIN

Members of the medical profession who have had to deal with severe brain injury are well aware that one of the great difficulties lies in making accurate prognosis in the critical initial period, before stabilization of the patient. In a brief article, "Reflections on Death," Dr. Denton Cooley gives an illustration of the case where decision is easy ([2], p. 507). The patient had suffered a severe blow to the head, which ripped away the skull, the brain coverings, "and a generous portion of the brain itself". In the emergency room, a dedicated young physician managed to keep the patient alive and even to achieve some stabilization of his condition. When the senior surgeon arrived, however, he discontinued treatment, disconnected support systems, and allowed the patient to die, because this was an instance of such massive damage that the outcome – if the patient had managed to survive – was inevitable and easily determinable.

Cooley offers a brief follow-up in which the senior surgeon is reported to have delivered "a sober lecture on the dignity of a quiet unmolested death – the necessary, fitting finale under such conditions." Certainly Jonas would most emphatically agree. But this is not the kind of case where the problems really lie, and notice, too, that this patient was not pronounced "dead" and then maintained, or even pronounced "dead" before treatment was discon-continued. The specific statement is that he was allowed to die, and he was not kept in a state of marginal life which was then redefined as "death" so that removal of tissues and organs could be made while he was, in fact, alive.

In many brain injuries the initial stages are confused, the damage is not precisely determinable, and even after stabilization of the patient it is not always possible to predict the outcome. One then faces the possibility of a comatose patient in a vegetative state, but no longer dependent upon support systems, as in the Karen Ann Quinlan case.

This does *not* mean that the decision to make a positive finding of "brain death" should be hurried. What it means is that it is often simply *impossible* to make the decision with certainty, for the very reason that we do *not* have precise scientific facts about many of the complexities of the brain, and the condition of coma is vague. After all, we are speaking of an area of the body in which some ten billion neurons are packed, in some places in densities of one hundred million to the cubic inch, each neuron connected with up to sixty thousand others, no two alike. It is not surprising that precision is impossible. We should acknowledge, as Aristotle advises: it is wrong to insist on greater precision in knowledge than the subject admits.

Perhaps some of our over-confidence is the result of our falling prey to what Jonas has termed the "seductive" language of the cyberneticians, a danger against which Sarkisov also warns. After quoting from Norbert Wiener's claims for the neuron model, Sarkisov points out that the model is fine for designing complex cybernetic systems, but is "extremely relative" in its application to the living cell, because it does not metabolize, does not exhibit the functional lability of living matter, and has no way of incorporating precise adaptations to changes in the environment of the living cell.

Besides this Wiener has underestimated the role of the stimulus, because model and living cells are so different, and "it should not be forgotten that a signal spreading along nervous structures (a nervous impulse) and a signal in an automatic apparatus differ qualitatively in their nature, for they carry different forms of energy".

Sarkisov believes that using principles of cybernetics to study biological processes, especially higher nervous activity, will lead inevitably to mechanicism, which results from "an unjustified extension of a concept from one field of science to another more complex field, where it is unsuitable or where its application is only partially permitted, taking into account the specific features of the problem under consideration".

Mechanicism is obviously an inheritance from the Cartesian machine-model of the body, and it still infects our thinking, and persuades us that in any area of study of the human anatomy we will find ways to simplify and to reduce problems to mechanical principles so that sooner or later all will be known to us.

One example of such over-simplification is the tendency to over-localization of function in the brain. Gaylin, in the article previously mentioned, notes that some members of the scientific community would like to see the definition of death narrowed even further to "cessation of cortical function". The problem, he adds, is that we do not yet know with certainty what the cortical functions are. Nevertheless, he is convinced that higher functions of personhood are seated in the cortex.

But Sarkisov is not. He writes

> Investigations into the action of psychotropic drugs on various structues of the brain showed that the intellectual processes and the affective behavior of human individuals depend not merely on the activity of the cerebral cortex, but also on complex processes taking place in the subcortical formations, the brain stem, the thalamus, the hypothalamus, the reticular formation, the rhinencephalon, the hippocampus, and the limbic structure [24].

We are hampered in our attempts to understand some of the complex processes and functions of the higher levels of the central nervous system by the absence of adequate axo-dendritic studies, which cannot be carried out because we have not yet developed methods which would make it possible to fix and stain these structures, yet "in the higher levels of the central nervous system, and especially in the cortex, the axo-dendritic transmission of impulses is predominant".

The fear the layman feels with respect to the medical profession and science in general, is often related to just such situations as this. We have not yet accepted the redefinition of death as "irreversible coma", and already Gaylin is pushing for acceptance of definition of death as "cessation of cortical functions", even while he admits that we do not know what these functions are. Now we learn from Sarkisov that we cannot even study the requisite functions adequately, so that we remain ignorant about the predominant form of impulse transmission in the central nervous system, especially in the cortex.

Sarkisov and his colleagues at the Moscow State Brain Research Institute, have done a great deal of work with the EEG in their study of brain processes and functions. Sarkisov feels strongly that the EEG should not be relied upon completely either as a diagnostic or prognostic tool, and says that it is "a mistaken notion that the biopotentials reflect all the distinct features and properties of the higher levels of the central nervous system in normal and pathological conditions."

In other words, the EEG, which is sometimes taken as the definitive feature in positive findings of "brain death", is not completely reliable for making diagnosis, or for making prognosis, either in normal or in pathological conditions. Even if it were, it is a mistake to assume that the biopotentials we are recording with the EEG "reflect all the distinct features and properties of the higher levels of the central nervous system." We are certainly justified in questioning whether the EEG tells us what we need to know in cases where "brain death" is being declared.

Even with all that we have learned about the brain we still do not know all of the properties and features of living matter, Sarkisov says, and this is especially true of "the physico-chemical properties of nerve tissues and the processes of its activity." He notes that there are new properties and substances being discovered, but they cannot be detected *in situ* "even by recording bio-electrical phenomena."

Although nerve tissue is composed of the same biochemical structures as other body tissues, and although the same metabolic processes found in other body tissues take place in nerve tissue, "against this common energy background, specific and highly complex neural functions develop in ways still to be discovered", and we do not have adequate studies of the biochemical compounds in individual brain structures or of the metabolic processes within these structures.

In warning against the illegitimate use of theoretical and practical cybernetics in the field of the biological sciences, Sarkisov quotes Engels: "we shall undoubtedly reduce thinking by some experimental means or other to molecular and chemical movements in the brain, but will these really exhaust the whole essence of thought?"

It is particularly interesting to read Sarkisov, because one of the most remarkable cases on which we will be reporting was treated in Russia. Sarkisov's book was written the following year, and is peppered with cautions about the limits of our knowledge and understanding of the brain and central nervous system, especially of the higher functions.

III. THE HARVARD CRITERIA: SUMMATION

Before we enter upon discussion of individual cases, it is well to review briefly the suggested guidelines for a positive finding of "brain death." The signs suggested as indicative of "brain death" are:

(1) unreceptiveness and unresponsiveness to external stimuli or inner needs;

(2) lack of spontaneous muscular movement or unassisted breathing – or effort to breathe – over a period of at least one hour;

(3) no reflexes (tendon jerks, pupillary reflexes, corneal reflex, for example).

In addition, if the EEG at maximal gain is flat, and if this state persists for twenty-four hours, then coma may be considered irreversible [25].

It is also suggested that if other signs are present and EEG waves appear, these waves may be ignored, because "spurious waves" are possible. If, on the other hand, all of the signs are present, then the absence of EEG waves is to be regarded as *confirmatory* evidence of "brain death". It is further recommended that the physician pronouncing death should not be involved in any way with later transplants of organs or tissues.

These are only guidelines, and they have no legal status. There are cases to be found in which not all of the signs were present, but attending physicians expressed their conviction that coma was irreversible. There are cases in which the real determinant was the EEG, and there are cases in which the suggested lapse of twenty-four hours was not observed.

IV. SPECIFIC CASES OF COMA

Suppose we take some specific cases and discuss them in relation to the suggested criteria.[1] The knowledge of medical details in these cases varies considerably, since most are not found in the literature, but appear in newspapers or magazines. We are limited to what some member of the medical profession has taken the time and trouble to report, or to what some editor regards as newsworthy. Coma was not a subject of interest in this country until the epidemic of "sleeping sickness" in 1914, and it faded from the news once the cause had been discovered and proper treatment determined.

Coma begins to reappear in *The New York Times Index* after World War II, when a reporter seems to have been struck by the possibility that a patient who had entered upon his two thousand and first day of coma might hold the record for longest coma. After this had been reported, other cases appear sporadically, but often we are not told of the outcome. We are also limited by the fact that local papers sometimes report cases that do not attract the attention of *The New York Times* or the national magazines.

From the material available the examples chosen offer more than name, place, and length of coma.

(1) Dr. Lev Landau, winner of the Nobel Prize for Physics in 1962, was critically injured in an automobile accident later in that same year [18]. This is a case in which there was severe brain damage, and an initial period in which the patient's condition was very unstable. The list of Landau's injuries is really appalling.

He suffered a fracture of the base of the skull, with laceration of the fornix cerebria; contusion of the frontal and temporal lobes; severe shock; fracture of nine ribs; pneumothorax and hemothorax; rupture of the pubic bone junction; fractures of the pubic, hip, and haunch bones; and of the head of the left thigh bone; severe contusions of abdominal organs; rupture of the urinary bladder; paralysis of both arms and both legs; gradual failing of circulation and breathing (apparently from brain damage) even while on a respirator. He was deaf, blind, speechless, without reflexes or pain reactions [29].

By all of the suggested guidelines, Landau was in "irreversible coma" and was therefore "brain dead". His skull was opened, but it was impossible to determine whether he was suffering from one large blood mass or from multiple hemorrhages. In other words, it was not possible to determine the extent of the brain injury, or its exact nature, and this is true of many such cases.

Although Landau was on a respirator, he was pulseless four times in one week, and his arterial blood pressure dropped to zero. One attending physician felt that it should be emphasized that Landau had never been clinically dead since his heart did not stop beating [21]. This is ironic. The redefinition is precisely for this kind of case – "brain death", but heart still beating, and the redefinition *refutes itself* here. Landau was *not* "brain dead" and his coma was *not* irreversible. Yet all the guidelines were met for positive "brain-death" finding. By the proposed redefinition, Landau was in "irreversible coma" for months.[2]

After seven weeks on the respirator, Landau was removed from support, but it was another seven weeks before he showed signs of returning brain function [19]. Director Boris Yegorov of Moscow's Institute of Neurosurgery noted that Landau's brain had had an insufficient supply of both blood and oxygen for well over one hundred days, and Yegorov felt that Landau's case compelled a reconsideration of "the whole of accumulated medical experience" in cases such as Landau's [29]. Specifically, Yegorov regarded the Landau case as a proof that it is not true that oxygen deficiency of the brain cells inevitably leads to their destruction. Landau was able to return to theoretical physics, but never fully recovered [19]. He lived another five years.

Neurologists from other countries did not agree with Yegorov, but they did concede that the Landau case undermined the argument – which was just then being advanced in the English medical journal, *Lancet* – "that doctors should not seek to prolong the lives of brain-damaged patients in 'irrevocable comas'" (the British equivalent of the American "irreversible coma"). But Landau had been in such a coma for months [29].

Yegorov's argument could be important to physicians who must make decisions in cases of brain damage with attendant coma. If it is *not* true that brain cells are inevitably (and rapidly) destroyed by oxygen deficiency, the prognosis that a brain-damaged patient will be in a vegetative state even if coma should reverse, might well be altered. Here is a case which again raises the question:

(2) Spc. 4/ Jacky Bayne, on routine patrol with his dog Bruno near Chu Lai, stepped on a land mine. At the field hospital Bayne was found to be pulseless, without heartbeat, not breathing, totally unresponsive, and with a flat EEG. After a forty-five minute attempt at resuscitation proved unsuccessful, Bayne was pronounced dead and sent to Graves Registration, where he lay for several hours unattended, until the embalmer reached him, opened the femoral artery to inject embalming fluid, and thought he discerned a very faint pulse.

Bayne was rushed back to the field hospital where a second attempt at resuscitation – this time with accompanying massive blood transfusions – succeeded. Bayne recovered with some speech impairment, and no other apparent brain damage [11].

This is a classic illustration of our ignorance of the borderline between life and death, and even of our inability to *detect* life. It is also an excellent example to support Yegorov's contention that it may not be true that oxygen deficiency inevitably destroys brain cells. Bayne was totally without support systems for several hours after "death" and had so faint a pulse that it could only be discerned by the opening of a major artery. There must have been acute cerebral anoxia, yet the resultant brain damage was minor.

(3) In January, 1966, Carol Dusold (now Rogman) was in an automobile accident, and was thrown from the vehicle, striking her head against a metal post and sustaining brain stem injury [23]. Attending physicians noted that the X-rays indicated the inside of Carol's head was "a jumble", and one physician later told Carol's religious advisor that if *he* could see the inside of her head he, too, would "just give up".

Carol was unable to swallow properly, remained totally silent, and exhibited no pain reactions or responses of any kind. She lay rigid, in deep coma. After six weeks, her eyes – which had been shut – popped open and remained staring fixedly, completely unresponsive to a bright light shined directly into them. When her physician was asked if there was not something more that could be done, he answered that it might be humane to discontinue the intravenous feeding that was sustaining her life. He did not believe that the coma would reverse, but feared that if it did the girl would be "a vegetable". Carol's mother refused to allow termination of treatment, and the girl remained in deep coma for a total of four months.

Quite inexplicably, coma gradually reversed to semi-coma, and Carol was sent to the Rehabilitation Institute in Chicago for a long and painful period of retraining. She is now married, has an associate's degree from a junior college, has borne a child, does most of her own housework, and continues to show improvement in coordination. She has a slight slurring of speech, but is perfectly intelligible, and occasionally she has some difficulty with memory. Don't we all.

Dr. Henry Betts, medical director of the Rehabilitation Institute, was asked by the journalist who was preparing a report on Carol's case, how he could explain her recovery, and he answered that he could not. He remarked that "You deal with millimeters when you deal with the brain". There had been doubt that Carol would recover the power of speech, and Betts pointed out that the speech area is so tiny "that insults just millimeters apart can make the difference in whether or not speech can come back". In other words, we have *some* "precise scientific facts", such as the fact that the speech center is tiny, but this does not make it possible to predict the outcome of damage in the area, for the very reason that it is so tiny that precision is impossible – just where it is most needed.

By the guidelines once again, Carol was "brain dead" for four months, and again it is not our intention to fault the attending physician, but to point out that in such cases as this the outcome *cannot be predicted* during the very period when accurate prediction is crucial in findings of "brain death".

(4) Lewis Sadowsky was critically injured when a truck collided with the family car. Here again we have an initial period of great instability and a totally unpredictable outcome. *The New York Times* reported:

> "It was a minute-by-minute struggle", said Marvin Klein, the physician in charge of pediatric neurology. "Fragments of his skull had been driven into his brain. His condition changed so quickly. He had endocrine problems, and there was danger of his dehydrating; he had seizures, convulsions, heart stoppage. He didn't breathe well. We did not have him medically stabilized for three weeks" [31].

Even after stabilization, Lewis remained in deep coma, and as the condition persisted, his doctors became increasingly pessimistic. He remained in coma for seventy days. One evening his father expressed the feeling that the family could not stand much more, and Lewis's sister asked the boy to touch a plastic cup in front of him if he could hear her. He touched the cup. She then directed him to touch their father's face, and he reached his hand in the correct direction, indicating that he could also see. Five months after the accident had occurred, Lewis was discharged from the hospital to complete his recovery at home. The reporter from *The New York Times* remarked on Lewis's "keen observations".

As far as Lewis's physicians were concerned he was never "brain dead". The case is included as a further indication of the inability to predict whether or not a brain-damaged patient's coma will reverse. The physicians *did* tell Lewis's family that reversal might *not* occur, and that the longer coma persisted the less likely reversal would be. That is, physicians indicated the possibility that "brain death" might result, and clearly they were unable to give an accurate prognosis, just at the time when such a prognosis was crucial, during the period of instability, when discontinuation of treatment allows nature to take its course.

Lewis's case illustrates another important aspect, that it is often difficult or impossible to determine whether or not a patient is conscious. Lewis was regaining consciousness when his sister asked him to touch the cup, but there was no indication of returning consciousness, and his father's expression of despair was made for the very reason that neither staff nor family could detect any change in the boy's condition.

Landau, too, exhibited no change in condition on the day a friend asked him to indicate by a blink of the eyes his awareness of the friend's presence, and Landau responded. Carol Dusold's mother daily pressed Carol's hand and asked her to return the pressure. One evening Carol seemed to respond. Carol's brother played a flashlight beam into her eyes every day. One evening he thought he saw a change in the pupils. But Carol's condition seemed to her physician to be unchanged, and he dismissed the family's reports as 'wishful thinking".

(5) Robert Spierer, eight years old, fractured his skull in a bicycle accident in July, 1957. After a second brain operation, he lapsed into deep coma, August second. For the first nine weeks, Robert's father remained at the hospital day and night. When he found it necessary to return to work, he still visited Robert twice a day, and whenever he was with the boy he spoke to him. On the evening of the seventy-ninth day of coma, Robert's father left the hospital without having elicited any response. The next morning he entered the room and found Robert in the same condition, but spoke to him because he *always* did. Robert answered him [16].

Again returning consciousness was indiscernible. The hospital staff also regarded Robert's condition as unchanged. This case took place in 1957, and the report is brief. There is no mention of EEG readings, but other signs of "brain death" do seem to have been present, and in that case the guidelines do not require confirmation by the EEG for a positive finding of "brain death".

(6) A case which combines a number of quite remarkable features is that of Roger Arntsen, a five-year-old Norwegian boy, who drowned in a fresh water river [28]. He was in the water for twenty-two minutes before being rescued, and for another two and a half hours had no heartbeat, even while on a respirator. His pupils were completely dilated, he had kidney failure, could not swallow or breathe, and required intravenous feeding for a week. His

physicians performed a blood exchange to rid Roger's system of toxins (the river was polluted), and he showed a gradual improvement over the first eleven days. Quite suddenly he went into crisis and lapsed into deep coma for a month.

Six weeks after the drowning, Roger underwent what his attending physicians referred to as an "inexplicable reversal." Apart from some loss of peripheral vision and a bit of clumsiness with his fingers, Roger recovered fully.

The length of time without heartbeat is noted in the *Guinness Book of Records* [8]. All of the other signs of clinical death (not just "brain death") were present for a time exceeding the suggested period. During the first three hours there must have been severe deprivation of oxygen for the brain, but damage was negligible. Again the prediction of outcome was impossible, and an apparently "irreversible coma" underwent "inexplicable" reversal.

Some of the cases of coma are very briefly reported, and apart from the statement that doctors did not expect recovery, and some mention of the extent of injury, we have very little information.

(7) Patricia Murphy, age 12, was in an automobile accident in 1950 which killed her grandparents and critically injured her mother [14]. Patricia received a fractured skull and was comatose for fifty-eight days. The prognosis was very pessimistic, but coma reversed. Her physician had no explanation.

(8) Clarellen Rible, twenty-two, was brought to the hospital in 1951 unconscious [15]. Physicians performed four brain operations in an unsuccessful attempt to diagnose her condition. The cause of her coma was never determined. She remained completely unresponsive for two months, and then quite suddenly awakened and was able to speak.

Physicians become increasingly pessimistic as coma is prolonged, but mere passage of time is no guarantee that a particular coma will not reverse.

(9) Pyrotr Vetrov, a Russian Soldier, received a severe concussion when a shell exploded close by him, early in World War II. He was comatose and totally unresponsive for eighteen years. When a nurse performing routine patient care thought she detected a faint response one day, he was given electric shock. The second treatment awakened him. He did not exhibit any diminished capacity [10, 17].

(10) We have very little information about the longest coma known to reverse. Karoline Karlsson, born in Monsteras, Sweden in 1862, lapsed into

coma (no cause given) on Christmas Day 1875, and remained comatose (no mention of depth of coma) for thirty-two years ninety-nine days. She awakened April 3, 1908, and lived to be eighty-eight years old, dying in 1950 [7].

Among children's diseases, one which often results in coma is Reye's Syndrome.

(11) Denise Hardin, six years old, was diagnosed as having Reye's Syndrome, and was in deep coma, totally unresponsive, for three and a half months [32]. Her physicians did not expect reversal and recovery, but removed two pieces of her skull (which were then frozen for later replacement) to allow the swollen brain tissues room to expand. The cerebral edema, which was assumed to have caused the coma, gradually dissipated. Coma reversed, and Denise returned home for the final stages of recovery wearing a football helmet to protect the still-exposed brain tissue. Doctors anticipated full recovery.

(12) Gerry Hay, also six years old and a victim of Reye's Syndrome, was in deep coma for a month, and had a flat EEG "for days" [27]. As a last resort, his physician put him on a respirator. Five days later he exhibited an independent heartbeat, coma reversed, and apart from a slight speech impediment which his physician regarded as temporary, the boy recovered fully.

In all of these cases there is deep coma, total unresponsiveness, and pessimism among the attending physicians. In every case, except that of Denise Hardin, the reversal of coma is regarded as inexplicable. For Denise, the relieving of pressure from the edematous brain tissue was probably the governing factor in reversal. Some of the cases, such as those of Gerry Hay, Lev Landau, Carol Rogman, show all of the signs of "brain death", but ended in recovery, and were certainly *not* examples of "irreversible coma" and "brain death." Two of the cases were apparently deaths of the whole organism. In several cases the return of consciousness was imperceptible to both staff and relatives.

Here, now, is a case that illustrates even more vividly the dangers inherent in redefinition of death as "irreversible coma." Unfortunately there are few details available, but the few we have are quite dramatic enough.

(13) A reporter whose primary interest was in the legal wranglings surrounding the Quinlan case, met on the steps of the Morris County (New Jersey) Courthouse, a young man named Guy Treadway, who was carrying a broadside describing his own case, and a clipping from a newspaper [31]. The journalist does not seem to have realized the significance of the case and gives a very brief report.

Treadway was pronounced "brain dead" after twenty-eight days in coma. A priest was called to administer last rites, and in the course of the ritual he pressed Treadway's hand and the "dead" man returned the pressure. The next part of the report is a bit disjointed, but it would appear that Treadway was then walked around the room, while still only marginally conscious, and fell on his face and woke up.

What if he had not been Catholic? No last rites, no hand pressure, and tragedy inevitable if that marginal consciousness had not been discovered *by sheer chance*.

Well, now, what member of the medical profession wants to be the first to practice some of the "exotic procedures" Gaylin mentions on a "neomort" such as this, who is "deposited" in the "bioemporium" on the twenty-eighth day? Who wants to be the first to administer, for example, viral leukemia to such a "thing" on the twenty-eighth day and find a conscious patient awaiting him on the twenty-ninth?

Make no mistake. It *can* happen. A cynic might well ask how we can be quite certain that it has *not* happened? We do not have the guarantee of "precise scientific facts" to *prevent* it from happening. It is just plain luck that Treadway was discovered to be alive before, perhaps, his still-beating heart was removed for transplant – without benefit of anesthesia, of course, since he was "brain dead".

Treadway obviously met the criteria for "brain death" but was decisively *not* "brain dead". He indicates that he was not even totally unconscious. He says his coma was like "a little pink womb". Yet certainly no consciousness was *detectable*, or he would not have been pronounced "brain dead".

V. THE EEG AS CRITERION

Even in 1968, when the guidelines were first suggested, it was known that temporary cerebral silence can result from hypothermia or from overdose of certain central nervous system depressants, and the advice was that in such cases the EEG tests should not be begun until the effects of such external causes had worn off. It was the knowledge that such temporary isoelectric EEGs can occur that reduced the EEG, in the guidelines, to the position of confirmatory evidence. But in many cases the EEG is, in fact, the determinant [13].

Dr. David J. Powner, writing in the September, 1976 issue of the *Journal of the American Medical Association*, offers a review of the literature, in a brief commentary "Drug-Associated Isoelectric EEGs: A Hazard in Brain-

Death Certification," and lists a variety of pharmaceutical agents and several disease-related conditions, all of which can effect cerebral silence [22].

As the sub-title of the article makes clear, Powner is concerned about "the care of patients with undiagnosed coma or drug overdose who may be candidates for organ donation if true 'brain death' can be certified." Powner has included even agents that produce relatively short periods of cerebral silence "because of the inherent tragedy of a 'false-positive' brain-death evaluation." He first lists the drugs formerly known to effect cerebral silences of any duration: barbiturates, methaqualone, diazepam, mecloqualone, meprobamate, and trichlorethylene. To these he adds phenothiazines, atropine sulfate, tricyclic antidepressants, nitrazepam, salicylates, heroin, insecticides, glutethimide, and Amanita phalloides (mushroom poisoning).

One of the listed articles reports a cerebral silence of twenty-eight days in a case of encephalitis, followed by return of cerebral activity. Another reports EEG evidence of cortical "death" followed by full recovery. Cerebral silence with subsequent return of cerebral electrical activity is also recounted in metabolic encephalopathy; with profound hypothermia and circulatory arrest in cardiac surgery; and after ischemic cerebral insults. (Powner notes that in the latter cases there have been some questions raised about certain technical aspects of the reports.) To these should be added two other listings of drug-associated comas. Succinylcholine sensitivity can simulate cerebral death [30] and an atropine-like coma can result from Ditran therapy [6].

Powner recommends that "quantitative assays of those drugs known to be associated with an isoelectric EEG be obtained in those patients comatose after a drug ingestion or in those patients with coma of undetermined origin who are being evaluated for certification of brain-death."

It is heartening to have such cautions coming from within the medical profession, but to the layman it is also disturbing that the warnings come so long after the pressures to redefine "death." This is only a report of what has been written, and we have no guarantee that the list of drugs or disease-related conditions which can effect temporary and reversible cerebral silence is complete. There is always the problem of idiosyncratic responses, too.

As Powner says, the EEG "is often regarded as the definitive feature of death," and we find that this "definitive feature" can be iatrogenic, temporary, reversible. How has it come about that a test which cannot distinguish between "dead brain, therefore silent" and "resting or drugged brain, therefore silent" has been accepted as definitive in determining "brain death"? How does one conclude that it is possible to distinguish between one silence and another silence?

It might be answered that in cases of severe brain injury the distinction can be made, but we have only to point to the cases of Lev Landau and Carol Dusold, to indicate that there can be serious brain damage and cerebral silence and quite obviously *not* "brain death," even when all of the signs of "brain death" were present.

When the guidelines were suggested in 1968, there was a warning that "spurious" EEG waves can occur and should be ignored if all other signs are present [13]. It is time to stress the numerous ways in which spurious *silences* can occur, and to recall Sarkisov's cautioning statement that the EEG is not completely reliable either diagnostically or prognostically.

VI. RESUSCITATION AND REVERSIBILITY

Several researchers have been compiling accounts of the experiences of patients who have been pronounced clinically dead and subsequently resuscitated. One of these researchers is Raymond A. Moody, Jr., whose credentials include both a PhD. in philosophy and an M.D. [9]. Moody has published a book with excerpts from a number of the accounts collected by him, and with some commentary of his own, and a Foreword by Dr. Elizabeth Kübler-Ross, who has been collecting acounts herself, has lectured on the subject, and is preparing her own book. She finds Moody's accounts to be strikingly similar to her own.

Kenneth A. Briggs, reporting in *The New York Times* says that although members of the medical profession who attended a lecture given by Kübler-Ross at Earlham College in the Summer of 1975, expressed polite scepticism at her conviction that these accounts prove that there is life after death, they were in agreement "that here are unresolved questions concerning such matters as clinical definition of death" [1].

Moody is a cautious researcher, and he himself raises critical questions, the most fundamental one being the question as to whether one can accept these accounts as relevant at all when death is, by definition, an irreversible state, and all of these patients were resuscitated. He also notes that the cases compiled by him do not include instances where cerebral silence was recorded, since most of the cases were emergencies, and there was no time to take an EEG.

Kübler-Ross, however, reports that some of her accounts include isoelectric EEGs, as well as all of the other signs of clinical death.

We will concentrate on just one feature of dozens of these accounts, i.e., the hearing and understanding of the pronouncement of their own deaths.

It is particularly interesting to note that the definition of "unconsciousness" in the *Merck Manual* is: "A state of insensibility in which the patient receives no sensory impressions and has no subjective experiences" ([4], p. 1276). Syncope, stupor, and coma are all considered to be states of unconsciousness. Death, of course, is not listed, but by this definition there are literally dozens of "dead" patients who have not even been unconscious.

VII. CONCLUDING REMARKS

Once again the point we wish to make is that qualified medical practitioners are sometimes simply *unable* to *detect* consciousness, and that this can be crucial. If we take the cases mentioned by Kübler-Ross in which there were confirmatory flat EEGs, we have the further contradiction of patients who were registering cerebral silence, and all of the other signs of clinical death, but who heard the pronouncement of death – and who, therefore, by definition were not even unconscious (since registering auditory data is a subjective experience).

The most important aspect of these cases, of course, is that they are evidence that it is possible for a patient to exhibit all of the signs of clinical death, including total unresponsiveness, fully dilated pupils, no reflexes, no pulse, no respiration, no heartbeat, and a flat EEG, and still hear and understand the pronouncement of death, be aware of the significance and react to the situation consciously.

If a patient without heartbeat and with a flat EEG can hear and understand in one case, what is the guarantee that the patient with beating heart and circulating blood, and a flat EEG is totally unconscious even when totally unresponsive?

Moody points out that a critic can always claim that there may be some residual biological activity in the supposedly "dead" bodies of the resuscitated patients to account for their experiences. This recalls Jonas' statement that we cannot be certain that there is no remaining sensitivity in the comatose patient, and it reinforces the point we are making here. First of all, this residual activity which is sufficient to render the patient conscious and aware is *undetectable.* Secondly, if the minimal amount of biological activity in patients pronounced clinically dead and not on support systems is enough to render them conscious and aware, we must revise our notions of the amount of energy required to render the brain functional, and we must ask what the possibilities are of undetectable consciousness in patients who are being maintained on life-support systems specifically designed to promote and sustain *far more* than residual biological activity.

That is, we know that the brain is greedy with relation to the rest of the body, and that it uses a disproportionate amount of blood and oxygen. If the rest of the body is being starved, the brain will take whatever nourishment is available. The assumption has been that this is an indication that the brain *requires* large amounts of blood and oxygen in order to function. The evidence of these resuscitated patients seems to indicate that we must revise this assumption. If the residual biological activity in the clinically dead body is sufficient to render the patient conscious and aware (sometimes even where there is confirmatory evidence of a flat EEG), then it would seem that far less consumption of blood and oxygen is required for a functioning brain than has been assumed.[3]

The belief that the EEG is sensitive enough to measure any remaining brain function will also have to be revised. Suppose that we take a case of cerebral silence in a coma of undiagnosed origin (not a case of massive brain injury) and assume that the patient has been pronounced "brain dead" and is then maintained on support systems which supply sufficient oxygen and nourishment to keep normal biological processes operative.

There is far more than residual biological activity, and we cannot take it for granted that this totally unresponsive patient is totally unconscious. Even if all of the evidence from resuscitated patients is discarded, there is always Guy Treadway to haunt us with the tragic possibility of the patient whose consciousness is not detected because *it is not detectable by our present methods*, perhaps because it requires far less energy than we have assumed it does to render the brain operative and the patient conscious and aware.

The medical profession does not know the precise borderline between life and death, or between consciousness and unconsciousness, and cannot detect the exact moment of loss of consciousness or regaining of consciousness. But the definition of death as "irreversible coma" *demands* that this borderline be detected with absolute certainty.

It is evident that the peripheral question of interruption of treatment without pronouncement of death is also raised here. If the patient is in coma as the result of some terminal disease such as cancer, then there does not seem to be a valid reason for not terminating treatment, especially if the patient has indicated this as a personal choice. Cases of massive brain damage, such as that cited by Dr. Cooley seem clear-cut.

As Jonas says, there is no need for a redefinition of death to deal with such cases. The California "Natural Death Act" is more to the point, enabling patients to discontinue treatment without endangering their physicians or next-of-kin, who might be open to murder charges without such legal protection.

But when Jonas says that in cases of truly irreversible coma he sees no need to continue treatment we are left with doubts. The Landau and Dusold cases indicate the gray areas, the cases where the predction of outcome of coma may *seem* clear, but accurate prediction is *not* made, and the "irreversible" coma reverses "inexplicably."

What we need, and do not have, are "precise scientific facts" about a very vague condition. It is likely that we are simply going to have to accept the fact that mistakes are made and sometimes lives that could be saved are lost. Physicians are not clairvoyant, and they are not gods, and it does them nothing but harm to make claims of precise knowledge that cannot be supported.

Until the precise scientific facts *are* available – and we must also live with the possibility that they may *never* be – the redefinition of death as "irreversible coma" is dangerous and should be opposed. We must agree with Jonas that "the only course to take is to lean over backward toward the side of possible life."

Cornwall, Connecticut

NOTES

1 The problem in acquiring details of cases such as those discussed in this paper is made evident by the Bibliography. Medical journals are devoted to articles reporting significant new findings, new techniques, new treatments. Reversal of coma by chance, or "benign neglect", or after a period of time determined, perhaps, by the next-of-kin's refusal to allow termination of treatment, is not a breakthrough and it doesn't make the journals. But it is sometimes what a local news editor thinks will make good copy, or an adequate "filler," and, of course, when it involves a Nobel Laureate it is news even to *The New York Times*. It is unfortunate that material so significant for the debate concerning the redefinition of death should depend upon an editor's determination of what may sell his publication. Occasionally the subject turns up on television, and then it may be entirely accidental whether the researcher stumbles upon the program or not. An example – I awoke in the middle of the night, Dec. 10, 1976 in time to hear a portion of an interview program with Jeanne Parr, in which a man was just stating the facts of his son's case. After severe brain injury the boy lapsed into coma Jan. 19, 1976, and remained totally unresponsive until June 6, 1976. Two and one half weeks before his inexplicable reversal, the attending physician informed the parents that the boy's condition was "irreversible." The parents refused to allow termination of treatment. The boy is well on the road to complete recovery.

2 A further irony – the Soviet Union has now accepted cessation of brain function as "death". Next-of-kin are not allowed to determine whether or not treatment will be terminated, and have no control over disposition of organs and tissues. *The New York Times* (3/2/77) has a very oversimplified statement of the issues involved, with a clear implication that physicians can, in fact, be quite certain when brain activity has ceased.

3 There are alternatives to this conclusion, of course. For example: (a) consciousness seated elsewhere than in the brain; (b) separable soul.

BIBLIOGRAPHY

1. Briggs, A.: 1976, *The New York Times*, Apr. 20, p. 15.
2. Cooley, D. A.: 1970, 'Reflections on Death', *The New York Times Encyclopedic Almanac*, New York Times Co., New York, pp. 507–508.
3. Gaylin, W.: 1974, 'Harvesting the Dead', *Harper's Monthly* **249**, 23–30.
4. Holvey, D. N. (ed.): 1972, *The Merck Manual of Diagnosis and Therapy*, Merck, Sharp & Dohme, Rahway, New Jersey.
5. Jonas, H.: 1974, *Philosophical Essays*, Prentice-Hall, Englewood Cliffs, New Jersey.
6. Kissel, W.A.: 1962, 'Atropine-like Coma with Ditran Therapy', *American Journal of Psychiatry* **118**, 847–848.
7. McWhirter, N. and McWhirter, Ross (eds.): 1971–72, *Guinness Book of World Records*, Bantam Books, New York, p. 31.
8. McWhirter, N. and McWhirter, Ross (eds.): 1976, *Guinness Book of World Records*, Bantam Books, New York, p. 40.
9. Moody, R. A.: 1975, *Life After Life*, Mockingbird Books, Covington, Georgia.
10. *Newsweek*, 1960, Aug. 13, p. 37.
11. *Newsweek*, 1967, Nov. 13, p. 99.
12. *Newsweek*, 1968, July 1, p. 61.
13. *Newsweek*, 1969, Aug. 19, p 54.
14. *New York Times*, 1950, July 12, p. 18.
15. *New York Times*, 1951, June 20, p. 29.
16. *New York Times*, 1957, Oct. 22, p. 21.
17. *New York Times*, 1960, Aug. 3, p. 9.
18. *New York Times*, 1962, Dec 5, p. 33.
19. *New York Times*, 1963, Mar. 2, p. 3.
20. *New York Times*, 1963, July 20, p. 2.
21. *New York Times*, 1963, Mar. 2, p. 3.
22. Powner, D. J.: Sept. 1976, 'Drug-Associated Isoelectric EEGs: A Hazard in Brain-Death Certification', *Journal of the American Medical Association* **236**, 1123.
23. Remsberg, B.: June, 1976, 'My Mother Wouldn't Let Me Die', *Good Housekeeping* **182**, 234–238.
24. Sarkisov, S. A.: 1966, *The Structure and Functions of the Brain* (ed. Naomi Raskin, transl. Basil Haigh), Indiana University Press, Bloomington, Indiana.
25. *Scientific American*, Sept. 1968 (anon.), 'What is Death?' 21, p. 85.
26. Spraggett, A.: 1976, 'The Unexplained', *The Western*, Winsted, Conn., July 21.
27. *The Star*, 1975, June 11, p. 47
28. *Time*, 1963, May 31, p. 66.
29. *Time*, 1963, Dec. 20, p. 66.
30. Tyson, R. N.: 1974, 'Simulation of Cerebral Death by Succinylcholine Sensitivity', *Archives of Neurology* **30**, 409–411.
31. Vecsey, G.: 1976, *The New York Times*, Sept. 29, p. 33.
32. Ward, R.: 1975, 'Blues for Karen Anne', *New York Times*, Nov. 28, p. 37.
33. Zimmerman, D.: Jan. 1975, 'Update on Last Year's Tragic Child Killer', *Today's Health* **53**, 43–45.

SECTION III

SCIENCE, INFIRMITY, AND METAPHYSICS

RICHARD KENNINGTON

DESCARTES AND MASTERY OF NATURE

I

The common judgment is that Francis Bacon is the originator of the concept of "mastery of nature," which is so indispensable in the technological crisis of this century. Attempts to trace the Baconian concept to anterior origins in Renaissance magic or Christian theology have touched upon but one or two aspects of his argument, which is moreover an indelibly philosophical one. On the other hand, the concept has never proved central to the interpretation of Descartes, the generally acknowledged "founder of modern philosophy." It is often noted, to be sure, that Descartes in a celebrated passage in *Discours* VI advocates the replacement of "the speculative philosophy of the schools" by "a practical [philosophy]" which will make us "like masters and owners of nature" ([6], pp. 61–62). But nowhere else does this phrase, or a similar one, occur in the Cartesian writings, whereas in Bacon's works the mastery concept is ubiquitous. The *Meditations*, Descartes' central philosophic writing, asserts in its title and in its preface, its place in the traditions of speculative "first philosophy" and Christian apologetics. We have no weighty efforts to bring its metaphysical doctrines into significant connection with the mastery concept of the *Discours*. Historians may observe the importance of *scientia proper potentiam* for Hobbes, but they do not stress mastery of nature as a thematic goal of modern philosophy. The influence of Baconian "mastery" remains dormant, it is usually held, until the sciences break away from their philosophic parentage, and develop their technological potential – most obviously in the nineteenth century. Among the philosophers of our century, Dewey is altogether rare in identifying the mastery concept as goal of philosophy and humanity; he traces it to Bacon and passes over Descartes. In sum, the power of the Baconian mastery goal is exerted only within the sciences, but only in a puzzling and delayed manner, and scarcely at all on the course of modern philosophy. Accordingly, there is scarcely any valuable research on the Baconian or seventeenth century origin of that modern technological humanitarianism, so powerful for good and evil, which has made the human future incalculable in unexampled degree.

"Mastery of nature" is today the enterprise of advanced technological

S. F. Spicker (ed.), Organism, Medicine, and Metaphysics, 201–223.

societies, singly and within the world scientific community, within which it is difficult to affix responsibility among the interconnected agencies and factors. Yet the one indispensable element must be located in its theoretical component – in what Hans Jonas has called "the new concept of ***nature*** [which] contained manipulability at its theoretical core" ([7], p. 48). Here we have to do with the mathematical and experimental theory of modern physics. Bacon indeed propounded the mastery goal, opposed the speculative tradition with argument unrivalled in scope and force, and recognized that experimentation was more precise than unaided sense perception, but the mathematical instrument it was to employ was unknown and even alien to his empiricism. It is rather Descartes who implants "mathematicism" in the heart of philosophy and science. To him we owe "analytic geometry" and "universal mathematics," the law of inertia, and the elaboration of a model of nature, more comprehensive than that of Hobbes, in which the "rules of mechanics are the same as those of nature" ([6], p. 52). Moreover, it has never proved possible to separate the metaphysical doctrine of Descartes, especially his substantial dualism, from close dependence on, or close complementarity with, his mathematicism and mechanistic physics. The question therefore must arise with which we are here concerned. Can mastery of nature be regarded as an intrinsic and guiding objective of Cartesian philosophy – and therewith of the problems and structure of the tradition of which he is the principal founder? The textual paucity of the phrase "mastery of nature" presents little obstacle. The one passage which turns on this phrase is the most explicit and complete treatment of the nature of philosophy in the Cartesian writings.

We must first indicate the measure in which Bacon established a framework within which to consider Cartesian mastery of nature. Bacon's starting point, often reiterated, is that "the greatest error of all the rest is the mistaking or misplacing of the last or farthest end of human knowledge" ([2], p. 34). His positive endeavor is that "contemplation and action may be more nearly and straightly conjoined and united together than they have been" ([2], p. 35). "Contemplation" (or "speculative philosophy") here means the quest for knowledge of the first or abiding beings or principles, whether further practical consequences are present or not. This conjoining, which suggests a synthesis in which contemplation and action are each given due weight, is a rejection of the contemplative goal simply, and a reinterpretation of it with a view to "action." The formulae of the new goal resound throughout the modern period. "Human knowledge and human power meet in one." "Dominion of the human race itself over the universe" for the alleviation of man's mortal condition ([3], pp. 39, 118). The repetitive sterility and barren disputatiousness of the contemplative tradition is contrasted with the steady progress of the arts. Even though some inventions have served human oppression – gunpowder, the compass are instanced, but not the printing press – and perhaps all contain that possibility, they are instruments and models for that comprehensive art of investigating nature in which alone the ultimate victory over evil can be found. Mastery of nature has both the character of an impetuous wager, and a somber, reasoned deliberation. Bacon's powerful formulae are rhetorical instruments of an appeal to humanity which has tended to conceal the underlying deliberation. No philosopher more openly declared ([1], p. 248) that philosophic speech must combine, in the same writing, open and public address with a less visible discourse addressed to those of philosophic capacity. His formulae are popular expressions of the conclusions of his deliberation. What is at stake in this wager concerns both the public and philosophy because it demands that humanity, using philosophy or science as its instrument, seizes control of its own destiny, instead of allowing philosophy to turn its back on humanity by contemplating the transhuman, eternal order. As regards the speculative tradition, if the "forms" discovered by the Platonic and Aristotelian philosophies were truly the efficient causes of the generation or destruction of at least some beings, some human works would have ensued. Their consequence has been barren. Hence no reflection on speculative philosophy from within its own intention is required. It seemed reasonable to conclude that a theory that has led, directly or indirectly, to no human production, while pretending to know first causes, is less genuinely knowledge than that which produces works as evidence for theory, even if not theory of first causes. It seemed reasonable to wager that a truly universal goal that conjoined theory and practice, the interests of philosophy and of humanity without exclusion, could be the engine that carried theory to new heights.

Bacon's endeavor is best understood as a critique of the anti-humanitarianism or non-humanitarianism of ancient political philosophy. The Platonic Socrates tells us in *Republic* V that unless philosophers rule as kings, or kings seriously cultivate philosophy, there is no rest from ills for the cities, nor for human-kind, nor will the best regime come forth from nature and see the light of the sun. This proposal is the peak of Greek philosophy's care for humanity. By the end of *Republic* VII it has been shown to be impossible, or extremely unlikely, and even undesirable or unnatural in certain respects. The goals of philosophy and humanity are separated by the greatest of differences; the philosophers are enamored of the contemplation of the whole or of the first or that which abides; the multitude do not philosophize. The coincidence of philosophy and political rule would require the coincidence of the perfections of virtue and knowledge, of natural philosophic endowment and social and material conditions, and such coincidences are always subject to chance. The supreme humanitarian solution must exist then only "in speech", or only "in imagination" – in the phrase of Machiavelli which Bacon takes over. "As for the philosophers, they make imaginary laws for imaginary commonwealths; and their discourses are as the stars, which give little light, because they are so high" ([2], p. 206). Bacon declares that he is "much beholden" to Machiavelli who teaches us to observe the low and universal, or almost universal: what men do and not what they ought to do. The mastery of "fortuna," nature, or chance of Machiavelli's *Prince* (especially Ch. XXV) is imitated by "the architecture of fortune" in *The Advancement of Learning*; what seems restricted to mastery of human nature is extended by Bacon to "lordship over the universe" in *Novum Organum* I, 129. The imaginary politics of Greek philosophy (here must be included the best regime of Aristotle's *Politics* which is also dependent on chance) is not only useless but dangerous: its non-humanitarian goal made it vulnerable to the equally or more "imaginary" politics of religion. It became the subordinate "handmaid" to revealed theology in the Christian Middle Ages. Non-humanitarian philosophy shows itself as anti-humanitarian philosophy: neutrality is impossible: if philosophy pretends to be the impartial spectator of the human fate within some eternal order it falls prey to the imperfect humanitarianism of religion. Bacon sought to conjoin the speculative quest for true nature of Greek philosophy with what he called "charity" – the care for humanity of the most universally received religion. This syncretism underlies the more fundamental syncretic union of philosophy and politics in Bacon's thought: a particular polity (e.g., the New Atlantis) or, in the best case, a universal politics that sanctions and propagates the mastery of nature, is the necessary condition of the fulfillment of its humanitarian promise.

What we have called syncretism is often misunderstood as "secularized Christianity." It is true that Bacon claims that mastery of nature is a privilege granted to Adam and to his posterity in the Garden. It is reasonable to conclude that that argument is directed to those who accept Biblical premises: his more sustained argument is independent of those premises. He contends that knowledge of nature is divinely sanctioned, so long as it is governed by charity, but that knowledge of good and evil was forbidden to Adam, as an object of inquiry. Good and evil in general, and charity in particular, are known by divine command, or by the Word of God; by the extension of charity Bacon purports to bring the fruits of the mastery of nature under divine command. By such a secularization, it is alleged, Bacon pours into the old bottles of Christian virtue the new wine of modern technology. Our wonder at this feat is aroused, however, when we consider the indefinitely long life, devoid of toil and replete with luxury, that Baconian mastery of nature holds out to mankind. But the secularization thesis collapses when knowledge of good and evil by divine command is openly replaced by knowledge of good and evil "nourished by natural philosophy" which furnishes principles to "moral and political philosophy" ([3], p. 77), although this is a promise for the future. As for the present, since Baconian philosophy is addressed to mankind and not Christianity or Christian Europe, its premises are laid not in Christian charity but in the universal human appeal of bread, freedom from natural necessity, and bodily longevity. Consequently, it is not Christian doctrine but Baconian philosophy that controls the mixture of syncretic elements in the *New Atlantis* in which Christianity and the Bible are combined with elements of Persian and Egyptian religion. Yet this religious syncretism is but a part of the underlying, "humanistic" syncretism of philosophy and politics.

That "human knowledge and human power meet in one" does not entail the literal coincidence of philosophy or science and political power. Bacon leaves it ambiguous whether or not the philosophic rulers of the research institute, Solomon's House, are identical with the political rulers of the *New Atlantis*. Bacon thereby respects the Platonic heterogeneity of philosophy and politics, even within their conjunction. Philosophy's domination employs the indirect means of rhetoric and "works" or technology. The political power of philosophic mastery of nature lies initially in its prophecy of irresistible benefits, and subsequently in the fulfilled prophecy of those same benefits, which permits further exercise of prophecy. Its power resides in the desires of "men as they are" and the rhetoric that addresses those desires. A distinction can be made in Bacon's writings between the preparatory rhetoric which gained such widespread European acceptance of the humanitarian goal, and the particular rhetoric required in the future which preserves some particular regime (e.g., the *New Atlantis*) devoted to that goal. The one leads insensibly to the other, as is easily confirmed by our "advanced technological societies." If all philosophic publication is political action of some kind, nonetheless Baconian philosophical publication, which seems to mold humanity to seek a society in which science is the principal benefactor of mankind, is political action of an unprecedented sort. After Bacon political rhetoric almost never constitutes the particular subject matter of philosophic writings, as in Aristotle's *Rhetoric* or the *Gorgias*. Political rhetoric becomes absorbed into the substance of philosophic speech because of the concern with the interests of humanity.

After Bacon scarcely any philosopher is exempt from the power of his humanitarian syncretism. What is properly called "Baconian" is the assumption by philosophy or science of the care and responsibility for the material and spiritual well-being of humanity, and not a narrow concern with material technology. Even those rare modern philosophers who seek to re-assert the ancient speculative goal, or who seek to combat the mastery of nature in one or all forms, typically show themselves members of "the party of humanity" by proposing political societies that are founded in universal rights and not in the rarity of human excellence; or in the "inevitable" progress of history towards just social forms; or in some redemption of humanity from the nihilisms of the age. Since all philosophic attempts to save humanity from the Baconian technological mastery goal are themselves humanitarian, it is hard to avoid the conclusion that they remain Baconian in principle. The reason for this power of Baconian humanitarianism is not often correctly located. It did not require, and never possessed, the new natural science, nor a new metaphysics, neither of which are found in Bacon's writings, except in nascent form. Its basis lies in no epistemology, but only in the human knowledge of misery, especially of death, and the conviction that it is ignoble not to avenge our subjugation at the hands of a niggardly nature, by the exaltation of the power of man.

II

The immediate sense of mastery of nature in Descartes must be drawn from the one passage of its assertion, in the *Discours de la méthode* (1637). We must first remove the misinterpretation of this passage which Gilson's *Commentaire* on the *Discours* has fathered upon two generations of scholarship. According to Gilson, mastery of nature and its potential for benevolence, is an accidental and unintended consequence of his philosophy which Descartes came upon after virtually all of his system was complete. This conclusion derives its plausibility from the opening paragraphs of *Discours* VI ([6], pp. 60–62). "Mastery and ownership of nature" is introduced expressly only after Descartes "acquired general notions concerning physics," and physics had followed method and metaphysics in the canonical sequence of parts of philosophy in the *Discours*. Descartes "notices" where these principles of physics might lead – to mastery of nature and its benefits – which obliged him to publish, since there is a "law" which obliges us to benefit other men, so far as it is in our power. Only at this juncture does Descartes for the first and only time expressly advocate the replacement of "speculative philosophy" by "practical philosophy" of mastery of nature. Accordingly, Gilson concludes that "that which inspires the philosophy of Bacon, inspires only the publication of the philosophy of Descartes" ([6], p. 444). Gilson's judgment would have force only if Descartes had indicated that his philosophy, prior to the discovery of the principles of physics, was directed to the speculative goal of the metaphysical tradition. But what Descartes asserts at this juncture is not the rejection of his own previous speculative philosophy, but "the speculative philosophy taught in the schools." There is also the possibility, to be sure, that Descartes had not hitherto reflected on the goal of philosophy, and so had hitherto belonged unwittingly to the speculative tradition. But the *Regulae* (1628) virtually begins by conceiving of "the ends of study" in terms of "the guidance of action" and *Discours* I employs the criterion of "the useful for life" to condemn the whole prior tradition of learning, philosophy and theology. "Utility" so pervasive at the inception and throughout the *Discours*, is assuredly not "mastery" but it prepares the turning to the perfection of various kinds of "masters" in *Discours* II, which in turn precedes method and metaphysics. Gilson failed to recognize that the utility criterion already excludes the speculative goal and thereby prepares the mastery objective. He did not envisage the possibility that discovery of the mastery potential of the physics is a realization of an original Cartesian intention, whose full disclosure was reserved until its realization became available. It would not be the first time that Descartes imputes to accident what is too untraditional to ascribe to his own deliberate design.

The initial as well as the ultimate problem we face in grasping Cartesian mastery of nature is the following. What is the measure of Descartes' opposition to the Church and its tradition, on the one hand, and to the tradition of speculative metaphysics on the other – and indeed how is the one opposition related to the other? The first opposition appears in Descartes' account of the most famous collision between the new science and the ecclesiastical tradition – the condemnation of Galileo by the Holy Office. Descartes does not name names, and does not indicate whether or not he too was a Copernican, but this reserve only supplies added force to an episode known far and wide to the learned public of the day. The question concerns the relation of religion and philosophy, not a particular scientific doctrine; he was unquestionably a Copernican, as even those who assert his invariable candor have acknowledged. He informs "the public" that he did not publish his earlier treatise on physics, either because he respected the authority of the Church over his actions, as distinct from the authority of his reason over his "thoughts," or because he feared a similar condemnation. However this may be, he replaces the authority of the Church over his actions by the authority of the "law" which commands benevolence, mentioned before, as interpreted by his reason. This transfer of authority is dictated by the discovery of the implications of his physics for benevolent mastery of nature. He does not suggest that the Church might be the proper interpreter of the law of benevolence, or of the subsumption of his physics under that law. He does not allude to the Bible as sanction for mastery of nature as Bacon frequently had done; he openly takes his departure from *Genesis* as Bacon had done implicitly. Mastery of nature will enable men to enjoy the fruits of the earth without pain "in this life." In the form of "medicine" it will conduce to health, the foundation of all human goods, and to the prolongation of life, or such victory over the grave as is available through human effort. It holds out the promise of an unprecedented practical wisdom, which we may reasonably identify with that knowledge of the use and enjoyment of the passions which Descartes calls "wisdom" at the close of the *Passions*, his final publication (1649). The significant departures from the orthodox religions of his time are apparent.

On the other hand, Descartes often speaks as a faithful son of the Church, and presents his metaphysical theology and dualism of substance in support of Christian apologetics. The "secular coloring" of Cartesian mastery of nature in the *Discours* seems emphatically rebutted by the metaphysics, especially when coupled with apologetic purposes in the *Meditations*. The measure of his opposition to the church and its tradition must be judged in terms of the measure of his opposition to the tradition of speculative metaphysics. In the regnant scholasticism which Descartes confronted, the Aristotelian speculative metaphysics was conjoined with revealed biblical theology. We must then ask some such questions as the following. In opposing "speculative philosophy," is Descartes rejecting its primary concern with the question of being or substance? In contrast with Bacon, Descartes expounds a substance doctrine of traditional type: is this legacy of the speculative tradition consistent with mastery of nature? Can Cartesian philosophy retain the traditional concern with being or substance while nonetheless regarding metaphysical knowledge as instrumental to the mastery of nature? In considering the elements of Descartes' double opposition to tradition, we take part in the contemporary discussion of the origins of modern "mastery of nature." One school, of vaguely Hegelian provenance, understands Descartes' effort as the secularizing effort of a child of his age; the other – Heidegger is its foremost spokesman – traces it to Greek speculative (especially Platonic) philosophy. These oppositions must be clarified before we can say in what measure Descartes' philosophy is a syncretic humanitarianism of Baconian type, but armed with the new instrument of a mathematicized physics.

III

The place of mastery of nature within the intention of Descartes must be determined within the *Discours* and not the *Meditations*. The authoritative position of the *Meditations* in Cartesian study appears assured because of the radicality of its beginning in universal doubt, its full-fledged articulation of the separation of mind from body, and its authoritative account of the perfect deity who is the ultimate guarantor of all human knowing. Yet none of these features touches upon the end of Cartesian philosophy, about which the *Meditations* preserves silence. Because of this silence the end is usually taken to be knowledge. But the quest of the *Meditations* is for "foundations" of the edifice of philosophy or science, as laid down in the first paragraph of *Meditations* I, and the goal of the edifice is not mentioned. The *Meditations* has the character of instrumental foundations for an unarticulated goal. On the other hand it also lacks the beginning of Cartesian philosophy. The universal doubt with which the quest for foundations commences has already presupposed the concept of the edifice to be founded, the ideal of certainty or self-evident indubitability which the foundations must possess, and above all the goal of the edifice. Moreover, in the *Meditations*, and in all its parts, the concept of nature as a corporeal world governed by the laws of mechanics is always present, often as a premise, a never doubted. Nowhere in the work is there a tincture of proof of the major ontological thesis that the essence of body is *extensio*. We must therefore allow the *Meditations* to recede from our view in any attempt to establish the structure and goal of Cartesian philosophy. The impression of autonomy, which the *Meditations* surely conveys, may be explained by Descartes' second or auxiliary purpose in presenting the work as a self-sufficient piece of Christian apologetics, whose ostensible themes are the existence of a perfect deity, and the independent – and therefore presumably immortal – existence of the soul.

The *Discours* is Descartes' only comprehensive statement, published or unpublished, on the pre-philosophic critique of the tradition, the necessity of beginning with a universal method, the parts of philosophy and their order, and its goal. Within the structure of the *Discours* the metaphysical argument, corresponding to the subject matter of the *Meditations*, is assigned a position in Part IV clearly posterior to the critique of the tradition (Part I), the method (Part II), and the provisional ethics (Part III). Our attention is therefore drawn to the criteria, utility and certainty, with which Descartes judges the prior tradition in Part I. But this double criterion has a pre-history: it develops as a criticism of the single criterion of certainty or self-evidentiality in the earlier fragment, the *Regulae* ([5], 1628, probable date), which suggests one reason for the incompleteness of that writing. Between the *Regulae* and the composition of the *Discours* in 1636 Descartes all but completed the physical treatise, *Le Monde*, which he withheld from publication because of the condemnation of Galileo. It is in this pre-*Discours* and post-*Regulae* interval that he most probably turned to the writings of Bacon in which he found the stress on utility, the arts as model of beneficence, and mastery of nature, all of which are absent in the *Regulae* and *Le Monde*, and thematic in the conception of philosophy in his first publication, the *Discours*. The structure of Cartesian philosophy is best understood as the attempt to unite two originally diverse lines of thinking, the mathematical science of nature of the *Regulae* and *Le Monde*, and the utility-mastery theme of Baconian origin. This fusion becomes possible because the non-teleological mathematical science, despite its earlier development, is assigned the status of means or instrument of the utility-mastery goal. The idea of this fusion is the very germ of Cartesian philosophy, and the heterogeneity of its elements is in continual tension with its unification. The elements of the fusion are discernible in the formula "certainty for the sake of utility" which functions as criterion of the tradition in *Discours* I.

Descartes employs the standard "clear and assured knowledge useful for life" ([6], p. 4) to judge the entire tradition of the arts and sciences, theology and philosophy in *Discours* I. Because certainty is given determinacy by the example of mathematics, whereas the "useful for life" remains opaque, it is often not recognized that the utility goal is a decisive emendation of Descartes' intention in earlier writings. *Regulae* II had begun with the character of knowledge, as distinct from its end: "all science is certain and evident cognition." *Regulae* I had indeed spoken, first of the pleasure of contemplation, and then of the guidance of the will in the contingencies of life, as "ends of study." But the universal science or method of the *Regulae* does not include good or evil or the ends of the will; the will is unrelated to the intellect; the goals of science and of the philosopher (or "life") fall apart. The "useful for life" of the *Discours* includes production ("the infinity of artifices") as well as guidance of will; it is aided by the model of the arts, which are on the whole disparaged in the *Regulae*. The *Regulae* in its quest for certainty may be regarded as pointing towards utility – to cessation of the state of uncertainty in many forms, which is a demand for contentment arising from the side of the "subject." This contentment, which Descartes calls by the Stoic term *bona mens* in the *Regulae* ([5], p. 2), is available only to those few capable of *humana sapientia*. Thus the *Regulae* remains "ancient" as regards the beneficiary of philosophy, and does not attain the standpoint of the useful for life which in the *Discours* (e.g, in the form of medicine) benefits man as man. Only with the *Discours* does Descartes become humanitarian and modern.

With the abbreviated formula "certainty for the sake of utility" it first appears that certainty is the decisive element. Because they are not certain, philosophy and theology, the arts and sciences, are useless. Only mathematics is certain but its present utility is trivial; it must supply "foundations" to philosophy which has hitherto been the source of uncertain foundations. Thus the certainty standard immediately points the way to the "mathematicization" of all branches of knowledge. Yet we must resist this conclusion, at least until certainty is brought into some relation with utility; indeed certainty and utility condition each other. "Theology" is useful as the way "to gain heaven," but it is uncertain because it lies beyond human reason, or comprises "revealed truths," and hence is useless to cultivate. Only certainty and utility in combination suffice to explain why the underlying standpoint is already "rationalism": the demand for the exclusive adherence to natural reason, subsequently expressed in the canonical sentence of *Discours* IV: "whether we are awake or asleep we ought never to be persuaded except by the evidence of our reason." Just as the demand for a certainty equivalent to mathematics veils the underlying rationalism, so also the utility demand tends to obscure the underlying agreement with the "realist" or "selfish" psychologies of Bacon and Hobbes. Utility excludes both "virtue" sought for its own sake – the virtue of "the writings of the ancient pagans which treat of morals," the only moral writings considered – as well as knowledge sought for its own sake – the "speculations" of "the men of letters" or the scholastic metaphysical tradition. Their common defect is their unawareness of the natural egoism of reason, or the pervasive power of the passions over all psychic life. Thus ancient virtue is often "insensibility" or "pride" – the attempt to be exempt from the passions which is itself passion. The "speculations" in turn are attributed to excessive self-esteem or "vanity" which plumes itself on a false superiority to the reason of "each" man, or "common sense" ([6], pp. 7–8, 9–10). The realism underlying utility thus fosters certainty, in so far as it seeks a psychological grounding in the universality of the passions, just as Hobbes and Locke found the basis of right in the allegedly universal passion for self-preservation; and the rationalism underlying the certainty requirement is broadly utilitarian in spirit.

"Inutile et incertain," Descartes' double charge against revealed Christian theology and Greek speculative philosophy, became in turn Pascal's indictment of Descartes ([8], p. 361). Pascal seems to have perceived that no means, however certain in itself, can be certain *as a means*, if the end is uncertain, and the goals of human life will always fall short of mathematically certain knowledge. Already in *Discours* I it becomes clear that the utility goal can never be brought within the charmed circle of certitude – it can never be treated by "clear and distinct ideas" or by self-evident intuition and deduction. Descartes is not guilty of the charge, often made, of universal mathematicism, of "the geometric prejudice," or of the inflexible demand for clear and distinct ideas. This mathematicism is only regional – applicable to the means, where, to be sure, it is already problematic. Yet the difficulty must be stated at a deeper level within the means – end structure: even if the means, e.g., a science of nature comprehensive of all natural qualities and actions, were as certain as arithmetic, it must nonetheless remain uncertain *as means* because the end lacks certitude. Mastery of nature in Descartes, precisely because it "improves" upon Bacon by making mathematical knowledge the instrument of utility, necessarily introduces an unprecedented gulf between its"means-knowledge" and its "end-knowledge."

IV

Descartes takes the step from utility to mastery in *Discours* II by laying down the precept that there is more "perfection" in works accomplished by one master than in those made by many masters ([6], p. 11). The precept is articulated by a series of examples which ascend from the pre-political arts (architecture, city planning and supervision) to the more comprehensive political art of the founder or legislator, culminating in the concept of a "pure and solid" reason whose "ouvrage" is a method which synthesizes logic, algebra and geometry. In this reflection, certainty is at first absent but is then generated out of the reflection on mastery in the arts, in the form of mathematical method. The reflection of perfect mastery in the arts determines the choice of "the ways I ought to follow," i.e., the choice of philosophy as the perfect life. For our theme it is of interest to observe that the mathematical method is made to develop out of the demands of philosophy as the most perfect mastery. This order can scarcely be other than deliberate, since it conflicts with the chronological order of Descartes' biographical development, as observed above, according to which the mathematical method of the *Regulae* was originally completed apart from concern with utility or mastery. The order of philosophy in the *Discours* employs the arts to articulate the goal of mastery of which the mathematical method becomes the instrument. The turning to the arts is determined negatively by the utility standard, which had excluded ancient virtue and speculative knowledge as ends in themselves. Whereas human nature as such, or the natural egoism of reason of man as man, sufficed to exclude these ancient conclusions about human perfection, only the arts supply a model for the perfection of the highest life.

This turning to the arts reflects the agreement with Bacon: the arts are the undeniably benevolent human activity, as men generally acknowledge. Yet only philosophic reflection on the arts lifts their benevolence above the level of opinion. Those art works which are not a plurality of diverse endeavors with diverse ends, or which proceed from one master and not several, have greater perfection. Moreover, arts which are more comprehensive are more perfect than those they include, as city planning is by contrast with the architect of a single building, or the legislator or founder of cities or peoples, by contrast with the city planner. Reason or philosophy, as the absolutely comprehensive art which provides for the goods of cities or peoples by the most comprehensive knowledge, is yet more perfect. This entire reflection largely repeats the argument of Bacon in *Novum Organum* I, 129, according to which arts and inventions are more beneficial and more peaceful, and therefore more worthy of honor than the political arts, and philosophy, the art of scientific mastery over the universe, comprehends all the arts, including those of political founders and saviors of humanity, and is deserving of the highest glory. Descartes, however, asks the further question: what guarantees that the master himself is a "one" and not a "many" of passions, faculties and goals? This question divides into the unicity of his knowing, on the one hand, and the singleness of his goal, on the other. The unicity of knowledge is satisfied by the purgation of reason of the heterogeneity of opinion, appetite and sensation through methodical doubt; and by the "pure and solid" element of its cognition, mathematics or mathematical laws of nature. But the singleness of the goal would only be satisfied if reason were to produce its own goal out of itself, and not to receive it from a distinct and alien source, i.e., from the opinion of mankind, religion, revealed theology, or inquiry into some source in the nature of man. Thus at this stage, neither the benevolence of philosophy in terms of its non-philosophic or public beneficiaries, nor the perfection of the philosopher dedicated to perfect mastery through mathematical science, has received justification.

In the first, or programmatic phase of Cartesian philosophy – which we may identify with Descartes' reflection in the celebrated South German stove, *Discours* II–III – these implications are stated with a certain reserve. From the same context one can easily get the impression that Descartes is unconcerned with the political, and that he thinks not of mastery but of acceptance of the world with Stoic passivity. He surely asserts that his reform is exclusively of his own private beliefs, and he denies that he seeks a "new reformation," i.e., the public reformation of the beliefs on which great states are maintained which might follow were he to publish the demand for the purgation of opinion. He therefore stresses that he does not invite followers to imitate a course of action, for which most of them lack the requisite credentials. In fact the world is almost wholly composed of men who lack the two essential qualifications of Cartesian philosophers, perseverance or resolution, and superior natural endowment of reason. By specifying the requirements in this exact manner, he tacitly invites the qualified to emulate him, and identifies his philosophic path as a "modèle" ([6], pp. 13–15). Moreover, this reservation to a purely private reformation of belief disappears in *Discours* VI when the new physics in its actuality discloses its benevolence to the public, and therewith the promise of glory to those who advance the cause of humanitarian science.

Descartes' Stoic passivity in the third moral rule ([6], pp. 25–26) also dissipates on a careful reading. He surely asserts that he will "try always to vanquish myself rather than fortune, and to change my desires rather than the order of the world." He concludes that everything but his "thoughts" is outside of his power, reminding us especially of Epictetus. However, the rule concerns those "actions" which are required while "cultivating my reason," i.e., in pursuing the application of method to nature. Hence, knowing "what is in our power and what is not" does not mean, as with the Stoics, "living in accordance with nature" which for Descartes remains as yet undiscovered. It is of importance that we recognize that the Stoic rule is part of a morality "par provision." Provisionally Descartes lived Stoically in accordance with the "order of the world" which here means only the human world of laws, customs and religion, even while prosecuting inquiry into the natural "world." Accordingly, his account of the Stoics, who believed they had final knowledge of the natural world, i.e., of what can ever become within man's power, has a decided element of satire. These ancient philosophers no more wished to be well, when they were ill, than they desired "to have wings to fly like the birds" or to be Emperor of China ([6], p. 26). By such absurdities they sought to "rival the gods in their felicity" while forgetting their own mortality. They preferred a spurious superiority to common men – we are reminded of the vanity of the speculations of the men of letters – to that acknowledgment of our common, human corporeality which could spur philosophy to develop the medicine that conquers disease and senescence. Even provisionally, therefore, Descartes asserts an opinion by no means in conflict with the demand for mastery of nature through medicine in *Discours* VI. The Stoics, among all ancient philosophers, asked the right question – what is within the range of human power? – to which only the future progress of science can supply an answer, which may always remain provisional.

V

This formative stage of Cartesian mastery of nature must now be confronted with the metaphysical doctrines of *Discours* IV and the *Meditations*. Descartes' dualistic metaphysics has much to commend it as the appropriate ontology for the mastery of nature. We may regard the heterogeneity of the subject and object, of mastering human ego and inert, objective nature, as requiring a ground in the diversity of thinking substance and extended substance. Even this initial conclusion, however, requires a drastic revision of the speculative form of metaphysics, as distinct from its content. As indicated above, the principles of metaphysics are now understood as "foundations" of the edifice of philosophy or science, and the edifice – or "the tree of philosophy" – exists only for the sake of the goal or the "fruits." Since the *Meditations* is silent regarding the goal, we may use the *Prefatory Letter* to the *Principles* to identify the "fruits" as the produce of the highest branches of the "tree of philosophy" – medicine, mechanics and practical, moral wisdom. Metaphysics, including knowledge of God, the highest principle, is instrumental to practice. Let us now assume that this novel, instrumental doctrine of substance could be regarded as compatible with its traditional content, doctrine of substance. Extended substance is devoid of life, self-moving organic compounds, purposes and final causes, and "secondary qualities" (in Locke's phrase) – all this precisely as demanded of the object of a mathematical physics. All such non-mathematicizable features of being are attributed to, or explained in terms of, the thinking substance, a separately existing mind or soul which has the exclusive privilege of being the autonomous source of its own activity. Furthermore, Descartes' theology is singularly appropriate to his mastery of nature goal. He knows that God is perfect, and the guarantor of the truth of clear and distinct ideas, but nonetheless cannot know the purposes of God, who is incomprehensible ([4], pp. 138, 166), in the natural world. Physics must henceforth abandon final causes, and neither from God or nature can be derived by rational procedures that knowledge of moral duties which might impede, or guide, the quest for mastery of nature. Rational theology is as agnostic as revealed theology with regard to human knowledge of right and wrong. The theological part of metaphysics thus furnishes the required negative sanction, the absence of suprahuman restraint on conquest of nature.

Descartes' metaphysics, summarily characterized with regard to our theme, has the following three functions. It claims to be Christian apologetics, as rational doctrine of God and the separate soul; "first philosophy" (sub-title of the *Meditations*) in the Aristotelian tradition, as doctrine of substance and first causes and principles; and "foundations" for the edifice of philosophy for the fruits of the mastery of nature. It almost goes without saying that no single metaphysical teaching can successfully perform these three heterogeneous functions without disclosing gross internal inconsistencies. The more clearly we grasp the foundations of Cartesian mastery of nature the less clearly does he stand in the traditions of Christian apologetics and Aristotelian "first philosophy". In *Meditations* VI Descartes recognized that the separate soul substance, so useful as a support for Christian apologetics, could not be brought into relations with extended substance. The benevolence or veracity of God is of no avail: there is no suggestion that the interaction of substances is or could be made intelligible. Still less does Descartes claim that the two substances, as substances, together comprise one being, in the critical case of human being. Only in private correspondence does he fall back on the hylemorphic solution to the unity of man – all too obviously a prudential, "Aristotelian," *ad hominem* argument, as is generally acknowledged. Nor is it reasonable to think that Descartes believed he could extricate himself from these difficulties by the pineal gland argument of *Passions* I, No. 30 ff., since the very term "substance" never occurs in that work. Moreover, recent Cartesian scholarship, seeking to penetrate to the origins of Cartesian dualism, has experienced acute difficulty in finding the elements of what one could call an argument for duality of substance; not finding them in the *Meditations*, it has looked for them in any and every Cartesian writing, published or unpublished, but without significant success. These agnostic conclusions will not surprise those who observe, and follow up their observation, that the *Meditations* is devoid of the characteristics of a metaphysical inquiry into being or substance. Descartes nowhere in the six *Meditations* addresses the questions, what is being, cause, substance, essence, accident, etc., and never defines any of these terms formally or informally. If he occasionally employs them in traditional meanings, he at the same time enjoins us to believe that he has made every effort to doubt and reject the tradition, and sometimes quite explicitly indicates that he is employing traditional terms as *ad hominem* premises with which to simulate an agreement with the tradition.

The presence of gross theoretical difficulties in the substance doctrine and the absence of even an attempt to resolve them on Descartes' part; the presence of prudential rhetoric to simulate agreement with tradition, which every scholar without exception grants in some degree; the unmistakable attempt to reject "the speculative philosophy of the schools" – all these conspire to compel us to seek in the Cartesian philosophy for solutions to its fundamental problems which abandon the metaphysics of substance. This will seem unnecessarily iconoclastic to those who believe that the consensus of twentieth century scholarship necessarily represents a progress over the findings and judgments of the philosophers and the learned public of the late seventeenth and eighteenth centuries. We need not fear the charge of iconoclasm if we are aware that a mind of the stature of Leibniz could declare, after four decades of assiduous study of the Cartesian writings, that Descartes dissimulated the agreement of his views on religion with those of Hobbes, and that the God of Descartes is virtually identical with the God of Spinoza. In the Cartesian writings the most sustained and searching inquiry into the problems of mind and body is the *Passions of the Soul* (1649), in which the abandonment of the substance doctrine has already been prepared by the latter half of *Meditations* VI.

Midway in *Meditations* VI, Descartes ceases to speak of the substance doctrine and begins a fresh inquiry into the nature of man. He draws a threefold distinction between (a) the things which only pertain to mind – the sphere of "thoughts" or consciousness, established by means of the radical doubt; (b) matters which only pertain to body, which prepares the first mention in the work of "laws of nature"; and (c) "those things given by God to me as a being composed of mind and body" ([4], p. 193). The concept of the compound here introduced is in no sense a metaphysical concept: it is not a compound of two substances, nor is it a substance itself. The compound has no metaphysical unity, nor any unity known by clear and distinct ideas, but only the experienced unity conferred on it by the sensation or feeling of pain and pleasure, especially pain. The judgment of unity inferred especially from pain is part of the "teaching of nature", and the compound, and only the compound, is the locus of the teaching of nature whose pedagogic instruments are pain and pleasure. Looking back upon the mind and body substances, it becomes clear that it is not only their theoretical incombinability that compels Descartes' step to the non-substantial compound; it is also the fact that the substances have been so defined that their combination would not permit but rather exclude an account of pleasure and pain as well as the passions. Since the *res cogitans* as thinking substance cannot, e.g., become angry, but only doubt, infer, judge, reflect, and the like; and since the *res extensa* by itself is exhaustively characterized by the law of inertia and the laws of mechanical interaction and so cannot become angry; and since the compounding of such substances does not yield the least possibility of accounting for such phenomena as pleasure or pain and passions such as anger, the ground for Descartes' turn to a non-substantial compound becomes apparent.

The natural teaching inherent in the compound furnishes him with the general foundation in human nature of judgments about good and bad. "These perceptions of sense [have] been placed in me by nature to signify to my mind what things are beneficial or hurtful to the whole of which it forms a part" ([4], p. 194). This introduction of natural purposiveness is not compatible with nature as *res extensa*, the mechanism of body, from which purpose is excluded. Purpose is limited to the region of the compound, which is conceivable, if only on experienced and not on "clear and distinct" evidence, independently of what is known clearly and distinctly to belong to mind by itself, or body by itself. This triad enables us to say that the *Meditations* concludes with what we may call "regional ontologies" which are "metaphysically neutral" as regards substance.

This triad, and especially the compound, should not be regarded as some lapse from the "rationalism" of "clear and distinct ideas" but rather as precisely that which furnishes the mastery of nature with the basis for its distinction of means and ends. In this context the example of the dropsical patient of *Meditations* VI is critical. From the standpoint of the mathematical science of nature, a body sick or a body healthy equally exhibit the laws of nature. But from the standpoint of the compound, i.e., of human experience, a sick body is naturally defective and bad. Descartes makes no effort whatever to unify theoretically the conflict between the scientific concept of nature which is neutral to good and bad, and the human experience of sickness which is bad by nature. Humanly experienced nature supplies the end, the goodness of health, to which only the scientific nature, in the form of medicine, conduces as the appropriate means. It is not the theoretical resolution of these antithetical concepts of nature, but the preservation of the antithesis which makes possible the mastery of nature and the Cartesian philosophy altogether. The necessity of the triad for the structure of Cartesian philosophy must be remembered when one considers the attempts of the *Passions* to reformulate the relations of mind and body, in a new version of the compound. The unification of the compound in terms of the operations or behavior of mind and body within "experience" is by no means the same as a theoretical unification of the compound with the other members of the triad. The disparateness or heterogeneity of the triad, it must be observed, appears only in the light of the demands of reason and is not an assertion about the nature of things. It is a conclusion fully in harmony with that rejection of the "speculative" quest for being already evident in the *humana sapientia* of the *Regulae*. "This knowledge is not the less science than that which exhibits the nature of the thing itself [*quae rei ipsum naturam exhibet*]" ([5], p. 37).

VI

With these clarifications of Descartes' critique of tradition we can return to *Discours* VI and identify the assertion of mastery of nature there as belonging to the second of the three phases into which the project typically divides: programmatic announcement of the goal, achievement of theoretical basis, and technical implementation. In the programmatic phase, the arts are treated in terms of their historic content, the opposition to the powerful doctrines of the past is presented as a modification which is continuous with the old intentions, and the syncretic character of philosophy – its union with politics – is presented with reserve. When the laws of physics are discovered, the program comes forth into the light of day as an actualized hope, and the program is revised to take account of more exact insight into its possibilities. The role of the arts as model becomes explicit at the same moment in which they begin to lose their autonomy by accepting principles from physics. "Practical philosophy" will know "the force and action" of "all the bodies that environ us as distinctly as we know the various métiers of our artisans" and "employ them in the same way" ([6], pp. 61–62). Medicine, mechanics and moral wisdom issue forth as branches from physics which is the trunk of the tree of philosophy (Pref. Letter to the *Principles*). This metaphor is nevertheless somewhat misleading. The laws of mechanics are identical with those of physics, which Descartes envisages as identical with those of medicine, and medicine in *Discours* VI includes study of the dependence of the mind ["l'esprit"] on the temperament and organs of the body, as well as of bodily senescence. The effort to unify medicine, in this enlarged meaning, with moral wisdom, would bring into harmony the means–end dualisms of certainty and utility, mathematical physics and mastery of nature.

The syncretic union of philosophy and politics, made explicit in *Discours* VI, may be divided into its elements: (1) the public as well as the secular rulers of states are informed of the potential benefits, if only for future generations, of the transformed goal of philosophy; and (2) are advised that these benefits will become available only if scientific research, especially experimentation, is fostered and financed, and freedom of communication is permitted; (3) both the public and the heads of states are advised of the identity of those who must necessarily oppose the humanitarian project – namely, those who would condemn a Galileo, and in general defend the subordination of humanity to the powers of tradition; (4) the project of humanitarian mastery of nature common to philosophy or science and society nonetheless divides into a two-sided relationship, each with its duties and rights to benefit; (5) the new relationship is brought under the sanction of the "law" of benevolence; and finally, (6) the benevolence law must be supported by a framework of belief that explains the possibility of happiness, now for the first time within human power – a kind of "theodicy" that replaces the old framework of belief.

The benevolence of Descartes' project only superficially receives its justification in the law that obliges us to benefit others to the extent that it is in us. Descartes may have regarded this law as the core of that moral belief required in society at all times; certainly it is the only categorical moral obligation advanced anywhere in his writings. Yet its apparent universality is weakened by its restriction to the context of the benefits of mastery of nature. It is not mentioned in the provisional morality of *Discours* III nor in the many other contexts in which duty to others would seem pertinent. Our skepticism is aroused by the lack of any supporting argument or clarification of the basis or source of the law. In the various passages of *Discours* VI which speak of obligation, the categorical status of the moral law is replaced by a hypothetical obligation, which may be summarized as follows. If the goal is desired – if humanity desires to receive the fruits of the mastery of nature, and if individuals desire to receive the honor and glory of benefiting humanity through that mastery of nature, then the means must be willed – the humanitarian project must be implemented. Although Descartes asserts that only those who accept this revised benevolence law have genuine virtue as distinct from mere seeming, it is better described as a precept of utilitarian hedonism than as a binding principle of moral virtue. Its force lies not in some known or received principle, but in the promise of satisfaction of desires and passions.

Mastery of nature promises those benefits that all or most men and societies have always desired: the alleviation of toil, the provision of necessities always so hard-won from the hands of a step-motherly nature, the accumulation of the means of comfort and luxury, the elimination of disease and the maximum possible postponement of mortality within human power. These benefits are as universally available to humanity as they are devoid of exacting duties or self-sacrifice. But mastery of nature promises as well a perfect moral wisdom which is peculiarly devoted to the requirements of the benefactor of humanity, the "strong and noble minds" who emulate Descartes and possess the passion or virtue of "generosité." Descartes apparently believed that he possessed a sufficient basis in his physics to supply the physiological basis for this virtue, the theme of *Passions* III. We must leave it at the assertion that he relies on a reformulation of ancient moral doctrines – Stoic, Aristotelian and Epicurean – and does not utilize his physical principles to explain this highest virtue. Cartesian mastery of nature culminates in a premature attempt to pass onto the third phase of implementation.

VII

The modern intention to master nature culminates in the problem of theodicy. This problem is invisible to many because they believe that mastery of nature is the work of science and scientific technology, as distinct from philosophy. But only seventeenth century philosophy gave the humanitarian justification for mastery of nature, and commenced the destruction of its antecedent traditions which is often misleadingly called "secularization." This action antedates the subsequent distinction of philosophy and science by more than a century. Only seventeenth century philosophy laid down the principles of the modern type of political society which became the agent of technological "progress" – by merging the goals of philosophy and politics in what we have called "syncretic philosophy." Just the utilitarian hedonism of Bacon and Descartes made their project easily combinable with the enlightened despotism and the liberal politics of Hobbes and Locke. But syncretic philosophy cannot take responsibility for the material and spiritual well-being of society without confronting the necessity of society for the over-arching framework of belief which is always, if in varying degree and form, a "theodicy." This responsibility is more clearly recognized if we divorce it from the classic theological formulations it received in the biblical tradition. "Theodicy" in the broad sense here employed means accounting for the relation between goodness and the expectation of happiness, and of evil and misfortune, in terms of human activity and the suprahuman whole, be it nature or history, in which man finds himself. Kant gave an almost perfect formulation of the problem of theodicy, when he addressed not philosophers but humanity with his tripartite formulation: What can I know? What ought I to do? What may I hope?

No pre-modern philosopher believed it possible or necessary to accept and discharge the responsibility for supplying a theodicy to humanity, as distinct from the sect of philosophers. He customarily supported the received religion while moderating certain of its features in the direction of the philosophic account of the whole, recognizing that religion supplies that account of the whole that sanctions morality, public and private, and satisfies an almost universal human need. For this reason, the published speech of the pre-modern philosophers is profoundly ambiguous: even a Lucretius, who taught that the gods of the *intermundia* do not care for humanity, praised Venus to the skies.

On the other hand, modern philosophies of the syncretic type, beginning with Bacon, and continuing most obviously with Leibniz, Kant, Hegel, Marx and Nietzsche, acknowledge, if in varying forms, the responsibility for theodicy. In the seventeenth century it is often difficult to distinguish their theodicies from the Christian apologetic theodicies they sought to replace, so that the ambiguity of their writing reminds us of pre-modern philosophic rhetoric, although it is nonetheless ruled by the new humanitarian goal. In Descartes the need for a new version of theodicy is by no means as clear as in Bacon's *New Atlantis*; yet we must not forget his praise of that writing in the prefatory letters to the *Passions*. In the *Passions* he seeks to defend all the passions as by nature good if rightly understood. The quest for the mastery of nature, which implies that nature is hostile, runs counter to that benevolence of nature, which even mastery of nature must premise if happiness is to be attainable through man's own natural powers, and within a natural whole which can never be eliminated. Only in the eighteenth century was it fully realized that modern scientific nature cannot be combined with an essentially pre-modern doctrine of the benevolence of nature, especially if nature in itself is not intelligible, as Kant perceived. After Kant's attempt to relocate the problem of theodicy to the plane of morality, it was translated onto the plane of history by Hegel and his successors, where it found a resolution so long as history could be understood as exhibiting an intelligible pattern or *telos*, or until Nietzsche. The "death of God" may be regarded not only in terms of Christianity but as the failure of modern syncretic philosophy to replace it with a new theodicy: what Zarathustra proposes is a new goal for humanity. The problem of nihilism, pervasive of all strata of society, cannot even be correctly stated unless it is recognized as the failure of modern syncretic philosophy which believed it possible to undertake the care of humanity through the mastery of nature.

The Catholic University of America,
Washington, D.C.

BIBLIOGRAPHY

1. Bacon, F.: 1870, *The Works of Francis Bacon*, R. L. Ellis and J. Spedding (eds.), Longmans and Co., London, Vol. III.
2. Bacon, F.: 1915, *The Advancement of Learning*, Everyman Edition, J. M. Dent and Sons, Ltd., London.
3. Bacon, F.: 1960, *The New Organon*, Library of Liberal Arts Press, New York.
4. Descartes, R.: 1931, *The Philosophical Works of Descartes*, Vol. I (transl. by E. S. Haldane and G. R. T. Ross), Cambridge University Press, Cambridge.
5. Descartes, R.: 1946, *Regulae ad directionem ingenii*, Henri Gouhier (ed.), J. Vrin, Paris.
6. Descartes, R.. 1962, *Discours de la méthode, texte et commentaire*, E. Gilson, J. Vrin, Paris.
7. Jonas, H.: 1974, *Philosophical Essays. From Ancient Creed to Technological Man*, Prentice-Hall, Englewood Cliffs, New Jersey.
8. Pascal, B.: 1951, *Pensées et opuscules*, L. Brunschvicg (ed.), Librairie Hachette, Paris.

WILHELM MAGNUS

THE PHILOSOPHER AND THE SCIENTISTS: COMMENTS ON THE PERCEPTION OF THE EXACT SCIENCES IN THE WORK OF HANS JONAS

It is difficult for a scientist to comment on the work of a philosopher. It may even seem to be improper for him to try since full competence is an acknowledged prerequisite for scientific publications. However, this is neither a philosophical essay nor a critical or hermeneutic evaluation of some of the writings of Hans Jonas. Rather, I shall try to formulate a response to the philosophical analysis of science and technology which appears in the work of Jonas, describing briefly its importance for scientists who are seeking an Archimedean standpoint for their activities. I assume that philosophy is not itself a specialized science but a discipline which can stay alive only through a never ending interaction with all human endeavors, and I propose to speak as a recipient, but not as a creator, of philosophical ideas. I hope that this will protect me against the reproach of going beyond my range of competence. I do not doubt that already my assumption about the nature of philosophy is open to criticism from at least some philosophers, but I feel certain that it is not objectionable to Jonas.

What I have to say here is, of necessity, of a very subjective nature, based on personal experience and on the observation of the experiences of fellow scientists. Nevertheless, I hope to contribute to a valid description of the relations of scientists to philosophy.

Of course, there is no unified attitude of scientists towards philosophy or towards any particular philosopher. There exist cases of scientists who are interested in philosophy as far as their time permits them to pursue this interest, which, however, has the same function for them as the interest in music, poetry, painting, history, psychology, economics, etc. has for others. To have interests of this type is the rule rather than the exception among scientists, a fact which should be kept in mind when using C. P. Snow's coinage of the "Two Cultures." I have yet to find someone who was brought up in the Western world and knows the second law of thermodynamics but has not read Hamlet.[1] There is only one human culture of which the exact sciences are a part. The question is how autonomous this part is and what relation it has with the whole. This is a philosophical question unless one chooses to consider it as a socioeconomic problem. The latter interpretation is widespread, but I do not see how it can cope with the following simple

S. F. Spicker (ed.), Organism, Medicine, and Metaphysics, 225–231.

observation. Research in the exact sciences and in technology uses the same ideas and methods in the United States and in Western Europe as it does in the Soviet Union. In particular, mathematical research is absolutely international. The understanding and, where external conditions are favorable, the collaboration between mathematicians throughout the world are perfect. Of course, economic conditions and the specific form of the organization of society can inhibit or favor the development of science. But apparently they cannot affect its essence.

The responses of scientists to the quest for a philosophical understanding of science vary widely. Many of them are not aware of the problem. Others consider the question as meaningless although they may never have read Wittgenstein. (It would be of some interest to know whether this answer has been given at all before Wittgenstein.) Still others have a positivistic or a pragmatistic position because these are considered as the views of "scientific" philosophers. The question after the motivation of scientific research is then usually answered by saying that it is a pleasing game, at least for the scientist. Especially the pure mathematician may then compare it with chess, and in this case even the competitive element of chess may be claimed as a motivation for research. ("We try to show that we are more clever than others.")

The less conventional responses from some leading scientists, especially physicists, are widely known. In particular, philosophical statements claiming a profound cultural role for science are due to Bohr, Einstein, and Heisenberg and have been well documented [6]. The following quotation is a statement made by I.I Rabi.

> It is only in science, I find, that we can get outside ourselves. It's realistic, and to a great degree verifiable, and it has this tremendous stage on which it plays. I have the same feeling – to a certain degree – about some religious expressions, such as the opening verses of the Bible and the story of the Creation. But only to a certain degree. For me, the proper study of mankind is science, which also means that the proper study of mankind is man [2]

Since the philosophical interests of Bohr, Einstein, and Heisenberg are sometimes explained by their European upbringing, it should be noted that Rabi, in interviews [1], emphasizes the fact that he received an American education.

Not only the very great scientists are interested in what may be called "the humanistic significance of science." An essay with this title, published in a philosophical journal by E. Cantore [3], drew nearly two hundred requests for reprints from scientists and engineers of all continents. It thus appears that there exists at least a substantial minority of scientists with a specific relationship to philosophy. Speaking for the members of this minority: What we expect from philosophy is first of all a communicable expression of our intuitive conviction that science is important beyond and above its own results. And here we can turn to Jonas. I quote:

> If we equate the realm of necessity with Plato's "cave," then scientific theory leads not out of the cave; nor is its practical application a return to the cave; it never left it in the first place. It is entirely of the cave and therefore not "theory" at all in the Platonic sense.
>
> Yet its very possibility implies, and its actuality testifies to a "transcendence" in man himself as the condition for it. A freedom beyond the necessities of the cave is manifest in the relation to truth, without which science could not be. This relation – a capacity, a commitment, a quest, in short, that which makes science humanly possible – is itself an extrascientific fact. As much, therefore, as science is of the cave by its objects and its uses, by its originating cause "in the soul" it is not. There is still "pure theory" as dedication to the discovery of truth and as devotion to Being, the content of truth: of that dedication science is the modern form.
>
> To philosophy as trans-scientific theory the human fact of science can provide a clue for the theory of man, so that we may know again about the essence of man – and through it, perhaps, even something about the essence of Being ([9], p. 210).

The title of the essay from which this quotation has been taken is simply, "The Practical Uses of Theory." To a pragmatist, this may raise hopes of finding, after all, a usable definition of "applied" versus "pure" science. I believe that the majority of scientists, like myself, will find Jonas' arguments convincing that such a definition cannot exist because there can be no theory of the practical uses of theory. However, this question is only a side issue in the essay. The very term "use" implies the question of "ends," and indeed Jonas provides a thorough analysis of the relation between "value-free" sciences and values. That this is an extremely important problem need not be emphasized. In fact, today's scientists have good reasons to pay attention to it since there is no lack of publications which describe science as a modern evil. However, I would have to write a philosophical essay myself if I wanted to discuss the specifics of Jonas' analysis of the problem. All I can do is try to explain the reasons why I believe that his essay meets the particular philosophical demands of scientists.

We do not expect scientific certainty from a philosophical investigation. Otherwise, we would stay within science, dealing with a specialty which is ours and which could, at best, arouse a passing interest. What we expect is not certainty but elucidation ('*Erhellung*', to borrow one of Jaspers' favorite terms). We will accept some speculation. After all, we use it freely in our own work, although it surfaces in our publications only once in a while in the form of conjectures. But most of us will demand restraint here, insisting on Occam's razor. [There are exceptions, for instance, the eminent nineteenth century mathematician Bernhard Riemann ([14], especially p. 511) and, to a considerably lesser degree, the contemporary astrophysicist Fred Hoyle in some of his popular writings ([8], p. 127).] In all cases we will insist on statements which are precise enough so that we can agree or disagree with them. We will expect information. This does not mean "facts" but an account of the thoughts of the great philosophers of the past and an analysis of their relations to each other. Also, we will demand some of the critical precautions which we have to take in our own work. Unqualified statements, sweeping judgments, and "nothing but" theories will almost automatically be suspect to the scientist. On the other hand, we will not be turned off by complexity.

Certainly, many philosophical investigations meet all of these requirements. What distinguishes the essay by Jonas is the fact that it also deals with a vital philosophical problem of the exact sciences. (I make haste to say that I do not call methodological or epistemological problems vital.) Unfortunately, today's scientists cannot turn to many philosophers in these matters. The quotation by Descartes in Jonas' essay – "Give me matter and motion and I shall make the world once more" – will at present be taken as an expression of exuberance which borders on arrogance but hardly as an expression of comprehension of twentieth century science. Ever since the mathematician C. F. Gauss abstained from publishing his results on non-Euclidian geometry because he was afraid of "the outcry of the blockheads"[2] (meaning the Kantians), mathematicians considered Kant's influence as at least potentially retarding. (Probably, he has had little actual influence on the development of the exact sciences, but this remark is not much more than a guess.) Hegel's characterization of Newton as "an absolute barbarian in conceptual matters" (5, p. 447) will not recommend him to scientists, at least not in matters concerning science. Nietzsche, who praises the "scientific method" (meaning both modern scholarship and the exact sciences) as the great contribution of his century to human achievements and who anticipated the mathematician Hilbert [7] by twenty years with his remark about the importance of relations rather than the definition of substance for our acquisition of truth ([13], end of aphorism 625), shows an astonishing lack of understanding when speaking about chemistry ([13], aphorism 623) or laws of nature ([13], aphorism 632).

Dewey, the great pragmatist, undoubtedly had a large amount of factual knowledge concerning the exact sciences at his command. It is rather strange to me and, as I believe, to many fellow scientists to see him denying the existence of any specific characteristics of research in the exact sciences. To quote:

> The marking off of certain conclusions as alone truly science, whether mathematical or physical, is a historical incident.

A few lines later, he continues:

> The temptation was practically irresistible to treat it (scientific research) as an exclusive and esoteric undertaking . . . . All the eulogistic connotations that gather about "truth" were called into play ([4], p. 220).

I do not see how these statements could be reconciled with those by Rabi (as quoted above).

Of course, I have to confess to ignorance with respect to a large body of philosophical thought, and I certainly do not claim that Jonas is the only philosopher to whom scientists can turn. However, he has to offer us much more than the essay mentioned above.

We are usually well aware of the fact that most of the exact sciences are a rather recent component of human knowledge. Especially in physics, it seems that only one quantitative "law of nature," discovered by Archimedes, was known in antiquity. But we rarely realize that certain tacit assumptions of a philosophical nature are indispensable for the emergence of the exact sciences. Least of all do we realize that these may have taken the form of theological arguments. In a detailed study [10] Jonas describes in particular the theological arguments of al-Ghazali (1058–1111) which anticipate the scepticism of Hume and the opposing arguments of Maimonides (1135–1204). He points out that any search for laws of nature would be hopeless according to Ghazali, and he traces the influence of the theological arguments on the conception of the world to modern times. In a second study [11] Jonas traces the relation between science and technology from the 16th Century to the present. The rather surprising historical facts have, of course, been established before, but a scientist will appreciate the interpretation of facts in the light of the emergence of new ideas.

The two volumes which contain the essays of Jonas quoted above contain also several other essays which touch on the work of the scientist. However, instead of commenting on them, I prefer to mention here an essay which deals with a problem preceding the specific problem of the nature of scientific research and which, nevertheless, exhibits one of its roots. This essay has the title "Image Making and the Freedom of Man" [12], and Jonas describes it as "an essay in philosophical anthropology concerned with determining man's 'specific difference' in the animal kingdom." The anthropological contents are summarized in the following quotation:

Former speculation demanded more concerning what should be regarded as conclusive evidence for Homo sapiens: at some time, nothing less than figures exemplifying geometrical propositions would suffice. This surely is an unfailing but also an overexacting criterion .... The criterion of attempted sensible likeness is more modest, but also more basic and comprehensive. It is full evidence for the transanimal freedom of the makers ([12], pp. 174–175).

But there is also the hint at the secondary form of image making:

The *adaquatio imaginis ad rem*, preceding the *adaequatio intellectus ad rem* is the first form of theoretical truth – the precursor of verbally descriptive truth, which is the precursor of scientific truth ([12], p. 172).

I believe that this passage contains the key to an understanding of the transcendental significance of mathematical theories and even of mathematics itself. Trying to formulate it more explicitly, I would say: We recognize in the mathematical models of particle physics and in the theorems of topology or number theory the latest metamorphosis of our distinctive human ability: They are images of elements of actual or potential order inherent in the universe.

Polytechnic Institute of New York,
Brooklyn, New York

NOTES

[1] "Hamlet Versus the Second Law of Thermodynamics" was the title of a review of a book by C. P. Snow in the *New York Times* some years ago.
[2] "Das Geschrei der Böotier." I have been unable to locate the exact reference (a letter by Gauss to a colleague). The quotation is widely known among mathematicians. It is generally assumed that G. means the Kantians.

BIBLIOGRAPHY

1. Bernstein, J.: October 13, 1975, 'Physicist I', *The New Yorker*, pp. 47–50, et passim.
2. Bernstein, J.: October 20, 1975, 'Physicist II', *The New Yorker*, pp. 47–50, et passim.
3. Cantore, E.: 1971, 'Humanistic Significance of Science: Some Methodological Consideration', *Philosophy of Science* **38**(3), 395–412.

4. Dewey, J.: 1960, *The Quest for Certainty*, Capricorn Books Edition, G. P. Putnam's Sons, New York.
5. Hegel, G. W. F.: 1959, 'Vollkommener Barbar an Begriffen', *Vorlesungen über die Geschichte der Philosophie*, Vol. 3, in the series Jubiläumsausgabe, Vol. 19, Fromann, Stuttgart.
6. Heisenberg, W.: 1969, *Physics and Beyond*, Harper & Row, New York.
7. Hilbert, D.: *Foundations of Geometry*, 2nd ed., Open Court Publishing Company, La Salle, Ill.; the original work appeared in German in 1901.
8. Hoyle, F.: 1957, *The Nature of the Universe*, a Mentor book, published by the New York American Library.
9. Jonas, H.: 1966, 'The Practical Uses of Theory', *The Phenomenon of Life*, Harper & Row, New York, pp. 188–210.
10. Jonas, H.: 1974, 'Jewish and Christian Elements in Philosophy: Their Share in the Emergence of the Modern Mind', *Philosophical Essays: From Ancient Creed to Technological Man*, Prentice-Hall, New Jersey, pp. 21–44.
11. Jonas, H.: 1974, 'Seventeenth Century and After: The Meaning of the Scientific and Technological Revolution', *Philosophical Essays: From Ancient Creed to Technological Man*, Prentice-Hall, New Jersey, pp. 45–80.
12. Jonas, H.: 1974, 'Image Making and the Freedom of Man', *Philosophical Essays: From Ancient Creed to Technological Man*, Prentice-Hall, New Jersey, pp. 157–182.
13. Nietzsche, F.: 1959, *Der Wille zur Macht*, Alfred Kröner Verlag, Stuttgart.
14. Riemann, B.: 1953, *Collected Works*, Dover Publications, New York.

MURRAY GREENE

LIFE, DISEASE, AND DEATH: A METAPHYSICAL VIEWPOINT

In our ordinary language and attitudes we tend to preserve a distinction between living and nonliving things. Unless we are speaking to a very young child we do not say the car cannot go because it is sick. We say a wrist watch needs repair but a wrist needs to be healed, an automobile is damaged but a person is injured. Although the crude reductive mechanism of an earlier era is past its high tide, a new mechanistic biology puts forward the claim that a living creature and a machine can be a model for one another ([11], p. 253). Where the old mechanism barred a thinking of purposiveness in nature, the new mechanism looks to cybernetics for a notion of purposiveness applicable equally to organism and machine.[1] Man-made machines are termed "self-regulating" insofar as they gather, store, and process information, monitor their own operations, and adjust their performance to a set goal. Is there any reason then why such machines should not be said to sense, remember, learn, adapt? A self-regulating machine performs like an organism. Why should not health be defined as successful performance, sickness as malfunction and death as irreparable breakdown of performance?

Two related problems seem to be involved in the neomechanist claim that a self-regulating machine can be a model for comprehending an organism. A machine by definition has a certain task or function, and a question for neomechanism is how to view a life in terms of a task or function. On first sight the neomechanist seems to be in the tradition of Aristotle, who employed for organism a term etymologically related to "work." But Aristotle set up a sharp distinction between the *energeia* of an artifact and that of a natural life, the very distinction neomechanism would collapse. Secondly there is the question of purposiveness or end, Aristotle's final cause: "for the sake of" what is the task or function? Purposiveness, the neomechanist claims, can be separated from finalistic thinking. Employing such cybernetic concepts as input, information, program, and feedback, the neomechanist substitutes "teleonomy" for teleology, the principle of natural selection for Aristotelian final cause.[2] In the perspective of evolution, organic purposiveness can be conceived in a manner that is no way consciously projective, providential, or based on a divinized nature. Heredity, for example, can be viewed as a "memory mechanism" governing the life process of the organism and survival

S. F. Spicker (ed.), Organism, Medicine, and Metaphysics, 233–263.

of the species. Logically it operates like any input that determines a particular mechanical performance. The programmer produces a plan, natural selection yields a genetic code, but in both cases we can speak of a goal-oriented performance and purposiveness. The concept of a program in this way has "made an honest woman of teleology" ([11], p. 9).

If organic purposiveness can be comprehended through a mechanical model, there seems no reason why capacities hitherto reserved for living creatures should not be attributed to machines, why disease should not be understood as disrepair, and death as total mechanical failure. What such a view of life, disease, and death would mean for our own humanity is a question I shall not pursue in detail. To suggest but one consideration, a Hippocratic oath to minister to a patient would be essentially a servicing warranty. But my main aim in this paper is theoretical.

Although feedback machines are of fairly recent vintage, the claim to comprehend organic life mechanistically goes back to Descartes in modern times. Capacities such as sensing and remembering, which neomechanism claims for its self-regulating machines, were reserved by Descartes for the thinking ego, which I shall deal with as "subjectivity". In opposition to some mechanists of his day, Kant sought to conceive organism teleologically by distinguishing "external" from "internal" purposiveness. Such a distinction had already been suggested by Aristotle in viewing a living thing as having its end in itself, as contrasted with an article of *techne* whose end lies in its use for another. Drawing upon concepts in Aristotle and Kant, Hegel presents his own teleology of organism as "the subjectivity of the Notion". Thus Hegel would comprehend natural life through a principle stemming from Descartes but denied by him to physical nature. In the present paper I would like to present Hegel's teleology of subjectivity, partly as a way of bringing out problematic aspects of the new mechanism, partly to contribute to a fuller historical perspective for comprehending organic life, disease, and death.

In the history of Western philosophy the concept of life has been of central importance metaphysically. Like Aristotle's divine *theos*, Hegel's Absolute Idea is an "imperishable life" ([6], p. 824). A perishable or natural life, which is for Aristotle a combination of form and matter, is conceived by Hegel as "immediate" Idea ([5], par. 216): an identity, though incomplete, of subject and object, Notion and actuality. The divine life for Aristotle is pure form and actuality, hence absolutely self-sufficient or self-relating activity conceived as a "knowing of knowing". Hegel's Absolute Idea is likewise a knowing of knowing, conceived by the German philosopher also as a self-manifesting and "the eternal vision of itself in the other" ([5], par. 214).

For both thinkers it is from the divine life that all natural life draws its being and obtains its notion. Hence for both thinkers the essential features of a natural life, such as self-sufficiency, are to be understood ultimately from the notion of the divine life. But while Aristotle conceives self-sufficiency as substance, Hegel conceives self-relation through subjectivity, whose determinations cannot be grasped as attributes merely but only as moments of its own self-differentiation. The movement of all self-differentiation is logically described in "the process of the Notion". Thus the Notion as "soul" is realized in a "body" which "expresses no other distinctions than follow from the characterizations of its notion" ([5], par. 216).

I

In its original meaning the English word "disease" meant a deprivation of ease, a deprivation expressed from the vantage point of the deprived. I need no one to tell me I feel a lack of ease; my condition is inseparable from my awareness of it. Such a situation, in a broader logical sense, in modern thinking came to be termed "subjective", and a subject in post-Cartesian philosophizing came to mean what can have an awareness of some sort or is capable of experience. For Descartes the subject could only be a thinker, whose every willing, desiring, feeling, etc., is a "cogitating," and this activity is in its very "being" different from anything in physical nature. Insofar as the thinker's being aware of any content is *eo ipso* a being aware of himself in some particular determination, the thinker in the first place always thinks himself: a situation, incidentally, evoking Kant's charge of "the scandal of philosophy". An animal eats, grows, preserves and reproduces itself, adapts to external conditions. In all this, however, there is only a motion of parts as in a clock, since the being of the animal is *partes extra partes* like everything else in nature. The animal's living activity can never be a *self*-relating but only a movement of parts. Hence the animal cannot possess awareness. According to the original meaning of the English word "disease", Descartes would have to say that an animal is not capable of disease but only malfunction, for the animal is a living machine.

Present-day mechanists who sometimes note approvingly the Cartesian concept of the animal automaton are generally unaware of a certain irony in their relation to Descartes. The exemplar for neomechanism is no longer the clock, which operates by a relatively simple action of parts upon parts, but the electronic computer, whose feedback loop keeps each component of the system "informed of the results of its own operation" and adjusts it in "the general interest" ([11], p. 252). The feeding back into the system "the results of its past activity" is a kind of sensing, remembering, experiencing. Thus with the new model of mechanism, organisms can be termed mechanisms not because organisms lack self-relation, as Descartes argued, but because mechanisms have it. For Descartes' subjective self-relation, neomechanism substitutes feedback, which renders a mechanism operationally a self-regulating "one". Armed with the concepts of feedback and operational purposiveness, neomechanism proceeds to challenge traditional notions of organism.

If we are to shake off all remnants of vitalism and the metaphysics of an "end in itself", argues the neomechanist, we must be open to new ways of considering anything "purposive", "self-regulating", a "one". Insofar as a machine's components operate "in coordination through a network of regulatory mechanisms", insofar as its materials and every feature of construction are dictated by the end to be accomplished, there is no reason to view a machine as less a self-regulating, purposive one than an organism. Operationally the homing activity of the automatic pilot and that of the pigeon are equally purposive and equally "in" each. Is it objected that the activity is but a "means" to an end that can never be in the machine? Granted, says the mechanist, but in what sense other than metaphysical can the pigeon be termed an end in itself? For evolutionary biology, which needs no metaphysics, the goal or end of an organism is to produce another like itself, an operation that in principle can be performed by a machine. In reproduction the end of the activity is a "product", distinguishable from the activity of bringing it forth. In organism the activity is the life of the individual creature. In thus distinguishing activity and product we see that the activity is as well "in" the machine as the organism, and the product as well "outside" both.

The new mechanists who employ teleonomy to make a statement about organic life and nature in general, also place organic purposiveness within a longer purview. An individual life is purposive not only as producing immediate offspring but also as transmitting a "program." This is no eternal form or essence but some pattern of life adapted to a given environment. The pattern, which has taken shape over eons through random mutation and natural selection, is more or less perpetuated in a species. But there is no transcendental necessity in nature for any particular species, and indeed "not only might the living world have been totally different; it might equally well never have existed at all" ([11], p. 172). The only purposiveness in nature is the performance of the individual creature in reproducing itself and continuing the species. Since the individual transmits the program through its progeny, Aristotle's final cause loses its meaning, for there is as much reason to say that the hen is the egg's way of producing another egg as the reverse.[3] But final cause is a metaphysical baggage. Once it is shed, biologists can use purposiveness without fear of violating scientific canons.

Before taking up again the problem of organism in modern philosophy, let us note an important feature of the Aristotelian metaphysics that is apparently discarded by neomechanism along with end in itself. In his notion of *energeia* Aristotle distinguishes between an activity like housebuilding which is a movement toward an end, and an activity that is complete and perfect at every moment ([1], 431a6–7), such as seeing, pleasure, happiness, and above all, knowing as *theoria.* The activity of a pure knowing is no *kinesis* like housebuilding, no overt performing or making, has no result or product other than itself, is never a means but always an end in itself. As a *noesis noeseos*, this absolutely self-contained activity is the divine life. As the natural creature's feeling of pleasure and well-being in the perfect play of its faculties, it is that which establishes a living being as an end in itself. Such a conception seems impossible for any *operational* definition of purposiveness, as put forward by neomechanism, and consequently of "artificial intelligence" in any of its aspects. The Aristotelian distinction between *energeia* and *kinesis* will prove decisive also for Hegel's approach to organism as subjectivity, for which we need to resume our account again briefly with Descartes.

The significant point in the Cartesian doctrine of the animal automaton is that awareness cannot consist in shifting combinations of parts but only in the activity of a self-related one. Later thinkers accepted this principle even while denying that an animal is but a combinative system of parts. Not only a thinking being but a living being needs to be comprehended in some way as a self-related one, through "appetition", for example. For Spinoza the identity of an organism consists in the "common conatus of the whole."[4] Insofar as the soul is the immediate idea of the body and the latter the immediate ideatum or object of the soul, the organism in its appetite remains a one in all its bodily parts. The Leibnizian thinking monad in apperception knows itself explicitly as an identity, while the organic monad is obscurely aware of itself in its immediate corporeal manifold. Both thinkers, through such concepts as conatus and nisus, seek to comprehend together ideality and materiality, to overcome the Cartesian separation of the self-identical activity of thinking and the mechanical movement of parts in organic life. Both seek to comprehend an organism as a self-related unity of its own bodily determinations directly and those of external nature indirectly. To the general problem of the unity of a manifold – not as a problem of organic life but rather of the certainty of knowledge – Kant contributed the notion of a *priori* synthesis. The thinking subjectivity is an identity not as a substance, as for Descartes and Leibniz, but an original synthetic unity of consciousness in whose pure forms the manifold of sensibility becomes an "objective" unity of experience.

Our thumbnail historical review is meant to call attention to several themes basic for Hegel's comprehension of organic life. Descartes' setting forth the concept of self-relating activity and at the same time denying this activity to animals, largely determines the shape of the problem of organism for later thinkers. The concept of self-relation on an organic level is advanced by Spinoza and Leibniz through the concept of an appetitive rather than purely cogitative self-relating activity. To the Cartesian self-relating subjectivity Kant adds objectivity through *a priori* synthesis. These major themes of modern thinking are retained and transformed by Hegel within a framework of "development". Development in Hegel is not an evolution in time, however, but according to the Notion, which means to say, in terms of subjective self-relation. In contrast to neomechanism's teleonomic purposiveness, Hegel's development of the Notion is a finalism that comprehends a natural life purposively, not as a means merely, but, as in Aristotle's notion of *energeia*, also an end in itself.

For Hegel, as for Kant, an organism cannot be comprehended through a logical model of mechanism. While the parts of a mechanism are related to one another as cause and effect, says Kant, an organism is a "self-end" [*Selbstzweck*], it is "cause and effect of itself" and its parts "are only possible by their relation to the whole" ([13], par. 64 ff.). Although Kant held in opposition to Descartes that an animal has representations, Kant did not seek to develop a notion of animal subjectivity. Hegel viewed Kant's notion of self-end as having "opened up the notion of life" ([6], p. 737), but for Hegel a self-end needed to be conceived as subjectivity, or according to the Notion. For this purpose Hegel employs the concept of "positive negativity" introduced by Fichte, which goes beyond the cause–effect relation of mechanism.

The animal subjectivity is termed by Hegel "soul," whose immediate object, as for Spinoza, is its body, and its mediate object the external physical world. But then Hegel draws upon transcendental thinking to provide the inner self-differentiation of the subjective one which psychophysical parallelism and pre-established harmony were unable to provide. As in Kant's subjectivity of consciousness, the soul for Hegel is a unifying activity of a manifold, namely, the body. This unifying activity, however, is not synthetic, in the Kantian sense, but a negating of particulars to "vanishing moments" of a concrete unity. In place of Kant's *a priori* forms of synthesis, which Hegel claims were presupposed and not philosophically derived, Hegel puts development according to the Notion. The Notion is for Hegel no inert identity but the logical archetype of all self-differentiating activity whose evolving categories articulate the rational structure of reality. In physical nature the emergence of organic life is a coming into existence of the Notion through the "inwardizing" of materiality's being-outside-self [*Aussersichsein*]. On succeeding stages of mechanism, chemism, and organism, physical nature's self-externality is sublated [*aufgehoben*] and the Notion comes forth from nature's protean forms as their "truth", which is mind or spirit [*Geist*]. Within nature itself, however, the inwardizing movement culminates in the animal, which "feels itself a one" in all its parts and members. The living *Gestalt*, inherently different from crystalline and other inorganic shapes, testifies to the fact that material externality is now but the being-there [*Dasein*] of an existent inwardness.

This is the *animal* nature which, in the actuality and externality of immediate singularity, is equally, on the other hand, the *inwardly reflected* self of singularity, inwardly present *subjective* universality ([7], par. 350).

The *Dasein* of the inner being, its shape, is a determinate embodying of animality in its notion.[5] But in this notion of the animal are contained the universal forms of nature sublated, in the course of the inwardizing movement, to moments of the animal subjectivity as self-end. In this way, for example, gravity, a universal form of nature, is sublated in the animal's faculty of locomotion.

The animal has freedom of *self-movement* because its subjectivity is, like light, ideality freed from gravity, a free time which, as removed from the real externality, *spontaneously determines its place*. Bound up with this is the animal's possession of a *voice*, for its *subjectivity* as *real* ideality (soul), dominates the abstract ideality of time and space and displays its self-movement as a free vibration *within itself* . . . . But above all, as the individuality which in determinateness is for itself, immediately *universal*, simply abiding with itself and preserving itself, it has feeling [*Gefühl*], – the *existent* ideality of being determined ([7], par. 351)

In these passages we are able to see how, by reconceiving transcendental synthesis as notional development, Hegel employs a Kantian epistemological formulation in order to comprehend organism as subjectivity. Where time and space are for Kant *a priori* forms of the unifying consciousness, they are for Hegel universal forms of nature's being-outside-self that have been inwardized in the organic subjectivity whose notion has emerged as nature's own inwardizing movement. Here too we see an example of the positive negativity. The animal's spontaneous determining of its place through the power of its limbs is a negating of an external physical power by employing the physical against the physical, like a seafarer's using seawater as ballast. The organic nature transmutes the inorganic, makes it over to its own life purposes, and this is the principle whereby the animal *Gestalt* derives from animality in its notion. From this principle we comprehend the organism in its determinate shape. Limbs, organs, systems comprising the organic configuration are the parts, which, in Kant's words, "owe their form and existence" to the whole. For Hegel they are embodiments, not of *a priori* forms of the unifying consciousness surely but of moments of the notion of the living creature as the self-end destined to employ inorganic nature as means. But this desiny of organism is encompassed in the universal purposiveness of the Notion: the self-manifesting of the divine Idea as absolute subjectivity.

As self-end, says Hegel, the living individuality "only is, in making itself what it is;" it is "the antecedent end which is itself only result" ([7], par. 352). The individual life as a self-reproduction is conceived again as a logical movement of the Notion. As in the case of the logical Notion the organism as identity inwardly differentiates itself and returns to itself in its particular determinations as concrete universality. In this way the organism actualizes itself as self-end, not as this singularity merely, but also as a moment of the universal Notion transcending its own perishable existence. In the latter perspective, animal life is a step in the spiritualization of materiality, a stage in the Idea's return to self from its other-being as nature. In the animal qua natural subjectivity emerge the first dim stirrings in nature of selfhood, awareness, and freedom. It is from this finalistic purposiveness of the Notion's actualization that the living creature obtains its own purposiveness as self-end.

Nature for Hegel is in general a realm of unfreedom, and the living creature's actualizing of itself as self-end is by far not the freedom of mind in its spiritual creations. But the animal's relative freedom in its body marks the "advance" of the animal nature beyond that of the plant. The plant is but "the first, immediate stage of subjective vitality [*Lebendigkeit*]," less a "subjective unity of members" than a differentiation into parts that are virtually separate individuals ([7], par. 343). The animal is the "veritable" organism that embodies organic life in its notion. In man the Notion is present as the subjectivity of consciousness, but the animal

exists as *subjectivity* in so far as the externality proper to shape is *idealized* into members, and the organism in its process outwards preserves inwardly the unity of the self ([7], par. 350).

In the first of the three syllogistic movements [*Schlüssen*] comprising the animal's "making itself" what it is, the organic "process outwards" is the articulation of the bodily organization as an "objectivity" in which the living subject "inwardly coalesces with itself." All features of structure derive from function: the animal's determinate way of actualizing itself as self-end; so that the objectivity is nothing other than that in which the living subjectivity is reflected into self as self-related one. These features, distinguished as notional moments, are: sensibility as abstract self-reference, or the animal's simple universal being-within-self in its corporeal manifold; irritability, in which the animal as particularity is stimulated by and reacts to externality; and reproduction as metabolism, wherein the animal posits itself as self-producing and self-preserving singularity. These three notional moments of organic life have their corporeal reality in the nervous, circulatory, and digestive systems, respectively, but in such a way that each system, embodying all three moments, in its own process passes over into the others. Of interest for our purposes is how the process of shape is conceived by Hegel as a process of subjectivity.

Living structure, says Hegel, is "essentially process." Kant said this cannot be conceived as the mechanical activity of part upon part, and today all hands are agreed. But can it be conceived on the model of the feedback mechanism, in which the performances of the components are coordinated in the interest of the whole? This "interest," as we know, is also a performance. For Hegel, coordination of performances for the sake of a performance is not the structural process of organism for it is not the activity of a self-end. The organic character of the process hinges on the relation of means and end. The organism as totality, says Hegel "converts its own members into its inorganic nature, into *means*, lives on and produces itself." Appetition, stressed by Spinoza and Leibniz, is conceived here by Hegel as "urge" [*Trieb*]. The conversion of means to end is different in principle from any relation of means and means where the end as "universality" is not implicit in the means as "particularity." In the organic process of shape, says Hegel, the urge of the subjective totality is

> the *urge* of each single, *specific moment* to produce itself and likewise raise its particularity to universality, to sublate the others external to it, engender itself at their cost, but similarly to sublate itself and make itself a means for the others ([6], pp. 766–767).

Thus the particular members, systems, etc., which derive from the organism in its notion, likewise derive their reciprocal negating activity as means from the urge of the subjective totality to actualize itself as "inwardly reflected self of singularity". The organic structural process is in this way a sublating of existent corporeal particularities to ideal moments.

It is the process that has as its result simple, immediate self-feeling [*Selbstgefühl*] ([7], par. 356).

That the end of the structural process is the inner reflectedness of self-feeling signifies a difference in principle from any process of a "self-regulating" machine whose component performances are coordinated in some one performance. In Hegel's view of the organic structural process in terms of subjectivity, the reciprocal negating of the particulars has as result no performance but "the existent ideality of being determined". Outer performance vanishes in inner ideality. While the one governing performance of a machine can be said to take up into itself the component performances, it remains but a movement in space. Even the sheer storage of "information" can be no more than a spatiotemporal arrangement and never an ideality. Hence the separation of activity and product in neomechanism's operational purposiveness. This is not the case for Hegel's process of shape. Insofar as subjectivity is its essence, the "product" is at the same time "producing agent," is product "only as the externality that equally posits itself as negative, or is product only in being the process of production" ([6], p. 767). In this sense the self-producing activity of the organism as self-end is an *energeia.*

II

In the process of shape, the first of Hegel's three processes of organic life according to its notion, the living creature has been considered as "enclosed" within itself, positing itself in reproduction as the permeating unity of its corporeal manifold and coalescing with itself as subjective one in self-feeling. Viewed thus as the organism's immediate being-with-self in its corporeality, Hegel's notional process of shape has prescinded from the outer world, whose immanent connection with the organic individuality is nevertheless implicitly present in all its structural features: e.g., that it has means of locomotion derives from the notion of the animal as ideality freed from gravity. For neomechanism structure is also important in comprehending organism. The macroscopic structure of an organism is evidence of an "autonomous determinism" that distinguishes organisms from most artifacts. But since "autonomous" can have no meaning of self-end, neomechanism's operational purposiveness cannot regard structure as inner self-reflectedness: an organism differs structurally from a crystal by "quantity of information" and can be conceived as a "self-constructing machine" (14], pp. 10, 12, 46).

The structural process is consummated notionally in self-feeling, according to Hegel, and it is as subjective one that the living creature encounters the outer world as an "other." Indeed it is only for an inner being that there can properly be an "outer." The external other is for the creature a "presupposed world already in existence." But from the notion of organism there is already a relation of the individual creature to its external world as its "opposite:" an objectivity in and through which it must actualize itself as self-end. As distinguished from a crystal or a mechanism, the living organism is an inwardness, a subjectivity

> which differentiates itself into members and which excludes from itself, as an objectivity confronting it, the merely *implicit* organism, physical nature in its universal and individual forms. But at the same time, it has in these natural powers the condition of its existence, the stimulus, and also the material of its process ([7], par. 342).

By virtue of the very emergence of organic life, the external world is at once an other "over against" the individual creature, and that to which the creature is bound in "absolute, indivisible, inner, and essential" connection.

Insofar as organic life has its notion in an opposition to inorganic nature, the opposition between the living creature and its external objectivity is simultaneously and necessarily an opposition within the creature itself. This is the case for all subjectivity. By virtue of the internal opposition the organism's relation to itself is a "contradiction" that sends it out beyond its abstract being-with-self in self-feeling. Therefore, in the process outwards that comprises the second *Schluss* of organic life in its notion, the animal is to be seen as determined by "need." Its inner tension is no less a relation of tension with the outer world, and it moves to overcome its own contradiction by sublating the otherness of the other to which it is bound by need. For this reason the animal's relation to external nature, for Hegel as for Aristotle, needs to be viewed as an "assimilation," a making of other like, where other is from the outset only "potentially" so for Aristotle and according to the Notion for Hegel.

In the process of shape we were occupied with the problem of the self-related one in its corporeal manifold, and in assimilation we need to see the organism as a self-related one in its physical environment. The relation of organism and external world, neomechanism often stresses, is an energy conversion marked by negentropy. In the face of the universal natural tendency toward decreasing order, the organism maintains its complex organization at the expense of its physical surroundings. But negentropy does not distinguish living from nonliving things, such as crystals, and organic utilization of energy already presupposes a principle of "limit" in the organization of the organism that is not contained in the notion as such of energy conversion.[6] For Hegel *Selbstzweck* is a principle of limit insofar as it signifies that the living creature transforms external nature in the manner of a living subjectivity, namely, that in the other the creature actualizes itself "for itself."

Making use of a distinction noteworthy in Aristotle and Kant in connection with problems other than organism, Hegel differentiates a "theoretical" and a "practical" assimilating activity of the animal. To be sure, this distinction in animal activity does not have to do with a knowing for the sake of knowing. One could say, however, that it is a first small step in that direction. The plant, as we noted, is not yet a veritable organism for it is not a true subjectivity. But the sentient animal subjectivity, which in its very process outwards "preserves inwardly the unity of the self," can retain its own freedom in the presence of the other, which too is left inviolate.[7] In this way the animal is capable of a theoretical relation to externality, a relation that can be comprehended neither as a performance nor in terms of exchange of energy but is nonetheless an "actual relation" where each of the two sides can be "free in the face of the other" ([7], par. 316 *Zusatz*, p. 180). Hegel's conception is not unlike that of the sensible form, where Aristotle says, not the stone but its form is in the sensitive soul ([1], 432al). As Aristotle had sought to relate the respective senses to the elements of physical nature, Hegel presents the senses as the subjective embodiment of the moments of materiality in its notion ([7], par. 358). Feeling as touch is the sense of the earthly element, the material individuality that possesses specific gravity, cohesion, resistance. Smell and taste are the senses of material dissolution, the chemical processes in which things are consumed. Sight is the sense of material ideality as outer spatial manifestation in light, and hearing is the sense of inner ideality manifested temporally in voice. In this way, Hegel claims to show, the senses comprise a system of the pure forms of materiality. From the perspective of the Hegelian finalism, or realization of the Notion, the "center" towards which selfless matter eternally "strives" ([7], par. 262 and Remark) is first attained in nature in the sentient animal as subjective one. From the viewpoint of Hegel's comprehension of organism, we should note that the "theoretical process" follows directly upon the organic *Gestaltungsprozess innerhalb ihrer selbst*.

While in its theorectical or "ideal" process outwards the animal subjectivity is "immediately reflected into self," the practical or "real" process with externality commences with a diremption of the creature within itself: its "feeling of lack [*Mangels*] and the urge to sublate it." For this reason, notes

Hegel, the role of externality in connection with organism is properly viewed as "excitation", not in terms of the cause–effect relation, which as Kant and pointed out, was adequate only for mechanism. For Hegel, as in transcendental philosophy's relation of externality and subjectivity generally,

> . . . nothing whatever can have a positive relation to the living creature if it is not in and for itself the possibility of this relation, i.e., if the relation is . . . not wholly immanent in the subject ([7], par. 359 Remark).

Like the theoretical process, the practical process of assimilation requires the notion of self-end, and is incomprehensible as a merely physical relation of organism and externality. Only the living can feel a lack, says Hegel, for as subjectivity it alone in nature is "the unity of its own self and its determinate opposite." A mechanism that meets an external obstacle or requires refueling could perhaps be said to be hampered. But such a limitation, says Hegel, is a negation "only for a third party, for an external comparison." There can be a lack only for that in which the overcoming is equally present, and "the contradiction as such is immanent and posited in it." But that which is capable of "containing and enduring its own contradiction" is a subject. Hence only through subjectivity can appetite, impulse, need, pain, and disease be conceived in their true character as "negations posited as contained in the affirmation of the subject itself."

The contradiction within the living creature that impels it outwards can only be resolved by negating the independence of the outer object, reducing it to a means. But that is not enough. If the creature has to do only with a means it never returns to itself as self-end. It must surmount the means-end relation or remain forever "entangled" with its opposite, retain ever the status of a particularity instead of coalescing with itself as subjective universality. Such entanglement, as we shall see shortly, comprises Hegel's concept of disease. Organic assimilation is a uniting of opposites, but as the activity of a self-end it is essentially different from the "neutral" unity of a physical system, such as in the chemical process of crystalline formation ([7], par. 201). Mechanical and chemical processes are necessary but not sufficient for organic assimilation. The process begins as a violence: mechanical seizure of the object, mastication, chemical dissolution, etc. But violence is directed against an "other." At what point is the organism no longer dealing with a means? As self-end, writes Hegel, the organism "makes this very process, which is on the point of lapsing into a mechanical and chemical one, into an object." The organism as self-relation relates itself negatively towards its own involvement with a means, not, however, in the manner of releasing waste products like the exhaust of a gasoline engine, but as a disdain towards its very dependence on externality, in Hegel's words, as a drawing away "from its anger towards the object" ([7], par. 365 Remark).[8] This negating of its involvement with externality is one with its "satisfaction" of need. It is the subjective return to self of a self-end but can find no place in the notion of a physical system merely.

While the initial process of shape was the animal's immediate being-with-self in its bodily manifold as abstract self-feeling, the process of assimilation is the mediated return through a negating of external difference. Having reproduced itself in its other, reduced to a "show" the independence of its opposite and revealed it to be but the organism's "own" inorganic nature, the organism no longer has the status of a particular face to face with another particular. In its theoretical or ideal process the sentient nature has shown itself to be the selfhood of selfless nature, and in its real process it has proved itself the truth of the mechanical and physical powers of external nature. The organism is now more than the singular "this" of shape, more than a particularity opposed to an other external to it. It is concrete universality, the third moment of the logical Notion, and its "process of kind" [*Gattungs-Prozess*] is the third and concluding *Schluss* of organic life in its notion.

III

Thus far we have seen how Hegel conceives the animal organism as self-end, the notional treatment proceeding from the process of shape to the process with externality and the process of kind. A self-end is for Hegel a subjectivity, and we have seen that it actualizes itself in the first two processes as self-feeling and satisfaction. In the third process it actualizes itself in sexual reproduction and passing away in the power of the kind. Insofar as this last and consummate self-actualization constitutes notionally the transition from nature to spirit, it represents the element of finalism in Hegel's conception of organic purposiveness.

The claim of neomechanism is to be able to comprehend organism purposively without finalism, and thereby according to the mechanical model. For neomechanism's teleonomic purposiveness, purposiveness in the individual organism is, as for Hegel, inseparably connected with the notion of the kind. But no finalism is here necessary, says the neomechanist, since purposiveness is operational. The purpose of the creature is to reproduce, and both performance and product can be regarded in terms of a program with two aspects, genotypical and phenotypical. Genotypically the program has developed through natural selection, is encoded chemically in the genes, and is transmitted in the germ cells from generation to generation through reproduction. At the same time, as contained in the somatic cells, the code governs the structure and function of the creature in its life-performance. It is through this performance, or the organism as phenotype, that natural selection makes itself felt. Random mutations are incorporated in the genetic program not when they abet survival *per se* but rather reproductive performance. Thus the role of the living creature in the ongoing program is that of a mediator like the feedback loop. A life is a feedback report from the phenotypical to the genotypical aspect of the program. In this way natural selection replaces finalism, and purposiveness can be salvaged for a mechanical comprehension of organism.

Hegel probably never dreamed of the discoveries in genetics and their importance for evolutionary theory. But his business as a philosopher was to think about matters like purposiveness; and the concept of mediation, inherent in any notion of purposiveness, is central to all his philosophizing. In distinguishing the "external" purposiveness of a machine from the "internal" purposiveness of an organism, the concept of the mediator provides the logical key. In external purposiveness, says Hegel, the means is broken up into two moments "external to one another": e.g., the machine and its performance, or, again, performance and product; the means is an "indifferent *medius terminus*" whose place can be taken by another.[9] By contrast, in the mediating process of organism, as in the process of shape, the living individual is directly the self-producing activity in which it is "for itself" the end. For this reason, a life for Hegel could never be viewed as purposive in the manner of a feedback loop.

But although the living creature is for Hegel not a means merely, its passing away in the *Gattungs-Prozess* is no less necessary for Hegelian finalism than for neomechanism's operational purposiveness. Yet there is an essential difference with regard to the notion of means in the respective concepts of purposiveness. The creature that is a means in the process of kind is for Hegel at the same time a self-end. But for teleonomy the creature is only purposive and a means through natural selection, the very factor that interposes itself between performance and product, thereby barring a notion of self-end.

That the living creature can be both end in itself and means is possible through the notion of the animal as subjectivity. Even while in reproduction it serves as a means the animal obtains a certain sense of itself in its universal being or kind. For Hegel, as we have seen, the animal individuality is for itself abstractly as self-feeling in its corporeal totality, and determinately in its satisfaction of appetite in its assimilating of externality. No doubt this being-for-self is connected with physical processes and performances. Unlike the situation in neomechanism's operational purposiveness, however, the performance of a subjectivity is a coming back to self as self-feeling. In the *Gattungs-Prozess*, according to Hegel, sexual reproduction, disease, and death are interrelated ways whereby the living individuality comes back to itself as self-end and is simultaneously a means to a transcendent end.

Gattung, which I translate as "kind," does not seem to be meant by Hegel as a particular taxon in a Linnaean or other classification.[10] Hegel speaks of the "universal type of animal determined through the Notion," and of man as the perfect animal. In the notional process of kind we are apparently to understand *Gattung* as some abiding variation of the universal type, exhibited in the different conditions of elemental nature.[11] In connection with reproduction Hegel speaks of the kind as the breeding group, though it is not species [*Art*] but divides itself into diverse specific forms occupying what later writers would call a niche.

The kind is for Hegel the "concrete substance" of the singular animal subject, its inherent nature and the power from which it never escapes. This universal substantiality has been in "implicit simple unity" with the singularity in the processes of shape, native capacities, appetites, and instincts attuned to its particular habitat. In realizing itself as self-end, the individual animal brings to determinate existence in itself this indwelling universality, and now the latter, in the process of kind, becomes for itself in its universality, i.e., posits itself as "subjective universality" ([7], par. 367). In the finalism of the Notion, this means the passing and transcendence of the singular animal life, an *Aufhebung* of purposiveness in the natural creature as self-end to a purposiveness of the Notion as spirit.

The connection between sexual reproduction and death consists for Hegel in the finalistic purposiveness of the Notion. For evolutionary biology the connection is empirical, and for the neomechanists it is understood in terms of feedback. In organisms that reproduce by fission or budding, death does not come as a natural limit of life. On this primitive level, as Hegel notes of the vegetable organism, one cannot yet speak properly of an individual life, for an individuality is a subjectivity, which first truly emerges with the sentient animal.[12] Evolutionist writers regard sexual reproduction as an advance over nonsexual reproduction in affording greater species adaptation through increased recombination of genetic materials and enhanced possibilities for evolutionary development. The distinction between somatic and germ lines, in the neomechanist view, enables the combination of sexual reproduction and death to afford a "reassortment of programmes at each generation"

([11], p. 309). In this way natural selection operates on the phenotype to produce an evolutionary feedback, and in the development of a hereditary "memory" individuals "have to disappear." Thus in teleonomic purposiveness the death of the individual serves the population group, and for this reason the hen can just as well be regarded as an egg's way of producing another egg, as the converse. This could never be the case in the finalistic purposiveness of Aristotle or Hegel.

In the *Gattungs-Prozess*, according to Hegel, the kind differentiates itself into subordinate animal groups and further determines itself to the immediate singularity, which, as "exclusive," is in hostile relation to others. By tooth and claw the animal distinguishes itself in reducing them to an inorganic nature. In this negative relation to others, however, its own fate remains one of violent death. But the kind is in the singularity no less in an affirmative way. While as exclusive the singularity pits itself in violence against the other, in the sex relation it "continues" itself in the other. The animal subjectivity possesses voice, and the mating call is an expression of the universal. In the other, not as consumed but subsisting, the animal attains to a feeling of itself.

Like all essential features of organic life, the sex relation is comprehended by Hegel through the dialectic of subject–object or the Notion. The subject's relating itself to the object is *eo ipso* its relating itself to itself. In shape the object is the immediate corporeality, in assimilation it is external nature, in the sex relation it is another subjectivity. In the sense of self attained in the sex relationship, according to Hegel, the living creature most fully actualizes itself as self-end even while it is a means that perishes in the service of the kind.

Sexual reproduction and death are to be understood together for Hegel not in terms of an evolutionary feedback but through the notion of subjectivity. In the plant, which is not yet truly a subjectivity, sexuality is more a relation of parts than individuals. But sexuality in its notion exists as an opposition of individuals. Lacking the self-feeling of subjectivity the plant feels neither an initial incompleteness in itself nor the satisfaction of completeness, and there can hardly be the choice of mate that occurs in some animals. The sexual process in the animal, says Hegel, is to be seen as beginning with need. But we cannot speak here of need in the sense that a machine needs fuel or even that a plant needs fertilization. The animal's is a felt need, a "contradiction" in it only insofar as it is at the same time "for" it. Viewed notionally the contradiction consists in that the animal singularity at one and the same time is: (a) not in accord with the universal or kind in it; and (b) the kind's "identical relation to itself" ([7], par. 368). In this way the singularity is a onesidedness that is implicitly totality, the kind in it making for the very "tension" of its incompleteness, hence

> the urge to attain its self-feeling in another of its kind, to integrate itself through uniting with it, and through this mediation to close itself together with the kind and bring it to existence, – *copulation* ([7], par. 368).

Unlike the notional processes of shape and assimilation, the opposites in the process of kind are "totalities of self-feeling," the urge of each being the production of itself as self-feeling. The nature of each permeates both, and in uniting they come within the sphere of the universal, the substantial relation of the kind. The latter, however, is no longer merely implicit, as in the process of shape, where the subjectivity was abstractly for itself as singularity in its corporeal *Gestalt*; or as in assimilation, where the subjectivity was determinately for itself as particularity in negating its opposite as its inorganic nature. Now each of the two sides attains its being-for-self as self-feeling. Insofar as each of the opposites is identical with the other and at the same time "infinite reflection" into self, the universal, or substantial relation of the kind, is "posited" as "subjective vitality." The creature, says Hegel, obtains the feeling of itself as universality, "the highest to which the animal can attain." But this is its limit, and the limit of physical nature generally. For the animal subjectivity the universal can never become a "theoretical object of intuition," which is possible only for the subjectivity of consciousness in spirit proper.

Thus sexual reproduction is viewed notionally as not merely a performance yielding a product,[13] as in operational purposiveness, but as a kind of self-knowing of the singular subjectivity in its universal substantial being. In some of the higher animals, notes Hegel, reproduction includes a kind of family tie and affectionate care of the young. But even here the identity of self and other can only be for the animal subjectivity in the form of another singularity, never in the manner of a universal that is for itself universal. The latter kind of self-relation needs to be grounded on the "I am I," the subject–object identity that is only possible for subjectivity as mind or spirit. The animal in reproduction is no doubt a means for the perpetuation of the kind, but in its self-feeling it is also an end in itself, never merely a phenotypical feedback loop or a "self-replicating machine" as in neomechanism.

As a performance, animal reproduction is indeed not a pure *energeia* but has a product other than sheer self-activity. Insofar as the product is another natural singularity destined, like its forebears, to grow, reproduce, and die, the universal is again only implicit. Hence propagation is termed by Hegel a "bad infinity," an endless succession of generations where each is a means bringing forth a product that is again a means, and so on without end. If the process of kind were no more than this, then Hegel's teleology would be like teleonomy, a purposiveness without finality. But the process of kind is for Hegel an "advance" of the Notion, and to understand this advance we need to see how Hegel's connecting of sexual reproduction and death is different from that of the evolutionary viewpoint, for which sexual reproduction and death are also necessary for advance.

The animal subjectivity in sexual reproduction feels the universal in the form of another singularity but cannot, in its self-feeling, become the universal that is for itself the universal. The animal can build nests and dams and fashion other works out of external nature instinctively, but it cannot detach itself from external objectivity like the introreflected "I" of consciousness, hence cannot objectify its own life activity as "experience," cannot attain to an objective knowing, to social institutions, traditions, and culture. It passes along its universal being not as a tradition but physically, in what evolutionist writers now call a genetic "memory," which makes possible evolutionary advance. Advance is not viewed empirically by Hegel. In the ongoing succession of the process of kind, says Hegel, the generations "fulfil their destiny in the process of copulation, and insofar as they have no higher one, go to their death" ([7], par. 369). But death is not viewed by Hegel as operationally purposive in the manner of teleonomy: e.g., as making possible greater organic complexity.[14] Death, like disease, is for Hegel the fate of the natural subjectivity which embodies the Notion physically but cannot exist for itself as Notion in the manner of spiritual subjectivity. The natural singularity, whose very being as subjectivity is infected with the being-outside-self of nature generally, is too constricted a vessel to embody the Notion as Notion. The animal is for Hegel a subjectivity, not, as for Descartes, *partes extra partes.* But the animal subjectivity has not entirely sublated the *partes extra partes* of materiality. The very coming to be, in the singularity, of the universality of the kind: the very actualization of the singularity as self-end, is simultaneously its dissolution and passing away as this singularity.

In Hegel's concept of organic life, disease cannot be comprehended as malfunction merely, as in the case of a mechanism, nor death as destruction beyond possibility of restoration. Disease is for Hegel a disrelation, not of parts to one another but of the animal subjectivity to itself as singularity and universality. It is because disease must be viewed in the context of the creature's substantial being that disease belongs notionally in the process of kind rather than the *Gestaltungsprozess* or assimilation. The possibility of disease lies in the inadequacy of the singularity to its own universal being as kind.

While the kind is the indwelling nature of the individual, it is also the mightier power that can come into conflict with it as an "external" universality or inorganic nature. The opposition between universality and singularity assumes the form of a "blockage" in the latter. The subjectivity no longer permeates its immediate objectivity, or corporeal manifold, as self-end. Provoked into conflict with the inorganic power, one of the organs or systems is deflected in its specific "urge" simultaneously to "produce itself" and make itself "a means for the others" (see above, p. 241). Rather it centers its activity on itself at the expense of the others, obstructing the fluid process whereby the organism comes back to itself in its objectivity as the subjective one of self-feeling. In mental illness the subjectivity's being "caught fast" in a particularity of itself takes the form of the *idée fixe*, the aggrandizement or distortion of a particular drive, in which the sick consciousness remains divided against itself ([8], par. 406 *Zusatz* p. 106, par. 408 and *Zusatz*). In physical and mental illness alike, the infirmity consists in the subjectivity's "remaining stuck" [*beharren bleiben*] in a particularity of itself which it is "unable to refine to ideality." This kind of disrelation of a self-relation is only possible for a subjective one, not for what can be put together out of parts.

Insofar as sickness consists in the organism's being caught fast in a particularity of itself, cure must lie in a dissolving of the blockage. The particularity that in disease has become active on its own account must be reduced once more to its proper role as self-negating moment of the fluid totality. Fever, a hyperactivity of the totality "against the isolated activity," is both symptom of disease and commencement of recovery. Medicine is compared by Hegel to a poison. It can serve as an excitation and challenge to the system to rouse itself in its defense against the external intruder, thus simultaneously ousting the usurper from within. In mental derangement, Hegel stresses, the physician needs to build on a "remainder of reason," wherever possible strengthening the universal selfhood by such means as work therapy and social intercourse. For mental illness, says Hegel, is not loss of reason but contradiction within reason, "just as physical disease is not an abstract, i.e., mere and total, loss of health (if it were that it would be death), but a contradiction in it" ([8], par. 408 remark). In physical disease the dominant note in Hegel's concept of therapy is that cure must come essentially from the organism itself. The whole medical art "does no more than assist the forces of nature" ([7], par. 272 *Zusatz* p. 434), and all therapy must rely on the organism's summoning of its own resources and reasserting itself in its natural vitality. Such a concept of therapy is evidently different in principle from any viewpoint of repair or readjustment.[15]

Like pain, which for Hegel is a "privilege" [*Vorrecht*] of sentient nature, sickness needs to be understood in terms of negative self-relation. A stone is incapable of disease but is destroyed in the negative of itself, as in chemical dissolution. The living subjectivity can endure its own contradiction because its negative is a negative "for it" and not its destruction merely. Disease is not merely the noxious effect in the organism of a foreign agency, for in the notion of life the animal is the microcosm in which "the whole of inorganic nature is recapitulated and idealized" ([7], par. 352 *Zusatz*). In health, where all members and organs are "fluid in the universal," there is "nothing inorganic for the organism that it cannot overcome." Only insofar as it becomes a house divided, i.e., when a particularity of itself "makes itself the centre," does it render itself vulnerable to a foreign power. But since the singularity is never adequate to its universal being or kind, it is never absolute master in its own house. What is ordinarily a part of its natural habitat, indeed a feature of its own body, can become for it an alien objectivity. This disparity, within the living creature, between singularity and universality means that an organism is not fully subject–object, not the perfect identity of Notion and reality. It is the possibility in the creature of disease, and the necessity of its death. The creature may recover from a particular disease, says Hegel, but finally succumbs to death, which is the disease of life.

IV

The disparity [*Unangemessenheit*] in the natural individual between its finitude and its universality, says Hegel, is its "original disease and the inborn germ of death" ([7], par. 375). Since this "death of nature" constitutes for Hegel the transition to spirit, – which transition connects purposiveness in organism with the finalism of the Notion, – we need to note briefly some considerations regarding nature generally. In the disparity between its reality and its notion consists the finitude, mortality, and "naturalness" of the living creature. To understand fully what Hegel means by naturalness, we need to go to the logical Idea for the purported emergence of nature as the Idea's "other-being." Because nature is the other of the Idea, it is an other to itself: i.e., its being is a being-outside-self. How Hegel conceives nature's main forms in terms of this *Aussersichsein* and its overcoming, we cannot go into further, except to note briefly how nature's being-outside-self means just the discrepancy, in every natural form, between its reality and its notion. This discrepancy, sometimes termed by Hegel "contradiction," is the Hegelian *dynamis* and *hyle* as the source of all movement. What does not have its notion in its reality does not have in itself its "center," so that the natural movement of all material bodies, for example, is their "striving" towards their center. But in bodies as in all natural forms, their center lies ever outside them and remains their eternal "ought." It is in this way that Hegel's notion of nature as being-outside-self takes on a dynamic character different from Descartes' concept of extension.

In the sentient animal, self-external nature first comes to have its center in itself. The living creature as subjectivity is for itself a self-end. Its life processes are its taking possession of its center, its actualizing within itself of its universal substantial being or kind. That the singularity is not able to bear its universality means that it cannot rise above its naturalness, cannot be wholly its own center. To be absolutely a center would mean to be, like the Notion, wholly self-determining. But the natural singularity, for example, cannot differentiate itself into the opposed sexes, and, like the Notion, return to itself in sublating the opposition. The sublation attained in sexual union is but the genesis of another singularity, another striving toward the center; and the procession of generations, like all movement in nature, remains purposive in the sense of an eternal ought.

As singularity, therefore, the natural creature cannot fully objectify its subjectivity, cannot render its reality commensurate with its notion. It can only attain to an "abstract objectivity," an objectivity that can never rise above a repeating round of physical processes which become devitalized and ossified as "habit." The natural organism, says Hegel, dies from the habit of life. The singularity achieves identity with its universality in habit, but insofar as opposition is essential to vitality, it is in the universal of habit that "the vitality itself" becomes extinguished. The oppositionless universal of habit is the quietude of death.

The process of kind is thus the culmination of the life of the natural individual and also its death. For our particular interest in Hegelian purposiveness we need to see more closely: (a) how this culmination is comprehended under the notion of *Selbstzweck*, and (b) how the passing of the individual constitutes an advance of the Notion from nature to spirit. Neither of these questions need burden teleonomic purposiveness, which rejects alike the notion of self-end and finalism, and conceives the living creature purposively only as a means. But they apply to Hegel's purposiveness inasmuch as it attempts to combine the purposiveness of the individual as self-end with the finalistic purposiveness of the Notion. We need to see first how in its very demise the individual can actualize itself as self-end, and then we need to consider briefly some problems concerning "advance" of organic life.

We have noted that for Hegelian and teleonomic purposiveness alike the connection of sexual reproduction and death serves the purpose of the kind. This is the case for Hegel, however, not because individuals "have to disappear" in order to make for a reassortment of genetic programs. The kind is for Hegel not merely some concatenation of organic traits suited to a given environment, but rather an embodiment of the Notion: although, mankind excepted, no particular animal kind is demanded by the logical Idea. The process of kind is a relation of identity in opposition between the creature as singularity and universality, a relation expressed in sexual reproduction, disease, and death. In reproduction the universal is implicitly present in the animal's feeling of its own self in another of its kind. Copulation is thus not a performance merely but an actualization of the creature as self-end and not only a means. At the same time the creature is a means, its performance has a product: the kind "come to be " [*die gewordene Gattung*], initially an "asexual life" in which the opposition between the sexes has been sublated. But as such a product the kind is again but implicit in an immediate singularity that passes away. In the succession of generations the kind is no introreflected, i.e., subjective, universality, but only a destiny of the singularities, at once their implicit nature and the power over against them in which they pass away. In disease the kind is also an implicit nature that can assume the role of an external power setting the singularity into opposition with itself. But by reactivating that very power as the power within it, the singularity can make itself whole again. In this way health, like the self-feeling in reproduction, is an identical actualization of the individual as *Selbstzweck* and of the universal within it. But since the singularity is in principle inadequate to its universal nature, health can only be a fending off of the inborn disease of life. Death posits the universal as the power over the individual, but not in the sense that death facilitates a "reassortment of programs" in the interest of the kind. Death is the *telos* of the natural individual, which, because it is natural, can have no higher destiny after reproduction than to live itself "into its body" ([7], par. 375 *Zusatz* p. 442). Unlike the lifeless planets that can endlessly pursue their rounds, the creature as subjectivity needs "interest," whose extinction in the "inertia of habit" means a repose of death. Thus death comes to the animal insofar as it is a subjectivity, and as an exhaustion and completion of the creature's possibilities as a natural subjectivity. It is in this way that organic purposiveness for Hegel would include in the notion of self-end the very demise of the natural creature.

But the living creature is not only an end in itself but also a means. It is used up. Its life activity is purposive as a performance yielding a product.

It is only in this fashion that the operational purposiveness of neomechanism can view a living creature as purposive, for the logic of operationalism bars a concept of self-end. The death of the individual is part of a package that includes sexual reproduction, genetic recombination, and the feedback effect of natural selection. The longterm result is the viability of the species, which is no end in itself but simply a consequence. In Hegelian finalism the death of the creature is the passing of the singular natural subjectivity in the power of the universal. This "negating" of the subjective singularity is at the same time a "positing" of subjective universality, or the Notion no longer as a natural life merely but a life of spirit. It is to this advance from nature to spirit, and to the problems of "advance" generally, that we turn in our concluding section.

V

Viewing nature and culture together, Dobzhansky writes:

> In producing the genetic basis of culture, biological evolution has transcended itself – it has produced the superorganic ([2], p. 20).

But how does the organic become the superorganic? Does the organic contain within itself the necessity of its own transcendence? The transcendence, says Dobzhansky, consists in the emergence of "self-awareness." Animals can be said to know, but man knows that he knows, and this is the essence of the superorganic ([2], pp. 337 ff.). Like most evolutionists, Dobzhansky rejects any a priori necessity or finalism. Animal knowing is purposive in the sense of adaptive, and it arises through natural selection. Natural selection sometimes brings about "ostensibly purposive results" but is in fact "automatic, mechanical, planless, opportunistic" ([2], p. 335). Although Dobzhansky's position is not that of neomechanism, his adaptive purposiveness is perforce an operational purposiveness. Evolutionary adaptation has to do with performance, conceived ultimately as reproduction. This is different in principle from an activity for its own sake, the activity of a self in being aware of itself. To expect a self-transcendence of operational activity to self-activity is like expecting a feedback mechanism, through sufficient corrective input, to elevate itself finally to self-awareness.

The difficulty of comprehending the development from organism to mind through operational purposiveness is part of the general incapacity of neomechanism to afford a concept of evolutionary "advance". How shall we speak of higher or lower, more or less advanced forms of life? Most criteria of advance are frankly quantitative: e.g., totality of biomass, more efficient energy utilization.[16] But why any such criteria should signify advance remains unclear. Indeed it seems that any criterion in purely physical terms must remain impossible, unless one takes the conclusion of a physical process, e.g., an oxidizing of a metal, as an "advance" compared with its beginning. Some writers speak of the increased organic complexity of structure and function, together with the increased freedom thereby afforded. But without a notion of self-end freedom can only be a meaningless word, and without freedom complexity as such can only revert to some other merely quantitative criterion. If survival *per se* is the criterion, mollusc is higher than man. Shall it be approximation to human intelligence? Apart from the problem of anthropocentrism, which remains an embarrassment for all parties to the discussion, some writers put forward the criterion of responsiveness to environment, perceptiveness, awareness. But either this is to be understood operationally, or through some concept of self-end. If operationally, then responsiveness becomes feedback, information in the service of performance, and the question only becomes, why is one performance, or more strictly its product, higher or more advanced than another?[17] Huxley, among the most humanist of writers and no neomechanist, would seem to avoid finalism by seeing progress as the pathway to "ever-fresh realizations of new possibilities for living substance." But "possibilities" can also be viewed merely quantitatively, so that in the end Huxley must point to the chief of possibilities as "mental capacity" and "organization of knowledge" ([10], p. 100).

For Hegel, purposiveness in the natural creature as self-end is not operational but grounded in the finalism of the Notion. The transition from nature to mind means that implicit purposiveness becomes explicit. On the level of mind the natural drives of the individual are satisfied in "universal," i.e., institutional ways, as in the family. But there are no institutions in nature, and the transition from nature to mind must show how insitutions are possible. This self-transcendence of the organic to the superorganic, to use Dobzhansky's terms, is conceived by Hegel through the notional moments of singularity, particularity, and universality. The creature has its *Selbstgefühl* in itself abstractly as singularity, determinately in an other of its kind as particularity: and now there remains "subjective universality," which is the condition for the possibility of institutional life, culture, the "superorganic." Subjective universality, as we noted, is the "end" of the *Gattungs-Prozess*: but in what sense? The tension of opposition has been sublated in sexual reproduction, the succession of generations proceeds, and the life of the creature turns into a "universal of habit." The universal of habit, to be sure, is a far cry from the universal qua institution, – but it is the utmost to which nature can attain, the closest it can come to embodying subjectivity as existent universality. The inertia of habit is for the singular subjectivity the quietude of death. The demise of the singularity is its passing away into the power of the kind. But such a universality is not yet the subjective universality of institutional life but only its abstract logical prefiguration. For this reason we need to say a word about Hegelian transition "according to the logical Idea."

Hegel's transition from nature to mind has been a stumbling block to many readers who seek to understand it in empirical fashion. This we are expressly told not to do ([8], par. 381 *Zusatz* p. 14). Hegel does not try to show a development of the sort that present-day evolutionists trace, for example, in the evolution of reptilian to mammalian forms. Hegel claims to show the rational necessity for the emergence of mind from nature, and no temporal sequence as such can provide rational necessity. The transition from nature to mind is indeed a self-transcendence of the organic to the super-organic, from knowing to self-knowing. But this can only be demonstrated through the forms of the Notion which enable animal self-feeling and human self-consciousness to be comprehended as notional moments connected through a notional necessity.

Nature's self-transcendence to mind is for Hegel an actualization of the Notion as subjectivity. The forms of physical nature embody the Notion "in itself" but not the Notion as subjectivity "for itself". The transition to mind is the emergence of the Notion as subjectivity "in and for itself," already prefigured in the animal subjectivity as self-end. The animal, which in sexual reproduction is for itself in an other of its kind, is implicitly Notion, i.e., in itself the identity of singularity and universality, subjectivity and objectivity. It is not yet the "achieved identity," for, still imprisoned in the being-outside-self of nature, the animal subjectivity cannot make itself for itself its own object. That its passing away is *eo ipso* the affirming of the universal as kind signifies that the Notion is no longer embodied in an "immediate" reality or objectivity whose determinations can hold themselves apart from their substantial unity like diseased bodily organs. With the passing of the singular animal subjectivity into the universal as kind, says Hegel, "the last self-externality of nature has been sublated" ([7], par. 376). In all of nature's forms and processes the Notion has been implicitly or in itself the unity. Now it is posited also as for itself: i.e., no longer, as in the processes of shape, assimilation, and kind, the subjective unity of determinations that need to be negated as self-external and thereby rendered ideal, but rather the unity of determinations of its own inwardness and as such already idealities. But in this way the Notion is posited as subjectivity whose reality or objectivity is suffused with its self-referring identity, a reality that is concrete universality. The Notion whose determinate being [*Dasein*] is thus commensurate with itself as subjectivity is no longer natural subjectivity but mind or spirit.

Here we see how there is no claim to show an empirical development, no claim to trace how mind emerged out of nature "in a natural manner." Could there indeed be such a tracing, would such an enterprise in principle be possible? No biological account can tell us when mind has arrived, for no such account can tell us what it is. Note, I leave aside the question of demonstration and ask here only about a tracing, but it shall be a tracing of knowing. Paleontology can trace sequence of structure, teleonomy show purposive function, and neomechanism rightly disclaim the need for self-end. But none of this can show how animal knowing becomes a knowing of knowing, or indeed how animal life can be comprehended as a knowing in the first place. Hegel's demonstration according to the logical Idea consists in showing that a certain logical structure, i.e., form of the Notion, embodied in physical nature preeminently in animal life as subjectivity, demands or implies another such logical structure, points to a higher embodiment of the Notion. Empirically we "find" this in mind, or rather, in the first instance, the subjective consciousness. But we can only find what we are looking for. This kind of demonstration applies also in the transition, earlier in the development of nature, from the vegetable to the animal organism ([7], par. 349). It is a mode of demonstrating an atemporal evolution that does not preclude a temporal one. Nevertheless plant and animal organism are necessitated as "eternal creations" as is the transition of organism as such to mind.

In contrast to the operational purposiveness of neomechanism, Hegel's purposiveness of the Notion is clear about its criterion of advance. Animal life is an advance beyond vegetable life because the animal is more truly subjectivity that is for itself in its determinations as its object, a more adequate realization of the Notion. A natural life is purposive not as a performance merely but as a self-relating activity. The logical model for such activity is no mechanical operation, whose purpose can never be for the mechanism, but a knowing, in which the knowing subject is for itself. The animal comes back to itself in its corporeality as self-feeling, "knows itself" in its body, its objevtivity and reality. The reality, on the level of subjective consciousness, is an external world whose objectivity is constituted by the very thought determinations wherein the subjectivity comes back to itself in its knowing. The divine life is absolutely self-identical activity, absolute subject-object as a transparent self-knowing that encompasses all reality. The Aristotelian distinction between an activity for its own sake and a *kinesis* may be taken as the essential distinction between the internal purposiveness of a self-end and the external purposiveness of neomechanism's mechanical model. In Hegel the Aristotelian *energia* is conceived as the self-relating activity of a knowing subject, and this provides Hegel's model for comprehending organic life.

For our "metaphysical" perspective on life, disease, and death, we have taken the Hegelian finalism of the Notion, for which nature as self-externality overcomes its being-outside-self and attains its "result" and its "truth" in the sentient animal life as self-end. Whether and to what extent this finalism can afford today a viable philosophical alternative, whether it can accommodate the empirical evidence for the evolution of living forms, this paper has not tried to say. Basic theses of the *Naturphilosophie*, such as the being-outside-self of physical nature, have been taken as given rather than claims to be established. Criticism has been muted in the effort to set forth the position on its own terms and for the issues it brings into view for a comprehension of organic life, disease, and death.

Hegel views a living creature as a natural subjectivity whose purposive life activity can be comprehended through the subject-object relation. Such a creature is a self-end, its life an activity for its own sake. In its "objectivity", meaning its structural process, assimilation of externality, relation to its kind, the living creature comes back to itself as subjective one, just as, in transcendental philosophy generally, the subjectivity's knowing of its object is ever also an implicit knowing of self. However essentially the objective processes of life are physical performances resulting in a product, they constitute ultimately an inwardness and an ideality.

Neomechanism claims to be able to conceive organic life purposively without finalism or a notion of self-end. Grounded in the notion of natural selection, which can only select among performances, purposiveness in an organism must needs be operational. A life activity is purposive in yielding a product, i.e., not as an activity for its own sake but for something other than itself, for example, the genetic pool. The creative can be operationally purposive as a means and only a means, never an end in itself. Its activity can be purposive never as a coming back to self, as Hegel sought to conceive through the subject-object relation. The "self-regulation" through feedback can be no true self-relating activity, and corrective adjustment of component performances in the "interest" of the whole – where this is but another performance – can be no different in principle from the action of parts upon parts in Descartes' animal automaton.

In operational purposiveness health can be conceived as optimum performance, disease as malfunction, but neither can be viewed in terms of self-relation. In speaking of evolutionary advance, operational purposiveness can perhaps offer a criterion such as efficiency of energy use but not perceptiveness or awareness or freedom, none of which can be comprehended operationally merely but only in some manner as self-activity. Such a concept Aristotle sought in *energia*, Hegel in subjectivity, and neomechanism claims it does not need at all.

Baruch College of the City University of New York,
New York City

NOTES

[1] For a trenchant critique of this claim, see Jonas, H., 1966, pp. 108–127.
[2] In the words of Williams, 1966, p. 259, the teleonomist seeks to explain evidence of design in an organism "as the inevitable consequence of the natural selection of alternative alleles in an environment described in relevant genetic, somatic, and ecological . . . terms."
[3] See the discussion in Simpson, 1966, pp. 63 ff. Insofar as natural selection acts more directly through the hen, one may give it a certain priority over the egg. But the priority can only be of one means over another.
[4] Jonas, H., 1965, "Spinoza and the Theory of Organism," *Journal of the History of Philosophy*, 3, p. 48.
[5] Cf. the discussion by Monod, 1971, pp. 10 ff., of the autonomous morphogenesis of living things. But this also characterizes crystalline formations, says Monod, and only a "purely quantitative" difference, to wit, the far greater "quantity of information" transmitted from generation to generation in living things distinguishes them from crystals.
[6] Aristotle rejects "fire" as the cause of nutrition and growth because "the growth of fire is without limit, so long as there is something to be burned" (*De Anima* 416a16).
[7] In sight, says Jonas, 1966, p. 145, the object "lets me be as I let it be."
[8] In assimilation, says Hegel, [7], par. 365 Remark, the organism takes from the outward process "nothing but chyle, its general animalization," and by ridding itself of this very process actualizes itself as "subjective" identity of its notion and reality, as compared with the merely "neutral" result in chemism, where self-end has not yet emerged in nature.
[9] See my article: "Hegel's Modern Teleology," in *Der Idealismus und seine Gegenwart: Festschrift fur Werner Marx*, ed. U. Guzzoni, B. Rang, and L. Siep, Felix Meiner, Hamburg, 1976, pp. 185 ff.
[10] The problems of taxonomy have no perfect solution, says Hegel. Since nature is the being-outside-self of the Idea, it is "impotent" to preserve the Notion in the protean and largely contingent forms of organic life.
[11] On several occasions in his Philosophy of Nature Hegel refers approvingly to Goethe's notion of archetypes. Evolution as an accounting for life in its main forms is rejected by

Hegel ([7], par. 249 remark and *Zusatz*). He does not regard species as eternal, however, and speaks to some extent of temporal development and atrophy of organic features. Unlike history, nature is not of philosophic importance in its temporal appearances, according to Hegel. Hence, while his lectures on nature contain much empirical material, it serves mainly to illustrate features of the "universal type" of vegetable organism and animal organism. It is these essential realizations of organic life in its notion that philosophy needs to comprehend, not in their temporal emergence but in their eternal necessity according to the logical Idea. Hegel's notion of "kind", I think, would in some ways correspond to what some writers on evolution speak of as "plan of structure" (e.g., Huxley, 1953, p. 69). Hegel's markedly holistic approach to the study of organisms, for which he cites the work of Cuvier and others, is congenial to many present-day writers (e.g., Harris, 1965, pp. 163 ff.), although much of what Hegel says is inacceptable in the light of later knowledge. More important than particular errors of fact is the challenge posed by evolution theory to Hegel's claim of eternal necessity for organic life and its main forms. Whether evolution, despite its express rejection by Hegel, could nevertheless be accommodated in the development of the Notion, perhaps as a "cunning of reason," as in the case of history, I cannot discuss in the present paper.

[12] For Huxley, 1953, pp. 105–106, 110, death is "the price life had to pay for individuality and the efficiency of its biological machinery"; what has a highly differentiated body organization is "dommed to die," See also Simpson, 1967, p. 260.

[13] The courtship ritual of certain birds, notes Huxley, 1953, p. 84, is "enjoyed for its own sake, as well as serving as a bond between members of the pair."

[14] For neomechanists the main, perhaps ultimately the sole, criterion for evolutionary advance is increase in organic complexity.

[15] See Bole, T.J., III, 1974, 'John Brown, Hegel and Speculative Concepts in Medicine,' *Texas Reports on Biology and Medicine* **32**, 287–297.

[16] See the discussion in Simpson, 1967, pp. 239 ff., of the concept of progress in evolution.

[17] See footnote 5, above.

BIBLIOGRAPHY

1. Aristotle, *De Anima*
2. Dobzhansky, T., 1962, *Mankind Evolving*, Yale Univ. Press, New Haven.
3. Dobzhansky, T., 1964, *Heredity and the Nature of Man*, New American Library, New York
4. Harris, E. E., 1965, *The Foundations of Metaphysics in Science*, George Allen and Unwin Ltd, London
5. Hegel, G. W. F., 1892, *The Logic of Hegel* (transl. by W. Wallace), Oxford Univ. Press, London.
6. Hegel, G. W. F., 1969, *Hegel's Science of Logic* (transl. by A. V. Miller), George Allen and Unwin Ltd, London.
7. Hegel, G. W. F., 1970, *Hegel's Philosophy of Nature* (transl. by A. V. Miller), Oxford Univ. Press, London.
8. Hegel, G. W. F., 1971, *Hegel's Philosophy of Mind* (transl. by W. Wallace and A. V. Miller), Oxford Univ. Press, London.
9. Hein, H. S., 1971, *On the Nature and Origin of Life*, McGraw-Hill, New York.

10. Huxley, J. S., 1953, *Evolution in Action*, New American Library, New York.
11. Jacob, F., 1976, *The Logic of Life* (transl. by B. E. Spillman), Vintage Books, New York.
12. Jonas, H., 1966, *The Phenomenon of Life*, Harper & Row, New York.
13. Kant, I., 1952, *The Critique of Judgement* (transl. by J. C. Meredith), Oxford Univ. Press, London.
14. Monod, J., 1971, *Chance and Necessity* (transl. by A. Wainhouse), Vintage Books, New York.
15. Simpson, G. G., 1964, *This View of Life*, Harcourt, Brace & World, New York.
16. Simpson, G. G., 1967, *The Meaning of Evolution*, 2nd ed., Yale Univ. Press, New Haven.
17. Waddington, C., 1976, *The Evolution of an Evolutionist*, Cornell Univ. Press, Ithaca.
18. Wiener, N., 1961, *Cybernetics*, 2nd ed., The M.I.T. Press, Cambridge.
19. Williams, G. C., 1966, *Adaptation and Natural Selection*, Princeton Univ. Press, Princeton.

RICHARD M. ZANER

ONTOLOGY AND THE BODY: A REFLECTION

> Between us and heaven or hell there is only life, which is the frailest thing in the world.
>
> Pascal

In the following brief, quite modest reflection, I try to come to an understanding of one of Hans Jonas' most fascinating and important discussions. I will also give it the respect and merit it is manifestly due – long overdue, it seems to me – by confronting it philosophically: i.e., critically.

I

Dualism, in whatever form – Gnostic ([9], p. xviii), Cartesian ([8], pp. 9–19, 54–63, passim), or even that frustration, epiphenomenalism ([10], pp. 146–54) – is not merely one of the passing points of view which pop in and out of the historical panorama of metaphysics, like some wacky Punch and Judy show. Nor is it simply an intractably persistent *bête noir*, forever dogging the tracks of monism's thrust to more pristine regions, where one can be ethereally innocent of the vexing questions of otherness. Dualism, Hans Jonas has brilliantly shown, is "so far the most momentous phase in the history of thought" ([8], p. 16), "the vehicle of the movement which carried the mind of man from the vitalistic monism of early times to the materialistic monism of our own . . . " ([8], p. 12). As thus historically bound, both idealism and materialism (in all their variations) are inherently "postdualistic" – partial products, both of them, of the dualism which by essence inhabits, shapes, and determines them, by leaving behind a "nature" (*res extensa*) bereft of life and quality and a "mind" (*res cogitans*) incorporeally stripped of concrete locus in the world ([8], pp. 17–18, 54).[1] It is thus no accident that, with dualism's positing that as matter can be without mind, so mind can be without matter, idealism and materialism are the two historically possible forms of monism. Both emerge from and presuppose dualism's ontological polarization, a bifurcation which cannot on principle be ignored or undone. Nor is it accidental that both, each in its way, leave something central out of joint, an incomprehensible which mutely but

S. F. Spicker (ed.), Organism, Medicine, and Metaphysics, 265–282.

insistently bespeaks its unaccountable presence; *the living body*. If taken as pure, extended matter, "then its being alive is not understood"; if taken as the sensible field of feeling of which it is one of a number of the ideas of consciousness, then its "being *my* body, my extended I and my share in the world of extension, is not understood" ([8], p. 18).

It is thus, Jonas writes, that

> the organic body signifies the latent crisis of every known ontology and the criterion of "any future one which will be able to come forward as science." As it was first the body on which, in the fact of *death*, that antithesis of life and nonlife became manifest whose relentless pressure on thought destroyed primitive panvitalism and caused the image of being to split, so it is conversely the concrete unity manifest in its *life* on which in turn the dualism of the two substances founders, and again this bi-unity which also brings to grief both alternatives branching off from dualism, whenever they – as they cannot help doing – enlarge themselves into total ontologies . . . . The living body that can die, that has world and itself belongs to the world, that feels and itself can be felt, whose outward form is organism and causality, and whose inward form is selfhood and finality; this body is the memento of the still unsolved question of ontology, "What is being?" and must be the canon of coming attempts to solve it ([8], p. 19).

An exact declaration-of-intent, and a telling formulation of the problematic which was[2] to occupy Jonas throughout his distinguished career: a philosophical biology whose major theme is "life" centered in the problematic of the living body. No mere partial monism, and certainly no dualism, can hope to solve it, much less even clearly articulate it. The surmounting of these "isms" can be accomplished only by a new, integral monism, one which must seek to "absorb" the radical polarity "into a higher unity of existence from which the opposites issue as faces of its being or phases of its becoming" ([8], p. 17). The "problem" at the root of dualism itself, that which gave rise to it in the first place, must be taken up, freed from the partiality and distortions of monism, as also from the intrinsic riddles of dualism: "the existence of feeling life in an unfeeling world of matter which in death triumphs over it" ([8], p. 17).

Descartes' bifurcation of reality is thus a major concern for Jonas. By no means an arbitrary invention, Cartesian dualism emerged as an apparently perfect response to difficulties encountered with early modern efforts to mathematicize Nature. To understand the "Book of Nature" as written by a hand mathematical, required that there be a way, in principle, of dividing off from her (soon to become "it") all elements or features which seemed not amenable to this expression. It also required that mathematics itself be conceived in a certain way, and of course that its applicability be secured (that things to be mathematicized be susceptible of this understanding, indeed were thus *truly* understood).

Without going into the necessarily difficult historical interpretations called for here (see [6], [1], [11], and [8], pp. 188–210), it seems clear enough that the key move (whatever its historical motives and expressions) involved making quantity, in the specific sense of spatial (later, temporal) spread-outness – *extension* – the sole category of Nature. Coupled with this, as Husserl has shown ([6], pp. 21–67, 353–78), mathematics (especially geometry) was taken to be the science of *measurement* (indeed, a "science" whose own origins were not questioned, but which was taken over as full-bodied and already-delivered as Minerva from Zeus). Nature, thus, as extended, or as a system of measurable particles in motion and defined by mathematical means, found its perfect match in mathematics. The difficulty (which Galileo had already seen) was that not everything was thus definable: the perceptual contact with Nature yielded "appearances," "secondary qualities," which, although merely apparent, resided in a subject whose "inside" consisted of these "ideas" (and the appertaining activities working on and with them) – and these could only be extrusions, unwanted because interferences, of nonmathematicizable things into the world of mathematicizable extension.

Just here, Jonas points out, Descartes' dualism came to the rescue: there are *two substances,* each of them self-existent, self-subsistent, and *sui generis. Matter*, defined as extension (*cum* motion), depends neither for its *being* nor its *being known* on *mind*, defined as thinking, and vice versa. What mattered here, of course, was the mutual *isolation* of the two: "the isolation of the *res cogitans* was the most effective way of securing the complete ontological detachment of external reality from what was not extended and measurable" ([8], p. 54).

And yet . . . : the stroke of genius (or ingenuity) became the mortal weakness. The theory "self-confidently *clashes* with experience" ([8], p.62), creating the enigma of how a person could possibly move even his own arm, or how an arm, struck by a physical thing, not only could be in its turn moved, but be *felt* as moving and painful. The living body of the thinker is the irascible and obdurate unaccountable: the mechanics of the body cannot comprehend its life, and the ideational theory of mind leaves its plain physical placement in the world the faintest shadow, an "idea" merely. Of course, for Descartes this crops up only in the case of man, for the rest of the "organic" world is but pieced in with *extensa*. This, though, is fully enough to undo, rudely but definitively, the whole cloth. For even if "the exemption of man was," as Jonas believes, "a mere inconsistency" ([8], p. 74), it is nonetheless decisive, as he also emphasizes.

And yet again: there are haunting vacillations in Descartes' words, curious clues which trickle through the otherwise sealed door between these substances. The reach of matter includes organisms, the human body too; hence, the human body as an organism within the material milieu, is *essentially* extension – and thereby is it barred from commerce with the human person so far as this signifies the *res cogitans*. Yet, and this reservation, "mere inconsistency" though it may be, is immense, as I see it: yet, the soul is *not* in the body in the way a boatman is in his boat – i.e., accidentally, removably, revocably. Indeed, so much the contrary is it that it seems a matter of *essence*, too: throughout, Descartes insists that the body is "intimately unioned" with the mind. Here, the stress is not at all on the *isolation* of extension from mind (whatever the benefits science, and later monisms, reaped from that), but rather on the "*intimacy*" (which was not grasped again until the late nineteenth and early twentieth centuries ([16], and [17]). But what are we, Descartes' explicitly invited readers, to make of this, if not that the "union" is essential? It seems almost as if Descartes let his better philosophic sense get the best of him, as if he blocked out the rigor of his otherwise tightly morticed system.

"Body" seems a wonderful enigma indeed (as Pascal acidly noted), within the very embrace of the Cartesian dualism: shifting, equivocal, at once *both* "mental" *and* "material," yet *neither* the one *nor* the other simpliciter. After all has been said, there remains the incredible *factum brutum: body, the human body, is the only affair within the entire "res extensa" which reveals this "intimacy," and thereby the only affair in the whole of reality which is "both . . . and . . . ," yet "neither . . . nor . . ."!*

This, surely, must make the reader pause. What is impossible here – an impasse of the first order – is that that "intimacy" can hardly be more than merely pronounced. While it is unavoidable that it be taken seriously, there simply is no way to do so. In a universe where everything is *either* matter *or* mind, nothing can be *both* matter *and* mind: the logic undergirding and giving life to the Cartesian metaphysic permits no devolving of the principle of non-contradiction into a logic of conjunction or disjunction. For that matter, even if it could, such a logic would be incapable of grasping the sense of "union" much less "intimacy" – neither of which are "both/and" or "either/or." Hence emerges the central problematic (Jonas is clearly correct) of subsequent philosophy: what is it we are called upon to think, when we meet – in Descartes' works or in our own experiences – this "intimate union"? What is it, to think "union"? And, to think it as "intimate" ? Descartes' ontology collapses under the weight of its own insight.

II

Just this question seems crucial for Jonas' quest, too; all I have suggested thus far is that it is also vitally within Descartes' as well. Clearly, too, it will not do merely to repeat metaphysical dualism, nor to smuggle in covertly any brand of monism. Whether Jonas is correct in calling for a "new" and "mediated" monism or not, we shall have to assess later. His more immediate concern is to pose the problem rightly, and begin its solution. There are, as I see it, several significant moments to this, and I want first to lay them out carefully, in order, second, to give them critical attention.

Unmistakably, there is a profound problem of "union." Still, as I read Jonas, that notion harbors serious difficulties unless it is purged of its dualistic overtones. The "two-ness" Descartes found is, to be sure, grounded in reality itself ([8], pp. 16–17), but even if my suggestion is right that Descartes seemed to be onto the main problem (the "living organism"), it is clear enough that he could do nothing with it. "Union" is, to put it bluntly, literally *unthinkable* in his final system. How, then, are we to conceive that "two-ness"?

In the first place, Jonas rightly stresses that it is only thanks to our being "living material things ourselves" ([8], p. 91), or on "the strength of the immediate testimony of our bodies" ([8], p. 79), that we can say anything at all about the body (I shall return to this point later). Without that source, "life itself would not even become visible, nor yet a blank in its place" ([8], p. 87). What we face with the genuine, "intimate union," is "the fact that the living body is the archetype of the concrete, and being *my* body it is, in its immediacy of inwardness and outwardness in one, the *only* fully given concrete of experience in general" ([8], p. 24).[3]

Secondly, this extensive outwardness and intensive inwardness "in one" are not only "genuine aspects of myself" ([8], p. 23), and thus give *concrete* significance to one side of the "two-ness," but also give the same significance to the other side. For it is not at all possible, on any traditional dualist or monist position, to find any legitimate ground for the causal interpretation of natural regularities. Hume was perfectly correct in his own way: there *is* no perception or impression of causality; nor is the nexus between data a perceptual content. Neither "force" nor "cause," hence neither "regularity" nor "connection," and hence *no idea of Nature at all* ([8], pp. 26–33). But, carefully consulting our own actual bodily life as embodied persons, the fact is that only our actual bodily life could possibly yield the very ideas which play such central rôles in the scientific account of nature: force, cause, connection, etc., are in no way *a priori bases of* experience, but are *themselves basic experiences*. Their root is the experience of "*effort*," resistance of things to my bodily acting on them, and undergoing the impact of things on myself as embodied ([8], pp. 22–23, 99–107). Thus, just as "causality" is a "finding of practical, not of the theoretical, self, of its activity, not of its perception" ([8], p. 23), so force "is not a datum, but an 'actum' humanly present in effort" ([8], p. 25).

Hence, if one renders life, especially bodily life, unintelligible, one renders both the world of nature and world of "mind" or "soul" unintelligible. "In the body, the knot of being is tied which dualism does not unravel but cut. Materialism and idealism, each from its end, try to smooth it out but get caught in it" ([8], p. 25). From this, as I see it, the conclusion seems evident: the "union" is ultimately an abortive notion, for its very sense effectively buys into a dualism, if only incipiently. Or, differently expressed, what Descartes had grasped in a sort of desperation – intimate union – and let slip through, was that there truly *is* no "union" but rather the "one" living body whose "extensiveness" locuses it within the actual world (and grounds subsequent knowledge of that world), and whose "intensiveness" is the "efforting" seat of subjectivity. The dual*ism* is wrong for it renders the dual aspects of reality, grounded in the *duality* of the organism, into ontological substances. *Both the aspects are myself* in the core sense of what alone "my life" can mean. There are not two *things*, then, to be somehow by some legerdemain welded together; there is rather the one and only embodied person, whose being is however *intrinsically complex*.

But in this complexus, what is it which answers to "life"? By virtue of what is it an "organism"? We have to do here with a "whole," but it is a "living" whole. This "selfhood with which 'organism' originally dared indifferent nature" ([8], p. 107), this "ontological surprise" nature springs with living things ([8], p. 79): what is it? The fundamental key, for Jonas, to unlock what must otherwise remain utterly closed, is *metabolism*.

In this remarkable mode of being, the material parts of which the organism consists at a given instant are . . . only temporary, passing contents whose joint material identity does not coincide with the identity of the whole which they enter and leave, and which sustains its own identity by the very act of foreign matter passing through its spatial system, the living *form*. It is never the same materially and yet persists as its same self, *by* not remaining the same matter. Once it really becomes the same with the sameness of its material contents . it ceases to live; it dies . . . ([8], pp. 75–76).

Anything else which occurs in and for the living organism – for human life: the emergence through motility, emotion, perception, image-making, etc. – could not occur without the functioning of metabolism. At the same time, the peculiar kinds of transcendence which appear at higher levels are already conditioned by, and prefigured within, metabolic activity. Thus, after recognizing and showing the differences between plant and animal life, Jonas insists that the dynamics of animal metabolism itself makes possible the distinctive sort of mediate actions in animal life ([8], pp. 100, 102, 103),[4] and on this ground become possible the still-richer levels of human life and freedom. The understanding of the complexus (extensive outwardness/intensive inwardness) will find its most *general* but also its most *fundamental* principles in the activity of metabolism, since this is all-pervasive in life as such.

Jonas thus contends, in effect, that this root of life answers the double question of life and organism qua "system." Metabolic activity is not a mere peripheral activity, but "is the total mode of continuity (self-continuation) of the subject of life itself" ([8], p. 76, n. 13). As against the machine, no matter how sophisticated cybernetically ([8], pp. 108–134), the organism is a system whose very character is at once the result and the agency of its metabolism, wholly and continuously. Hence, the organism is the peculiar self-referential center it is solely by means and on the basis of its continuous interchanges with its material environs. The "unity" here is one totally innocent of "synthesizing activities" of thought or perception, as also of the concurrence of forces binding its parts into a manifold: it is strictly "self-unifying, by means of changing multiplicity. Sameness, while it lasts . . . is perpetual self-renewal through process, borne on the shift of otherness" ([8], p. 79). This conception alone can give significance to the ontological notion of "individual," a conception beyond the reach of phenomenology ([8], p. 79).

Here, we find a striking reversal in the ontological relationship of "matter" and "form": where inert, purely physical "wholes" are "identities" thanks to their *matter*, for organisms, *form* is the essence, matter the accident. Thus, "the organic form stands in a dialectical relation of *needful freedom* to matter" ([8], p. 80). Form in living things is thus efficacious (i.e., "real"), and even though enjoying a certain *independence* from its material – precisely by preserving self-identity through continuous metabolic change of matter – it yet, just thereby, is ineluctably *bound* to its matter. The "reality" of the organism's *form* is thus an "ontological revolution" in respect to matter: while the latter's "identity" is fixed, the organism's is mediate and functional, revealing always changing and shifting substrata: an *act* incessantly accomplished, a genuine "self"-continuation.

This complexus, then, is the ontologically basic sense of "self": emergent self-identity of form which thereby is singled out as distinct from the environs, from all the rest of reality. "Profound singleness and heterogeneousness within a universe of homogeneously interrelated existence mark the selfhood of organisms" ([8], p. 83). An identity whose being is to continue, maintain, assert *itself* is always within a milieu set off against it, polarized by otherness: "self" is, too, a polarized identity within the context of otherness.

Ceasing to be *thus* is tantamount to ceasing to be *at all*. The very freedom of the living form *from* matter is also the (dialectical) necessity of being *bound to* matter. To be self-continuing, the living form *stands in need of* the material world, and thus assumes the richer forms of "outwardness," being-at-world, while yet that very being-turned-outward is an "inwardness," a "for-itself" whose form is thus complexly inner/outer.[5] The *turning-outward*, however, is of necessity (for the form *to be* what it is) a *taking-into-inwardness* of its other, foreign matter. That rudimentary activity is what ultimately gives sense to higher-level modes of interaction (pursuing food, fleeing danger, etc.) The "primary antinomy of freedom and necessity," at all levels of life as such, are "inherent in organism as such" ([8], p. 84).

For this self-continuation of efficacious form to be, some modality of sensitivity, awareness, interest is requisite. Self-continuation of the metabolic process purports the presence, however minimal, of a *needful self-concern*, active by way of *selective* modes of relation to world. Affection by the foreign matter brings about the sense of being-affected, impinged upon, hence of *self-feeling*, hence of sensitivity to what is affecting this "self" – viz., the other. Thereby emerge the germinal senses of the "within-ness" and "without-ness" (the self of organism), in space (without) and time (within, as stretched to what "will" be brought into the metabolic process). The incorporating of what is other is already an encountering of a "there" (meshing it with the "here"), but it is a "there" which also signifies a *time* necessary to bring it to the "here." Self-transcendence, thus, is "life facing forward as well as outward and extends 'beyond' its own immediacy in both directions at once" ([8], p. 85). In different terms, the very fact of metabolism, that always changing and always precarious relation of organic form to its matter, signifies that *purposiveness* is essential to the being of this form. Jonas thus sums his view: "there is no organism without teleology: there is no teleology without inwardness; and: life can be known only by life" ([8], p. 91). All higher modes of life, which open up peculiarly animal forms of "distance" and thus ground specifically human forms of relatedness to the world, even theory itself, are constitutively grounded in this elemental activity of metabolism. With their emergence in motility and emotion ([8], pp. 99–107), perception ([8], pp. 135–156), and image-making ([8], pp. 157–182), there emerge as well the richer modes of the dialectic of freedom and necessity, risk and dependency, which single out the human world itself.

There is thus a far richer world beyond what has been indicated thus far. Enough is at hand, though, to give pause for reflection – indeed precisely *before* one goes any further with Jonas. It is imperative here to take seriously Jonas' claim that what he has maintained stems from "the evidence of each one's own organic awareness," which "evidence we find in ourselves is an integral part of the evidence concerning life which experience puts at our disposal" ([8], p.91). It is *this evidence*, indeed, which is continually, albeit most often silently, operative in every biologist's and behaviorist's *actual* work and study – without this, they would, like Sir James Jeans' "mathematical God," utterly miss the very "life" and "behavior" they seek to claim to know.

III

A brief rehearsal of the issue is in order, to secure the terrain of our reflection. Within the panvitalism of classical times, seeing life everywhere rendered death enigmatic; but to acknowledge death was effectively to negate it by making it a transformation, a stage along life's way. In modern materialism precisely the reverse holds: not death – a merely "natural" event among others – but life becomes the riddle. But materialism, and its modern counterpart idealism, are creatures born by the Cartesian sectioning – if not abortive quite, these new monisms, then surely severely handicapped as neonates, carrying always within themselves, each in its own way, the primal marks of their parentage. What stands as the riveted impasse of dualism, becomes in its postdualistic offspring an egregious misfit, a cunning disguise to be unmasked: the living body.

Hence, within the imperative to think from within one's own historical context, it is inevitable that one situate oneself with respect to that history – to the origins of Cartesian dualism. The major theme of our thinking today is perforce the theme of life, for that dualism, stripping the material world of human presence, forcefully deprived the realm of life of any *effective* status (hence, reality) in the scheme of things. By cluttering up the soon-to-be irrelevancy of "subjectivity" with all the leavings of that sweeping democratization of nature *à la bas* (to "extension" and "motion" only) – sensory qualities, feelings, values, goals, and life itself – in short, by directly clashing with the evidence of concrete experience, dualism necessarily created a needful, demanding *problēmā*. Man, in the terms of that theoretical posture, becomes a puppety caricature; but so, too, does the wealth of animate life, its marvel of complex organization, observable opulence, and plain display of difference. "Theory," Jonas reminds us, became as it hitherto had never been: an ingenious, inventive, even sportive effort to solve "a difficulty never faced before and itself an invention of theory" ([8], p. 63). However much one must admire the inventors of those theories, one must nevertheless be centrally suspect of their problem.

"The problem": what is it, for Jonas? What is the "evidence" I am supposed to attend to, and what does it disclose? On the one hand, I am asked to focus my unprejudiced regard on my own experience more especially on my own living organism. Only life can understand life; but more than that: for, this "evidence" I have in my own case is "an integral part of the evidence concerning life which experience puts at our disposal." Thus, on the other hand, I am asked to take this evidence I obtain in my own case and presumably make use of it "critically": i.e., to avoid the pitfalls of possible anthropomorphism ([8], p. 91), but also to enable me to know what *life itself*, beyond *my own* life, truly is. Here, several puzzles make themselves felt.

First, as for possible anthropomorphism: why not? Or, rather: within Jonas' own "new monism," what else could there be? While not exactly ruling it out in the passage cited, in his "Note on Anthropomorphism"[6] on the other hand, it seems clear that it is most definitely ruled *in*. The history of anthropomorphism is the same as the history of dualism/postdualistic monism, a history of struggle against teleology (final causes). In modern materialism, as we know now, we are left in the same dilemmas and caricatures: alienation of man and obfuscation of the reality of life. The other alternative mentioned in this Appendix (presumably the one Jonas is disposed toward, and in any case the one implied in this emphasis on "evidence"), is "to take the presence of purposive inwardness in one part of the physical order, *viz.*, in man, as a valid testimony to the nature of that wider reality that lets it emerge, and to accept what it reveals in itself as part of the general evidence . . . " ([8], p. 37). I gather that it is with an aim to clarifying this, and related issues, that Jonas points out the two major tasks this study points to: a philosophical biology and a philosophy of nature. It is, he states here, *solely on the basis of* the philosophical biology that the philosophy of man is at all possible. Thus, the "charge"of anthropomorphism is turned into a positive acceptance by Jonas, but solely in the context of philosophically grounding it in philosophical biology ([8], p. 92).

But here things quickly become obscure. On the one hand, Jonas appeals to the evidence we find in ourselves, the evidence which gives us (presumably "purposive inwardness" in the most direct and incontrovertible way, on the basis of which we shall then move to "life itself": it is "valid testimony" to the wider reality, and is the sole basis and condition for biology and behavioristics being able to do what they do ([8], p. 91).

On the other hand, however, the philosophy of man is said to be strictly dependent on philosophical biology. Anthropomorphism is effectively embraced; but then it is submerged with what may as well be called biologism. In the first case, the "evidence" can only come from within *reflection*, for in no other way could one possibly become cognizant of "my own organism" as mine, "my own self," and the like. In the second, it is at least uncertain what rôle reflection might play, if any, and thus uncertain, too, precisely what "evidence" comes into play. If we wonder about this latter issue, so far as I can tell, Jonas refers us back to the evidence mentioned: I am able to know "life" because, in the strongest sense, I myself am "alive," an organism. But *this* evidence seems clearly to imply the precisely reverse relation between the anthropology and the biology!

This seems not only full circle; it seems on the face of it vicious. For if philosophical anthropology is itself not possible except on the basis of philosophical biology, then neither can the evidence for the first figure in the securing of the latter; it must rather presuppose the latter. Thus, only the reverse relation could possibly apply: the biology is what can alone inform and ground the anthropology, *and therefore the evidence in question*. It cannot be had both ways, and this is nonetheless the proposal; one is left wondering about the ingenuity ([8], pp. 62–63) of *this* "new" monism.

This is no small quibble: neither irrelevant nor irreverant. It concerns not only the possibility of Jonas' ontology; it concerns more immediately the question of that "evidence" and the "method" for grounding judgments made on behalf of the philosophical biology and the anthropology. All that I have suggested thus far is that, given the difficulty mentioned, the problem of method must loom large in this project, precisely to the extent that the sense of the evidence appealed to here remains uncertain, at the least, in respect of what it is supposed to establish, or give access to. (I return to this in a moment.)

The point has another relevance. Might it not be that the very history Jonas unravels so well has a quite different significance? For him, it signifies the necessity of a "new, integral monism" whose mission is to "absorb" the conflicts of dualism, idealism and materialism "into a higher unity of existence." Thus do equally integral and purportedly comprehensive visions of the real become "stages," chapters merely, in the future history of metaphysics. But is that step really necessary? Might it not be, rather, that the history has the significance, negatively, of precisely *not* repeating the urge to comprehensiveness? That "a new monism," whatever it may purport to do, is yet but *another* monism – and thus suspect from the very beginning? It may be that the yearning for totality, i.e., is what is so awry in the "issue" Jonas critiques, their "partiality" not merely an historical mark of particular parentage but also of the very effort itself! Perhaps partiality is the inevitable result of totalism; perhaps, too, *any* subsequent effort to conceive and write the "oneness" within "twoness" bears the inevitable mark of that very same parentage. I only note that at the most rudimentary levels of life, for Jonas, there is yet the radically alien: ephemeral or not, "matter" stands not just as "other" but radical stranger to "form"; and "form," however efficacious, however needful of "matter," is yet its radical opposite. What is different here – that "form" rides the crest of "matter," and that "nature" springs forth with "form" – may not make all that much difference for resolving the dilemmas of dualism. In the effort to construct the new monism, i.e., the duality of matter and form seems but to return, assuming only a new kind of tension and opposition. Jonas' dialectical move may thus not at all escape the hold of any totality's grip, wherein "having it both ways" seems but native air. But the doing of that skillful feat may well be the very thing our recent history, from Descartes on, has made impossible: the urge to comprehensiveness may well require far stronger curbs than even Kant envisioned. I thus rephrase the "crisis of every human ontology" in, not dialectical, but reflexive terms: every future ontology which will be able to come forth as a science ([8], p. 19) must be capable of *accounting for and justifying itself* first and foremost (cf. [2], p. 47). This reflexive inquiry, however, may only show the impossibility of making good on the wager of comprehensiveness. However that may turn out, the main thing is that there cannot on principle be such efforts in the absence of *continuous and radical, i.e., transcendental self-criticism*. Taking the appeal to "evidence" which Jonas makes with philosophical seriousness, thus, leads directly to the transcendental grounds for "one's own awareness of organism," and in this specific sense to the *transcendental-phenomenological problematic*.

This returns me to my initial concern about evidence, and to the shifting locus of Jonas' problematic. To what am I directed in the case of "evidence" in my own case? One would think that front and center would be, precisely, this *body which is mine*. And here, as I have elsewhere suggested[7] ([16], [17], [18],) it would seem that this "front and center" is the issue: by virtue of what is this one organism uniquely singled out, from the universe of objects and organism, *as mine* (cf. [5], pp. 60–61)? I "want" to eat, and *I* "move" *my* arm . . . .

While insisting that just this daily action is bequeathed to us as a profound riddle by Descartes ([8], pp. 60–62, e.g.), we are in the meantime left similarly wondering by Jonas' analysis. That central question is quickly submerged, for what he presents is not at all a study meeting and redressing the Cartesian legacy, uncovering critically why it is that *this* body is so "intimately unioned" to me, but rather an exercize in philosophical biology: *metabolism* and its implications for understanding "life" and "organism."

Fascinating, profound, and subtle, his analysis however can hardly be warranted on the grounds of evidence he appeals to. For it is in this study that he appeals to "my own living organism" and, so it seems to me, the evidence here can in no way penetrate to the metabolic activities *of my own* organism, nor does this instruct us on the grounds for my experiencing *this* one body as uniquely *mine*. Our attention is deflected from the latter to the former, and we no more follow up that move than we find the sense of "self" and of "my body" understood, not on the basis of the evidence in one's own case, but on the scientific-biological model of metabolism. Indeed, either that is the case, or Jonas has all along presupposed that we *already know* the answer to the question, that we have, to express it better, *already done* the philosophic work necessary to understand the very categories of his analysis of metabolism: "self," "other," "foreign," "freedom," "necessity," and the like.

To be clear about this, we need to ask: what is the status of the inquiry into metabolism and its results? It may be that its significance lies in its *illustrative* character – *viz*., that, already having some knowledge ("evidence") about the sense of "body" and "self" (for instance), one can better understand the kind of activity metabolism is or reveals. It may be, on the other hand, that the inquiry's status is found in this, that there is a *regression* from the "evidence" in my own case to its roots in metabolic activity (in which case, as far as I can see, there is a crucial departure from the "evidence" of one's own embodiment, to what can be garnered from the highly sophisticated sciences of biology, biomedicine, and the like).

But, Jonas would endorse neither of these, I think. It seems rather that he is maintaining that the inquiry into metabolism is *foundational* for an understanding of "life," and therefore for understanding *one's own* "life": the disclosure of the fundamental sense of self and body is to be found first at this level, and helped by that one can then move to the specifically human life (philosophical anthropology). In this case, however, it must needs be emphasized that the appeal to evidence has no formal status in the way claimed; at best, it can enter only subsequent to the philosophical biology. *Either*. then, the latter inquiry grounds that which would be indicated by the "evidence of each one's own organic awareness"; *or* the latter (the anthropology) grounds and guides the former, in which case, the status of the claims made about metabolism must be radically altered.

In fact, however, it seems to me that Jonas' arguments, while expressly departing from "mere phenomenology" ([8], p. 79), *trade* on it – or, if not on it, then on our common understanding of the central terms. As it were, I "know" what he means when he writes about the subtle workings and weavings of metabolism, *just because I "know" ahead of time about "self," "continuation," "transcendence," and the like, from my own experience.* Or, perhaps more accurately, thanks to the "evidence of my own organic awareness," the inquiry into metabolism makes sense, can be checked out, confirmed, or disconfirmed, and the like. But Jonas just does not give what is promised in his appeal to the "evidence." The evidence, taken seriously, concerns phenomena which the inquiry takes for granted have already been delivered to us.

If we step back a bit to assess what has transpired, it might be suggested, without appearing in the least bizarre, that the logic which had trapped Descartes, despite his insight, may well be at work in Jonas' effort to set up a "new monism." I mean: Jonas' strenuous and frequently brilliant double-edged effort to refute any materialist interpretation of "life" and to secure the "different nature altogether" ([8], p. 82) of "organic identity," while it unquestionably succeeds, may do so at the substantial cost of reintroducing a kind of dualism. Now dubbed "form/matter," there seems little doubt that Jonas' analysis reinstates a gap between the two, hardly less formidable than the one Descartes had made. Not only is the sense of "identity" radically different, not only is the source of that identity radically different, but so too is the principle or essence radically different: for the physical world, matter is the principle; for life, it is form. And, throughout the analysis of life, matter is taken as the "alien," the "foreign," the "accidental." Indeed, so profound is the chasm that Jonas is led to argue that no mere analysis of the physical record could ever possibly yield "life"; and, on the other hand, only a being itself alive could possibly know life, even in the indirect ways of inference ([8], p. 82).

Of course, what Jonas calls "form" in organic life *does* take "matter" (of some sort, but not all sorts) into itself; it literally "in-corporates" its other, it bodily transforms matter. Granted for the moment that this occurs, Jonas' subsequent analysis follows. But, I fail to see how *in principle*, granting this *description* of the emergence of self and world from the rudimentary reaches of metabolic activity, there could be the initial and continuous contact between two such radically opposed "things." That is, within Jonas' study, at its initiating point, there seems clearly the very "dualism" he has in other contexts rigorously refuted. And, as Descartes had indeed had a glimmering of the significant issue – "intimate union" – so Jonas has, with much greater force and clarity, seen the issue – the "living body" which is *mine*. The turn to metabolism, with the ontological categories of "form" and "matter" in hand, however, seems to generate or to include an equally enrooted dualism. In a most telling footnote to the analysis, Jonas writes:

> . . . when we call a living body a "metabolizing system," we must include in the term that the system itself is wholly and continuously a result of its metabolizing activity, and further that none of the "result" ceases to be an object of metabolism while it is also an agent for it. For this reason alone, it is inappropriate to liken the organism to a machine . . . . But metabolism is more than a method of power generation, or, food is more than fuel: in addition to, and more basic than, providing a kinetic energy for the running of the machine (a cause anyway not applying to plants), its role is to build up originally and replace continually the very parts of the machine itself – and this becoming itself is a performance of the machine: but for performance there is no analogue in the world of machines ([8], p. 78, n. 13).

Here, independently of the problems raised earlier about the formal status of "evidence" (which can be taken as a quite different issue), Jonas seems to me perfectly correct. But, as it is inappropriate to liken the organism to a machine, so does it seem inappropriate to take this systemic activity as a dialectical relation between form and matter. No matter how dialectically conceived, a form of dualism seems inevitable with that. The above passage, however, does not use that distinction; instead, the living body is a "metabolizing system" whose "parts," continually changing and being replaced, are integral to the body itself as a consequence, and as the agency of, metabolism. It is thus not so much a relation of form to matter which is revealed here, but of "part" to integral "whole"; or, "form" is expressive of the *integrity of the system* itself, "matter" of the *constituents of that system.* And, the "life" of this integral whole is expressed by the specific and distinctive "activity" which continuously exhibits that complex and reflexive self-maintaining and self-enhancement. In these terms, it is not so much that the "form" rides on the "crest of a wave" composed of "matter"; nor, that the latter comprise merely a "passing collection" whose individual "unity" (unity of matter) is of no particular consequence as regards the sustaining (self-sustaining) unity of the "form," the organism. It is rather that the "whole" is itself precisely nothing other than the system of mutually self-referent constituents, which *are* nothing themselves outside their functional role (or significance) within that whole (cf. [3], esp. pp. 318–25). To take the "parts" of the organism, or of the metabolizing activity, as somehow "different altogether," as forming a "collection" all their own, and the like, is to vitiate the very "system" itself, to fragment it into a dualism easily as profound and consequential as any Descartes constructed.

V

It is thus not so much that "the" organic body signals the "latent crisis of every known ontology." It is rather that "my" organic body does. The ease with which Jonas glides from the one to the other is striking, and suggestive, I think, of the temptations of ontological thinking itself: the urge to comprehensiveness, so deeply rooted and so apparently endemic to us all, may nevertheless be profoundly inimical to what we (with Jonas) so mightily seek to know and be. The breakdown of traditional ontologies – dualistic and monistic alike – is signaled preeminently by "the *only* fully given concrete of experience" ([8], p. 24), *my* own living body, which forms an integral, living, and incredibly complex whole – a veritable *contexture* achieved and sustained continuously by its own unique activity.

But having once posed the issue, one cannot let it slip away in the embrace of whatever totalism it may be – surely not, at the least until its concreteness and fully presented contexture has been explicated. Only that evidentially guided and grounded task can possibly be what is "used" in the understanding of life more generally and thence be "testimony" for the wider reality beyond it. And, it seems to me, that critical, phenomenological explication of the "evidence of each one's own organic awareness" is exactly what lies beneath Jonas' otherwise penetrating analysis of metabolism – what stands under and makes it possible to understand it. Philosophical biology is thus based in philosophical anthropology, and that, in turn, finds its ground and continual critical accompaniment in phenomenological explication of the "evidence" at hand.

Southern Methodist University,
Dallas, Texas

NOTES

[1] Jonas: 1966, *The Phenomenon of Life*, Essay I, Dell Publishing Co., Inc., New York. He emphasizes: " . . . since the point of departure in either case is partial with respect to integral reality, they severally embody the internal contradiction of a partial monism," i.e., the failure to reduce mind to matter, or matter to mind (p.17).

[2] "Was:" the passage is from Jonas' opening lecture course on "The Problem of Life in the Theory of Being," delivered at Hebrew University in Jerusalem, in 1946/47 (see [8], p. 156, n. 3).

[3] This is preciselv the point, and almost the formulation, of Gabriel Marcel, as early as the 1920's ([12], and [13]). So far as I know, it was only during my MA thesis writing, directed by Professor Jonas, in 1958–59, that Jonas became aware of Marcel's seminal writings on the "body-qua-mine." Indeed, without in the least faulting Jonas (or Marcel; so far as I know, Marcel was unaware of Jonas' work as well), there are serious convergences here with the work of Marcel; and, also with the then unpublished work of Husserl ([4]), as well as with Sartre ([15]), and Merleau-Ponty ([14]), and others, which, as I read it, not only testifies to deeply-seated historical trends and convergences, but as well to the significance of these issues. There simply is no way for them to be avoided, if one is serious about the problems of life and the body.

[4] "Only living things have needs and act on needs. Need is based both on the necessity for the continuous self-renewal of the organism by the metabolic process, and on the organism's elemental urge thus precariously to continue itself. This basic concern of all life. . .manifests itself on the level of animality as appetite, fear, and all the rest of the emotions" ([8], p. 126).

[5] Although neither Husserl nor Merleau-Ponty are mentioned here nor elsewhere in *The Phenomenon of Life*, there is a notable parallel here, too. See, e.g., Husserl ([7], pp. 72–86, 381–390), and Merleau-Ponty ([14], p. 363), on "*Urpräsenz*" and "*Entgegenwartigung*."

[6] Apparently this "Note" was written after the original essay; no mention of this is given, nor dating of the Appendices.

[7] While Jonas was a member of my doctoral committee, as well as director of my master's thesis, we did not discuss these views – to my real disappointment and loss.

BIBLIOGRAPHY

1. De Santillana, G.: 1955, *The Crime of Galileo*, The University of Chicago Press, Chicago.
2. Gurwitsch, A.: 1966, 'An Apparent Paradox in Leibnizianism', *Social Research* **33**, 47–64.
3. Gurwitsch, A.: 1964, *The Field of Consciousness*, Duquesne University Press, Pittsburgh.
4. Husserl, E.: 1952, *Ideen zu einer reinen Phänomenologie und phänomenologischen Philosophie*, Zweites Buch, Hrsg. M. Biemel, Husserliana Bd. IV, Martinus Nijhoff, The Hague.
5. Husserl, E.: 1959, *Erste Philosophie (1923/24)*, Zweiter Teil, Hrsg. R. Boehm, Husserliana Bd. VIII, Martinus Nijhoff, The Hague.
6. Husserl, E.: 1970, *The Crisis of European Science and Transcendental Phenome-*

nology, tr. and intro. David Carr, Northwestern University Press, Evanston.
7. Husserl, E.: 1973, *Experience and Judgment*, rev. and ed. L. Landgrebe, tr. J.S. Churchill and K. Ameriks, Northwestern University Press, Evanston.
8. Jonas, H.: 1966, *The Phenomenon of Life*, Delta Books, Dell Publishing Company, New York.
9. Jonas, H.: 1974, *Philosophical Essays: From Ancient Creed to Technological Man*, Prentice-Hall, New Jersey.
10. Jonas, H.: 1976, 'On the Power or Impotence of Subjectivity', in S. F. Spicker and H. T. Engelhardt, Jr. (eds.), *Philosophical Dimensions of the Neuro-Medical Sciences*, D. Reidel Publishing Company, Dordrecht-Holland, Boston, pp. 143–61.
11. Koyré, A.: 1957, *From the Closed World to the Infinite Universe*, The Johns Hopkins Press, Baltimore.
12. Marcel, G.: 1935, *Être et avoir*, Éditions Montaigne, Fernand Aubier, Paris.
13. Marcel, G.: 1952, *Metaphysical Journal*, tr. B. Wall, Henry Regnery Company, Chicago.
14. Merleau-Ponty, M.: 1962, *Phenomenology of Perception*, tr. C. Smith, The Humanities Press, New York.
15. Sartre, J. P.: 1956, *Being and Nothingness*, tr. H. Barnes, Philosophical Library, New York.
16. Zaner, R.: 1964, *The Problem of Embodiment*, Martinus Nijhoff, The Hague.
17. Zaner, R.: 1974, 'The Alternating Reed: Embodiment as Problematic Unity', in J. Y. Fenton (ed.), *Theology and the Body*, The Westminster Press, Philadelphia, pp. 53–71.
18. Zaner, R.: 1975, 'Context and Reflexivity: The Genealogy of Self', in H. T. Engelhardt, Jr. and S. F. Spicker (ed.), *Evaluation and Explanation in the Biomedical Sciences*, D. Reidel, Dordrecht-Holland/Boston, pp. 153–74.

JITENDRA NATH MOHANTY

INTENTIONALITY AND THE MIND/BODY PROBLEM

In this paper, I want to further develop a line of reasoning which I first sought to articulate in an earlier paper bearing a similar title[1]. The basic problem in that paper was: does the theory of intentionality commit us to a dualistic ontology? Is the defense of intentionality a defense of Cartesian dualism or of mentalism? The prevailing attitudes toward this and various allied issues are pretty sharply divided, but there is also a basic confusion owing to: (a) a narrower concept of intentionality with which one generally operates which may roughly he defined in terms of Franz Brentano's thesis along with Roderick Chisholm's criteria, superadded, and (b) a methodological belief that the critical problem is the dispensability or indispensability of intensional logic, as though the problem of intentionality is reducible to this issue, whereas in my view it should rather be the other way around. I sought to show that the intentionality thesis did not commit one to Cartesian dualism, that a certain form of pre-theoretical identity thesis is phenomenologically justified and also supported by the intentionality thesis ([14], pp. 133–154). This pre-theoretical identity was contrasted with the theoretical identity posited by objectifying thought of a philosophical theory. In this paper I want to pursue those reflections by emphasizing how a phenomenological conception of body helps us both to overcome the traditional mind body problem *and* even to trace the latter back to its genesis within the phenomena themselves. I will assume the intentionality thesis to be well-known, but will here add this much: the intentional act is not only directed toward an object (which is Brentano's form of the thesis), but also has its own *sense*, meaning or correlative *noema* which is the object precisely as it is intended in that act (which is the Husserlian addition to the basic Brentano thesis). Furthermore, intentional acts owe their sense structure to the temporal structure in which they are embedded; they are, *quâ* intentional, not merely occurrences in objective time, concurrent with other physical events, but in fact constitute, by virtue of their retentional and protentional context-dependence, an inner temporal horizon of their own [8]. It is of this richer intentionality thesis that I will make use. The basic problem will be formulated in the particular context of Wilfrid Sellars' "Empiricism and Philosophy of Mind" [17], especially to one crucial move in that paper. I will also appeal to P. F. Straw-

S. F. Spicker (ed.), Organism, Medicine, and Metaphysics, 283–300.

son's concept of 'P-predicate' and the very promising way it seeks to overcome the conceptual hiatus that vitiates philosophy of mind. Thus the thoughts of Sellars and Strawson will serve to provide the foil against which I will expound the basic features of a phenomenology of body with particular reference to Husserl. Finally, I will outline a brief sketch of the way the insights of phenomenology may be brought to bear upon the traditional body/mind problem.

I

Wilfird Sellars imagines an "anthropological science fiction" ([17], pp. 197–213), where the linguistic resources of pre-historical humans were restricted to descriptive vocabulary for public properties of public objects in space and time and contained elementary truth functional operations together with the subjunctive conditional and "over-textured" and vague concepts. The problem that Sellars sets for himself is: what linguistic resources are to be added to this apparatus such that our ancestors could come to recognize themselves as thinking, feeling and sensing animals? The first important step is to enrich the language with semantical categories so that the humans could stay "'Rot' means red" and "'*Es regnet*' means it is raining". Now does this semantic discourse enable the users to talk about *thoughts* together with the intentionality, reference and aboutness of the latter? According to a classical schema, thoughts are inner episodes other than overt behavior and verbal imagery. According to a modern orthodoxy to which Sellars is affiliated, all talk about 'thoughts' is reducible to semantical talk about verbal components. Sellars' problem is to reconcile these two views: his attempt owes its philosophical importance to the fact that he does not outright reject all classical talk about 'inner episodes' but seeks to find a place for its possibility in his own scheme of things. This he does by the enrichment of language, by the addition of theoretical discourse such that overt verbal behavior is regarded as the result of "inner speech" conceived analogously to overt speech. The already available categories are applicable to this inner speech as much as to overt speech. These inner speech episodes which are posited as theoretical entities, not as immediate experiences, may be called "thoughts." But the primary use of "'   ' means that" is in connection with overt speech; the application of this category to inner speech is modelled on overt speech. The next step in this thought experiment is that once the ancestors are taught to interpret each other's behavior by referring to these "inner episodes," what was originally introduced for purely theoretical use acquires a "reporting rôle," and men begin to introspect with the aid of these concepts. This last step that "What began as a language with a purely theoretical use has gained a reporting role" is crucial for Sellars' thesis.

I will not, for my present purpose, question any of the steps of the "science fiction." In fact, the last crucial step seems to me to embody an important philosophical insight: namely, that in some cases, if not in all, theoretical concepts sedimented into language, constitute the world as perceived. What I propose to do is to question the basic presupposition of Sellars' thought experiment: unproblematic acceptance of the thought that "we are not puzzled by how people acquire a language for referring to public properties of public objects, but are very puzzled indeed about how we learn to speak of inner episodes and immediate experiences" ([17], pp. 199–200). If we take seriously Austin's advice that "In philosophy it is often good policy, where one member of a putative pair falls under suspicion, to view the more innocent-seeming party suspiciously as well" ([1], p.4 fn.), it would be philosophically naive to let the concept of 'public outer objects' pass as unproblematic and to look upon that of 'private, inner entities' as problematic and in need of explanation and analysis. For, after all, the two concepts entail each other and are parasitical on each other; and radical philosophizing should cast suspicion on both, not in order to reject either or both, but to trace them back to their origins in the sense in which phenomenological philosophy speaks of "origin." For this purpose, let me suggest another imaginative variation, closer perhaps to reality and so less of a fiction. Let us suppose that the conception of a purely outer, public object deprived of all interiority and also the conception of a purely inner and private domain, are both late achievements of human culture. Our mythical ancestors were not Cartesian dualists. Rather, they "saw" spirits in stones, trees and human bodies (not ghosts in machines – for stones, trees and human bodies were not for them mere machines.) For these ancestors, let us suppose, language referred to a world that had not yet been bifurcated into outer and inner, matter and mind, machine and spirit. The philosophical problem, then, is: how from such discourse there emerged discourse about public, outer objects on the one hand and private, inner entities on the other? Thus Sellars' fictive variation serves for us as a foil for formulating what seems to be a more radical philosophical problem, which does not begin with Cartesian dualism, not even with one element of that pair. It is curious that some suppose that the Cartesian bug is only removed by getting rid of the private and the inner; but the *res extensa* is equally well a Cartesian achievement, and a philosophy which begins with the inner and the private alone as problematic and finds no problems with regard to the outer and the public, is equally not aware of its own presuppositions.

II

Our philosophical situation is post-Cartesian, and our task is both to overcome the Cartesian hiatus and to trace it back to its phenomenological origin. One most striking attempt to overcome the hiatus is made by Strawson, from whom we may learn a great deal ([18], Ch. III). The problem which Strawson seeks to resolve may be formulated as follows: I ascribe mental states not only to myself ("I am depressed") but also to others ("He is depressed") – in both cases "depressed" having the same sense. Any satisfactory theory of mind should be able to account for the possibility of both sorts of ascriptions. The Cartesian allows for the first possibility, not for the second; the behaviorist, for whom mental states are bodily states, can account for the second, not for the first. One cannot say that the meanings of the mental predicate are different in the two cases – which indeed would be an absurd consequence of the verificationist principle that meaning is identical to the method of verification, for certainly the method of verification in each case is different (if we can at all speak of 'verification' in the case of first-person ascription). To say that they have different meanings is not only counter-intuitive, but renders the possibility of a conjunction such as "He and I, both are depressed" absurd. And yet the bases on which two sorts of ascriptions are made are strikingly different: one says "I am depressed" without appealing to observation (even self-observation), and one says "He is depressed" on the basis of observed data. Both are equally original possibilities grounded in the nature of mental predicates, and it will not do to attempt to reduce the one to the other, to say – for example – that my saying "I am depressed" is reducible to "J.N.M. is depressed" or that your saying about me "He is depressed" is reducible to my saying "I am depressed," for the sort of evidence on which each is based is so different. Confronted by this hiatus, what Strawson does is to take up into the structure of his "P-predicates" both these possibilities along with their radical difference, such that to understand or acquire a mental concept is to be able to do both these two kinds of things: to ascribe it to one self independent of observation and to ascribe it to the other on the basis of observation. Both the Cartesian and the behaviorist see one half of the total situation, Strawson wants to accommodate them both.

And yet Strawson would not let these two aspects be retained – even if within the structure of the same concept – independently of each other. For he also argues that I cannot ascribe a state of consciousness to myself unless I could also ascribe it to the other, and in order to ascribe it to the other I should be able to identify the other, who therefore cannot be a Cartesian ego but must of necessity be an embodied person. He does not, however, likewise argue for the correlative thesis which would require that ascription of a state of consciousness to an other necessitates the possibility of self-ascription. The former dependence – namely, that self-ascription requires the possibility of ascription of an other – was maintained on a purely logical ground, but that logical ground cannot be used to justify the latter dependence, for obvious reasons. But he could have equally well argued that unless I did or could ascribe a state of consciousness to myself, a state of consciousness *quâ* state of consciousness could make no sense to me; and when I ascribe it to the other, I ascribe it not to the other's body but to him as a person. I ascribe it to him as one who is also a self-ascriber. If identifying an ego requires, as Strawson argues, identifying a body, it is also true, as he also argues, that identifying a body requires identifying he who owns the body.

It should be clear that underlying this laudable attempt there is a conception of the relation between body and states of consciousness that is novel. But this conception is also hard to formulate with precision. The important element of that conception is that the notion of 'P-predicate' *spans* the Cartesian dualism within its own structure, for it is neither a bodily predicate nor a pure egological predicate. Thus we have a domain in which the dualism does not apply, and a range of concepts whose applicability enables us to overcome the alternatives of Cartesianism and behaviorism.

However, closer examination reveals that despite this attempt to break new grounds, Strawson continues to remain under the influence of Cartesian dualism. The two metaphysical theories which he argues against so effectively – namely, Cartesianism and the so-called no-ownership theory – share a common concept of body, i.e., the Cartesian, so that:

> for Cartesianism, the person = body + soul;
> for the no-ownership theory, the person = body + nothing.

The two concepts of body, however, are the same. In his attempt to avoid and improve upon both, Strawson does not overcome the Cartesian concept of body as *res extensa*, merely assimilates both terms of the dualism within the concept of 'P-predicate'. It is this vestige of Cartesianism, then, that needs to be overcome. Strawson appreciates this point when he requires for ascribing P-predicates to others that one not only observe the other's body but also see the other as a self-ascriber. Thus, again, self-ascription and other-ascription are in fact inseparable.

The merit of Strawson's theory is that it helps us to avoid both the analogical-inferential theory as well as all purely behaviorist theories of our knowledge of other minds, and makes it conceptually tenable to speak of non-inferential (i.e., direct) apprehension of the other's mental life. Since the possibility of ascription of mental life on the basis of the observation of the other's behavior is a constitutive element of the logic of such predicates, such ascription to the other may be as much non-inferential and direct as is self-ascription although the latter is made independently of observation.

The loose ends of the Strawsonian theory may then be tied: its overcoming of Cartesian dualism made secure and its rootedness in the everyday life world (or, 'ordinary language') less open to challenge. We need only reject the Cartesian concept of the body as *res extensa* which Strawson does not question and which, in fact, is a constitutive element of his theory.

III

It is now time to consider the concept of body in phenomenology. Brentano's formulation of the intentionality thesis was meant to distinguish between mental and physical phenomena; consequently, intentional directedness was put forward as a defining character of all mental acts. Although Brentano defended the idea of a descriptive psychology, he nevertheless aimed at finding a place for that within the total body of knowledge and in fact did not question the validity of physiological explanations of mental phenomena. Husserl began, already in the *Logical Investigations*, by restricting Brentano's thesis in various ways, but it seems he did continue to operate with the conception of body as *res extensa.* The resulting attitude towards body finds its best expression in Husserl's *Ideen II* ([7], sections 39, 53): ascribing to myself in this world a body is said to be possible within the naturalistic standpoint, this latter standpoint is founded upon or "draws its nourishment from" sensory experience which is the experience of things and of our body within it. Husserl asks: "How does consciousness, so to speak, enter into the real world, how can that which is absolute in itself abandon its immanence and put on the character of transcendence?" and answers:

> We see at once that it can do this only in virtue of a certain participation in transcendence in its first and primordial sense, and that obviously is the transcendence of material Nature. Only through the empirical relation to the body does consciousness become real in a human and animal sense, and only thereby does it win a place in Nature's space and time . . ([7], p. 164).

The relation between consciousness and body is then said to be "a natural unity," and Husserl recognizes that:

only thereby can every subject that knows find before it a full world containing itself and other subjects, and at the same time know it for one and the same world about us belonging in common to itself and all other subjects ([7], p. 165).

It is also called "realization of consciousness," its "linking-on" to the corporeal – which gives rise to "a peculiar type of appreception" – or, rather "apperceptive interweavings." The "linking-on", however, does not affect the essential nature of consciousness: consciousness cannot assimilate anything that is foreign to its own essence, which would indeed be absurd. Consciousness might very well be incorporeal. A stream of experience in which no coherent unity gets constituted would not be constitutive of "nature." Such a consciousness would not "apperceive" itself as human or animal.

Even in this formulation, according to which body is a part of the world of things and the relation of consciousness to body is an external linking-on (in spite of the fact that the two are said to constitute a *natural* unity), the body is nevertheless assigned a most unique function: namely, that the world-constituting function of consciousness is mediated by the body, so that were it not for body there would be no intersubjective world. In other words, the body *quâ* body has a constituting role -- even if it is also true that *quâ res extensa* it is constituted. But right at this point we confront the following paradoxical situation. Body, according to this formulation in *Ideen II*, plays its rôle in making possible the constitution of an inter-subjective world only in so far as it is itself a part of 'nature,' so that by being 'embodied consciousness' it is 'linked-on' to nature. How can that which is a part of nature also be responsible, in whatever measure, for the constitution of that same nature, in so far as 'nature' signifies the intersubjective world around us? The paradox is further intensified by the consideration that the constitution of my lived body [*Leib*] as a material object [*Körper*] is made possible by the other: as Husserl writes in a note of a later date: "*wie sein Leib für mich, so mein Leib für ihn*" ([6], p. 34). And yet this transformation of my lived body to a material object is never quite complete.

This paradoxical situation may help us to understand Husserl's growing awareness of the uniqueness of body on the one hand, and the *intrinsicality of corporeality* to the life of consciousness, even in its reduced or purified form, on the other.

The phenomena which support the thesis of the uniqueness of the body among other things, all cluster around the concept of lived body. The concept is of course negatively defined by contrast with body as a material object. Its positive "nourishment" derives from phenomenological evidence. It would be an error to identify this concept with that of an animal organism or a living body which is what Husserl and many others often tend to do. My body is at once my lived body, an animal organism, and a material object, in that order of constitution. The phenomena which lend support to the concept of lived body are well-known and need not all be recounted here. However, the crucial point is that the lived body is body as pre-reflectively lived from within, yet prior to being objectified either by observation or by reflection. It is my body in its pre-reflective givenness, as such it is no doubt a noematic unity of meaning as much as are the higher order constituted unities such as 'animal organism' and 'material object.' But it is a noematic unity, a unity of sense that is phenomenologically a more primitive stratum, founding rather than founded in relation to the latter two. As lived *from within* ('within' here has no Cartesian implications) my body is not presented [*vorgestellt*] to me, it is not an object of a *Vorstellung*. I do not take, as I necessarily do in the case of outer objects, perspectival attitudes towards my body. That I do not perceive my body as I perceive outer objects is corroborated by the fact that I may not be able in photographs to recognize my own hands and feet as *mine* and may even find them strange. I do not also, in perceiving things around me, localize my body in a region of space in the environment. The body is always 'pregiven' along with the world ([10], p. 57; [5], p. 94), but it is not necessarily given as located in that world. It is not localized in the environment, for I effect all localization, in the long run, with reference to my body, which is therefore the "zero point" for such purposes. If all outer perception is perspectival, all such perspectives refer back to my body as the "center of orientation." Likewise in relation to movement: if I move all things with my body or with my body and tools, I move my body immediately. In fact, moving my body is moving myself. (It is possible that paralysis of an arm leads a subject to experience *his* arm as foreign, as an other, as a *mere thing* that needs to be raised as one lifts an inanimate weight.) Moreover, the lived body's behavior is intentional. As lived from within, my hand's reaching out to a glass of water is directed towards the glass of water in the 'how' of its givenness, and not the physico-chemical process described by physiology. Nor is this piece of behavior intentional only in the derivative sense of being caused by mental acts that are intentional. As Merleau-Ponty has shown, it is a form of intentionality that is *sui generis*, that is not analysable into the intentionality of a mental act and a non-intentional physiological process. Finally, the ownership (mineness) of my body is radically different from my (or anybody else's) ownership of anything else. This radical difference in the very meaning of 'ownership' also affects the nature of that which is owned. Strawson seeks to single out this difference by characterizing this ownership as 'non-transferrable' not merely in fact but in principle, i.e., logically.

Most of these phenomena Husserl came to recognize, some more explicitly, some less so. However, his primary interest in *Ideen II* seems to have been the constitution of body as animal organism and as the substratum of mental life. That the body, for Husserl, is constituted has led some to suppose that the conception of body as intentional is neither Husserlian nor compatible with Husserlian philosophy. To me this seems to be an over-hasty judgement. Mental acts are intentional, and yet they are, for Husserl, constituted in the stream of inner time consciousness. For any transcendental phenomenology, a mental act *quâ* mundane, i.e., as belonging to a real human's inner life and as localized in objective time, is constituted. Thus it is not a true proposition that what is intentional cannot be an object of another intentional act. In reflection, my own mental act is made the object of another intentional act. In so far as my own as well as the other's body is an object of my consciousness, there is a noematic *sense,* 'body,' whose constitution refers back to the appropriate structures of consciousness. But nothing in this constitution analysis rebels against the thesis of the intentionality of lived body ([9], p. 197).

Thus one may have to distinguish between:

(1) the lived body as a mode of pre-reflective, pre-objective experience;
(2) my consciousness *of* my body and *of* the other's body;
(3) 'body' as a noematic unity of *sense* ([9], p. 198); and
(4) the transcendental, constituting life of consciousness.

I will briefly comment on (3) and (4) – only to the extent that they directly bear on the theme of this paper.

Human body, in its fully developed noematic sense,[2] is a many-layered structure. At least four such strata may be distinguished:

(a) the body as physical object;
(b) the body as living organism;
(c) the body as expressive object; and
(d) the body as cultural object.

In *Ideen II*, Husserl is primarily concerned with the constitution of (b), but here, as well as in other works, he is also concerned with the constitution of (a). A serious inadequacy of his philosophy of body lies in a paucity, not absence, of concern with (c) and (d) – although by the very nature of things and in consonance with many of his other ideas, (c) should have acquired prominence in his thinking. One such idea may, in particular, be recalled here: there is, in Husserl's thinking, a close analogy between one's understanding of the other's linguistic utterance and one's knowledge of the other's mind. In both cases, the presentation (of the utterance or of the other's body) provides the basis for an interpretive appresentation (of the intended meaning or of the other's mental state). Now certainly if this is to be more than an analogy, the other's *voice* and the other's *face* have to be brought under the common concept of 'expression.' Husserl refused to take this step; already in the First Logical Investigation, he restricts 'expression' to linguistic acts and excludes, for example, gestures from its scope. There is both linguistic and phenomenological ground for bringing both linguistic and non-linguistic behavior under 'expression' – without having to deny the deep abyss that separates them.[3] In both, to recall Helmuth Plessner's words, "the "inner' becomes visible, it moves outward," "the outer is no mere containing wall which encloses something inner, but it is incorporated into the inner and, conversely, implicates the inner" ([16], pp. 43, 45). It is not surprising that Merleau-Ponty who takes this step should also move on to (d) the body as cultural object: to the extent that expressivity of an expression is culturally conditioned, body is also a cultural object.[4]

Any transcendental account of the constitution of body as a noematic unity has to exhibit the constitution of all these strata within that unity. Given this task, one may proceed in either of two ways: one may begin with body as physical object, as *res extensa,* then proceed to the constitution of body as animal organism, as living (not lived) body; the next step would be to show how something like 'psychic reality' is constituted on the foregoing foundation whereby body – living body and also *res extensa* – acquires an 'expressive' significance; and finally, how with historicity and intersubjectivity achieved, this expressive dimension acquires the significance of a cultural object. The other route that one may take is to begin with body as a cultural object, which is how body, as an object, is most primordially given. As with all cultural objects, the inner and the outer are not yet separated. In an expression *quâ* expression – which is, if one follows this route, the next in order of constitution – the distinction is made in thought, but no separation in fact. The expression 'expresses' what is *other* than it, i.e., the inner, but the inner is not there concealed or at most to be inferred, but rather is revealed through the outer. In the living organism *quâ* living, the inner is completely absorbed into the outer, the living body with all its phenomenal features such as self-locomotion *is* its life, there being nothing 'psychical,' or even 'historical,' behind it, either concealed or unconcealed. The body as physical object, as *res extensa,* is the limiting case of this process of exteriorization and deinteriorization: it is the merely outer with no inner. Once the body is constituted as 'mere thing,' as *Körper*, the hypothetical inner can only be – in the other's case – 'indicated,' 'analogously inferred,' 'indirectly apperceived' but not expressed; it is (as most discussions on the existence of other minds reveal) finally regarded as a theoretical posit and eventually as a useless fiction.

Husserl, it seems to me, has followed the first route. But the second route appears to be sounder from a strictly phenomenological point of view. For if the life-world is phenomenologically primitive in order of constitution, it is not yet the world of physics. Animal and human bodies, in the life-world, are not yet sheer physical objects. They rather carry with them their expressivity. Human bodies and bodily behavior, in particular, carry with them their cultural significance – which is not to say that bodies do not move in space or are not extended, but they are not yet constituted as *merely* physical objects or as mere *res extensa*.

There is one more aspect of Husserlian thinking that interests me here. This is the conception of transcendental subjectivity. One is tempted to think that in a transcendental philosophy for which the constituting principle is a disembodied consciousness in general, body has to be one object among others. Merleau-Ponty has emphasized a close connection between objective thinking, objectification of the body, and the thesis of a universal constituting consciousness. Not infrequently, Husserl has been regarded as a paradigmatic case of a philosopher in whose thought this nexus is pre-eminently exemplified. It seems to me that Merleau-Ponty was right in emphasizing the connection between what he calls objective thinking and the understanding of body as one object among others. In this, he has, quite unexpectedly, Husserl's support. Constitution of the pure in-itself presupposes, according to Husserl, intersubjectivity, and intersubjectivity is mediated by body, but not alone by *Leib* but by the constitution of *Leib* as *Körper* ([6], p. 34). What is not Husserlian is the thesis that the body is but a physical object; in fact, it would seem that though, according to Husserl, *Leib* gets constituted as *Körper*, it never quite becomes *Körper*. Furthermore, neither is the Husserlian constituting consciousness or transcendental subjectivity the same as the neo-Kantian or the Cartesian *cogito*; it is neither the purely logical principle of *Bewusstsein überhaupt* nor the self-translucent domain of acts of reflective thought. As Ludwig Landgrebe rightly points out, the constitutive condition of the possibility of experience of nature contains a passive structural moment which announces itself in the corporeality (*Leiblichkeit*), i.e., in the kinaesthetic sensations, in the "I can move myself" – where, according to Landgrebe, the traditional distinction between the inner and the outer, the immanent and the transcendent breaks down.[5] Without belaboring this point further, let me sum up by saying that the constituting life of subjectivity, even in its transcendentally purified form, contains a stratum of corporeality in which the lived body itself is constituted, but which is prior to the Cartesian distinction between body and mind, outer and inner.

IV

The mind–body problem which has gained notoriety in the tradition of modern philosophy has its genesis within the horizon of Cartesian dualism. If this problem, or something like it, antedates modern thought, it does, in any case, presuppose a similar dualism between: body and soul, outer and inner, objectivity and subjectivity. The problem thus presupposes a historically accomplished framework of thinking. Once this is recognized, it would seem as though by *overcoming* this tradition, – at one stroke, as it were – one gets rid of the problem. However, this is a move which, because of its radicalness and yet because of the ease with which one desires to eliminate a nagging problem, has proved tempting to many philosophers. But it is precisely the ease with which so radical an amputation is to be achieved, which makes it suspect. What is needed, most importantly, is to exhibit how the problem as well as the framework, i.e., the Cartesian distinction, come to be constituted, how they come to emerge *from within phenomena* themselves. The mind–body dualism is not merely a philosophical theory. Rather, the philosophical theory lays its hold upon and radicalizes modes of interpretation, i.e., noematic senses under which we apperceive ourselves and others. If Cartesian dualism is grounded in phenomena themselves, it is not for that reason a phenomenological theory, for it does not see the genesis and the limits of those phenomena and ascribes to them a primordiality that does not belong to them. If this be so, then to characterize Cartesian dualism – as I have done – as a framework that is a historical accomplishment cannot be the whole truth about it. To the extent it is also grounded in phenomena, it transcends history

In our pursuit of the problem of the constitution of the body, we have distinguished between: body as a material object, body as a living organism, body as expressive, body as a cultural object, the lived body in its pre-reflective self-experience, and the constitutive transcendental subjectivity with its structural moment of corporeality. (The last does not concern us here, its introduction earlier was meant only to counteract an erroneous but widely held view about Husserlian transcendental philosophy.)

(a) Lived body in its pre-reflective, pre-objective self-experience is a directedness towards the world, the world being intended as having its center around the body and the things belonging to it as well as the dimensions and relations (space, time and causality) in which they are ordered being characterized in terms of the powers and abilities constituting the lived body's self-understanding. This world is not a totality of objects, but a structure of references. At this level, body and mind are not yet distinguished. It is not the case that body is lived as something outer and extended while mind is lived through as something inner and unextended. There is rather one, undivided experience of being in the world and with others. If the Cartesian dualism has not yet constituted itself, the origins of that fissure nevertheless lie already there within the texture of lived experience, a fissure that is conceptually solidified by later objectifying thought. The lived nondistinction between the outer and the inner is threatened by objectivating intentions, so that the body, or rather parts of it, tend to emerge as distinct objects, though always against the background of anonymity, in which role the body as a whole functions. The body functions in this unstable manner of anonymous background and objective figure, though the totality of one's body never quite abandons the former role. Plessner sought to capture this nature of body by the expression "*Leib in Körper*," suggesting an ambiguity which is also 'an actual break in his way of existing," which accounts for both the locutions: a man *is* his body and also *has* it ([16], pp. 32–38). I am trying to isolate these two as two distinct strata in the constitution of body, and the constitution of body as thing, it seems to me, has its phenomenological basis right within this unstable texture of lived body. Reflective thought lays its exclusive hold on this fissure and objectifies the body as a whole (which presupposes, as Husserl has said on various occasions, the mediation of the other: "*wie sein Leib für mich, so mein Leib für ihn*" [6]). Except in theory, however, the body – even the other's body – never quite becomes a *Körper.* The other's body, perceived as a thing, continues to be apperceived *as* what the other lives as his *Leib*. Even the corpse is perceived as what *was* another's lived body but is no more.

(b) If lived body is pre-Cartesian, so also is body as expressive. To be sure, expression expresses something. The body 'mirrors' the soul. Verbal utterances and also non-verbal gestures express thoughts and intentions. But in so far as body is expressive, the inner is not the Cartesian inner, the *res cogitans*, the invisible substance or process, but the 'sense' or meaning (*Sinn*) of the expression. The outer–inner distinction here is pre-Cartesian, it is the distinction between the sense and its concretization. One often suspects that the talk of 'meaning' is founded on the Cartesian myth of the 'ghost in machine.' This suspicion is unfounded, for the sense of expressive behavior is 'expressed,' realized, concretized, so much so that the two – the sense and its concretization – are inseparable, their unity is the expression. It is here that Strawson's concept of 'P-predicate' becomes particularly helpful: for example, the concept of 'depression' spans within its structure, both the inner and the outer in one – with the addendum that the outer that is so inseparable from *its* inner is not yet the Cartesian *res extensa*.

(c) The otherness of the sense or meaning which still haunts an expressive behavior is apparently overcome in the body as a cultural object. As belonging to a cultural world, a look, a waving of the hand, nodding of the head, a gait, facial gesture as also the voice – each becomes a significant entity by itself; *quâ* expressive, it still points to a sense, but *quâ* cultural object the sense has been transformed into the style. If expressive behavior was 'natural,' style becomes conventional, needing intersubjective recognition.

(d) Body as living organism is freed from conventional, historically accomplished, intersubjectively acknowledged features; it is natural, but again, unlike expressive behavior, the behavior, the behavioral marks of life do not point to any sense or meaning beyond them. The inner, the life, is here nothing but the organization of the behavior. Life is the form of the behavior.

(e) Abstracted from this form, from that conventional style and also from the sense or thought expressed, the behavior and the expression tend to become merely physical processes bereft of intentionality; the body tends to become the mere *res extensa*. The constitution of body as *res extensa* is not the product of a philosophical theory, it is not the result of the physical sciences. It rather takes place prior to philosophical and scientific theories, and defines the horizon within which the traditional formulation of the mind/body problem first becomes possible. With the constitution of body as *res extensa*, the other member of the pair, mind, as *res cogitans* also constitutes itself. The two are indeed correlatives. Only after these terms are constituted, is such a philosophical theory as identity theory (of whatever variety) first meaningful, but then one also realizes that it cannot be valid as asserting the identity of two terms which are distinct but correlative. The more sophisticated version of the identity theory makes use of the sense reference distinction and asserts the identity of referent but does not want to deny the difference between the two concepts 'bodily' and 'mental.' From the phenomenological point of view, this presupposes that one can lay one's finger on the referent without the mediation of some sense or other, whereas the referent is identifiable only through some sense or other. The pure, unadulterated, bare particular which is the identical referent is a fiction, if neither of the senses 'bodily' and 'mental' makes any difference to it. The identity of referent through variations in sense is a basic theme for phenomenology, but that identity is not indifference to sense variations, but is constituted by them. The modern identity theory has no place for this notion of constitution, for then the senses would be constitutive of the reference, and the ontology of bare particulars would have to go.

Whereas, the body as *res extensa* lacks intentionality, the Cartesian inner, even when objectified, is still intentional. The inner mental state (event, act, or process) is directed towards something, and has a sense or content as its intentional correlate. This sense or content – e.g., in the case of an act of thinking, its propositional content, that s is p – is the recalcitrant element that refuses to be absorbed into the identity of the referent. The brain state theory in this respect is from one point of view weaker, but from another point of view stronger than behaviorism. It is weaker, for the concept of behavior (as I have argued [3]) is more amenable to and in fact is in need of an intentional interpretation; it is stronger in the sense that the most viable models, generally information-theoretical, of neural processes seem to require an intensional logic.[6]

Even Strawson's thesis, illuminating up to a point, founders. To have the concept of 'depression,' according to this thesis, is to be able to ascribe depression to oneself independent of observation *and* to ascribe it to the other on the basis of observation (of the other's behavior). Although this does span the gap between first person ascription and third person ascription, between Cartesianism and behaviorism, what one can ascribe to the other, on the basis of observation, is depression but not what he is depressed about, i.e., not the noematic content of depression. In other words, it is only the possession of a state such as depression or anger that one can ascribe to the other on the basis of observation of his behavior, but one cannot therewith ascribe to him that his mental state has such and such noematic content. It seems then that an entire dimension of the other's mental life remains inaccessible except through his appropriate verbal utterance or through inference on the basis of circumstantial evidence.

The empirical *theory* which, ideally considered, succeeds in eliminating all intentionality by reducing body to *res extensa* and replacing all talk of the mental by talk of appropriate bodily states, in so far as it is a scientific theory, is, or shall be (when it is completed), a historical achievement of intentional acts of a community of scientists. Thus the denial of intentionality is its own refutation and itself an accomplishment of intentionality. At the same time, there is no reason to cast doubt on the success of scientific theory in achieving its own projects; nor is there any reason to doubt the validity of its theses, for the only criteria of validity to which it can be submitted are internal to the project itself.

V

The preceding, cursory account of a phenomenology of body, projected against the background of two contemporary analytic discussions of the mind/body problem – those of Wilfrid Sellars and P. F. Strawson – is intended both to show the conceptual limitations from which analytic philosophy's discussions suffer as well as to make it possible to appreciate the contribution that phenomenology can make toward the understanding and resolution of this classical problem.

I conclude with one last remark to ward off a possible misunderstanding. One of the central theses underlying the phenomenology of body is that of bodily *subjectivity*. It might appear as though by speaking of 'bodily subjectivity' one is in effect assimilating 'body' into 'consciousness,' so that instead of the real body one is talking about the thought or idea of the body. Nothing could be farther removed from our intentions. Body is neither a modality of consciousness, nor is subjectivity coextensive with consciousness. In fact, one of the implications of the concept of bodily subjectivity is that the concept of subjectivity is wider than the concept of consciousness. It also entails that intentionality is a distinguishing feature, not of the domain of consciousness, but of the larger domain of subjectivity. The concept of subjectivity should also be dissociated from the epistemological concept of 'subject.' Nor do the concepts of subjectivity or consciousness necessarily hang together with the concept of 'representation' (of reality) and/or the priority of the temporal dimension of *presence* over the other modalities of time as Heidegger would have us believe. Released from these historical and metaphysical preconceptions, the phenomenological concept of subjectivity is multifaceted: it is both pre-reflective and reflective, both bodily and intellectual, aesthetic as well as logical, non-temporal as well as temporal. We cannot pursue that line of thought longer, only this much needs to be said now: the phenomenological subjectivity is not the Cartesian inner, but that domain of experience within which the Cartesian dualism has its genesis.

The University of Oklahoma,
Norman, Oklahoma

NOTES

[1] See [2], [3], and [4]. For an exposition and discussion of the Chisholm criteria from the phenomenological point of view, see [14].

[2] The Noema is correlative to the specific act of objectification, and so what the noematic sense is depends upon *how* precisely, i.e., under what description, a body is being apprehended.

[3] This is what Husserl seems to be driving at in Beilage XIV to his *Zur Phanomenologie der Intersubjektivitat.* See [10]. Also see Beilage XVI where he distinguishes between different levels of "*Ausdruck*". Also on p. 65: "*Jeder Ausdruck ist Leib für einen Sinn und ordnet sich damit der allgemeinen Leiblichkeit ein als ausgezeichneter Sinntrager* ...." In fn. 2, on p. 65, Husserl compares '*Leib*' with the written text as "objective spirit."

[4] "The very first of all cultural objects, and the one by which all the rest exist, is the body of the other person as the vehicle of a form of behavior." See ([13], p. 348).

[5] The place of *Leiblichkeit* in the structure of Husserlian transcendental subjectivity has been ably worked out by A. Lingis. See ([12]; [11], pp. 143–144).

[6] The relation between inten*t*ionality and inten*s*ionality is in a hopelessly muddled state of discussion. I can only console myself at this point with the thought that the clarification of this muddle is not vital for the central concern of *this* paper.

BIBLIOGRAPHY

1. Austin, J. L.: 1962, *Sense and Sensibilia*, Oxford University Press, Oxford.
2. Chisholm, R. M.: 1957, *Perceiving A Philosophical Study*, Cornell University Press, Ithaca, New York.
3. Chisholm, R. M.: 1958, 'Sentences about Believing', *Minnesota Studies in the Philosophy of Science* II.
4. Chisholm, R. M.: 1967, 'On Some Psychological Concepts and the "Logic" of Intentionality', in H. N. Castañeda (ed.), *Intentionality, Minds and Perception*, Wayne State University Press, Detroit.
5. Claesges, U.: 1964, *Edmund Husserls Theorie der Raumkonstitution*, Martinus Nijhoff, The Hague
6. Claesges, U (ed.) 1972, *Perspektiven transzendental – phanomenologischer Forschung, Fur Ludwig Landgrebe zum 70 Geburtstag*, Martinus Nijhoff, The Hague.
7. Husserl, E.: 1913, *Ideas*. (trans. W. R. Boyce Gibson, 1931), George Allen and Unwin Ltd., London. The original German, *Ideen zu einer reinen Phänomenologie und phanomenologische Philosophie*. Band II, Martinus Nijhoff, The Hague.

8. Husserl, E.: 1928, *Vorlesungen zur Phänomenologie des inneren Zeitbewusstseins*, (trans. J. S. Churchill 1966), Max Niemayer, Halle.
9. Husserl, E.: 1962, *Phänomenologische Psychologie*, W. Biemel (ed.), Martinus Nijhoff, The Hague.
10. Husserl, E.: 1973, *Zur Phänomenologie der Intersubjektivitat*, Texte aus dem Nachlass. Zweiter Teil: 1921–28, Iso Kern (ed.), Martinus Nijhoff, The Hague.
11. Landgrebe, L.: 1967, *Phanomenologie und Geschichte*, Guters Loher Verlaghaus, Gerd Mohn.
12. Lingis, A.: 1971, 'Intentionality and Corporeity', in A. T. Tymieniecka (ed.), *Analecta Husserliana* I, D. Reidel Publishing Company, Dordrecht, Holland.
13. Merleau-Ponty, M.: 1962, *Phenomenology of Perception*, (transl. Colin Smith), The Humanities Press, New York.
14. Mohanty, J. N.: 1972, *The Concept of Intentionality*, Warren Green, St. Louis, Mo., Part One, Chapter 2.
15. Mohanty, J. N.: 1974, 'Intentionality and the Body-Mind Problem', in M. Chatterjee (ed.), *Contemporary Indian Philosophy*, Series Two, George Allen & Unwin, London.
16. Plessner, H.: 1970, *Laughing and Crying*, (transl. J. S. Churchill and M. Grene), Northwestern University Press, Evanston, Illinois.
17. Sellars, W.: 1956, 'Empiricism and the Philosphy of Mind', *Minnesota Studies in the Philosophy of Science* I; reprinted in 1972, A. Marras (ed.), *Intentionality, Mind and Language*, University of Illinois Press, Urbana, Illinois. References are to this volume.
18. Strawson, P. F.: 1959, *Individuals: An Essay in Descriptive Metaphysics*, Methuen, London.

EPILOGUE

HANNAH ARENDT

METAPHOR AND THE INEFFABLE: ILLUMINATION ON 'THE NOBILITY OF SIGHT'

Mental activities, driven to language as the only medium for their manifestation, each draw their metaphors from a different bodily sense, and their plausibility depends upon an innate affinity between certain mental and certain sensory data. Thus, from the outset in formal philosophy, thinking has been thought of in terms of seeing, and since thinking is the most fundamental and the most radical of mental activities, it is quite true that vision "has tended to serve as the model of perception in general and thus as the measure of the other senses."[1] The predominance of sight is so deeply embedded in Greek speech and therefore in our conceptual language that we seldom find any consideration bestowed on it, as though it belonged among things too obvious to be noticed. A passing remark by Heraclitus, "The eyes are more exact witnesses than the ears,"[2] is an exception, and not a very helpful one. On the contrary, if one considers how easy it is for sight, unlike the other senses, to shut out the outside world and if one examines the early notion of the blind bard, whose stories are being listened to, one may wonder why hearing did not develop into the guiding metaphor for thinking.[3] Still, it is not altogether true that, in the words of Hans Jonas, "the mind has gone where vision pointed."[4] The metaphors used by the theoreticians of the Will are hardly ever taken from the visual sphere; their model is either desire as the quintessential property of all our senses – in that they serve the general appetitiveness of a needy and wanting being – or they are drawn from hearing, in line with the Jewish tradition of a God who is heard but not seen. (Metaphors drawn from hearing are very rare in the history of philosophy, the most notable modern exception being the late writings of Heidegger, where the thinking ego "hears" the call of Being. Medieval efforts to reconcile Biblical teaching with Greek philosophy testify to a complete victory of intuition or contemplation over every form of audition, and this victory was, as it were, foreshadowed by the early attempt of Philo of Alexandria to attune his Jewish creed to his Platonizing philosophy. He was still aware of the distinction between a Hebrew truth, which was heard, and the Greek *vision* of the true, and transformed the former into a mere preparation for the latter, to be achieved by divine intervention that had made man's ears into eyes to permit greater perfection of human cognition.)[5]

The following pages are excerpted from The Life of the Mind, Vol. 1: Thinking, Chap. II, Part 13, pp. 110–125, *published by Harcourt Brace Jovanovich, Inc.*

Judgment, finally, in terms of discovery the late-comer of our mental abilities, draws, as Kant knew so well, its metaphorical language from the sense of *taste* (the *Critique of Judgment* was originally conceived as a "Critique of Taste"), the most intimate, private, and idiosyncratic of the senses, somehow the opposite of sight, with its "noble" distance. The chief problem of the *Critique of Judgment* therefore became the question of how propositions of judgment could possibly claim, as they indeed do, general agreement.

Jonas enumerates all the advantages of sight as the guiding metaphor and model for the thinking mind. There is first of all the indisputable fact that no other sense establishes such a safe distance between subject and object; distance is the most basic condition for the functioning of vision. "The gain is the concept of objectivity, of the thing as it is in itself as distinct from the thing as it affects me, and from this distinction arises the whole idea of *theōria* and theoretical truth." Moreover, sight provides us with a "co-temporaneous manifold," whereas all the other senses, and especially hearing, "construct their perceptual 'unities of a manifold' out of a temporal sequence of sensations." Sight permits "freedom of choice . . . dependent . . . on . . . the fact that in seeing I am not yet engaged by the seen object. . . . [The seen object] lets me be as I let it be," whereas the other senses affect me directly. This is especially important for hearing, the only possible competitor sight might have for pre-eminence but which finds itself disqualified because it "intrudes upon a passive subject." In hearing, the percipient is at the mercy of something or somebody else. (This, incidentally, may be why the German language derived a whole cluster of words indicating a position of non-freedom from *hören*, to hear: *gehorchen, hörig, gehören,* to obey, be in bondage, belong.) Most important in our context is the fact brought out by Jonas that seeing necessarily "introduces the beholder," and for the beholder, in contrast to the auditor, the "*present* [is not] the point-experience of the passing *now*," but is transformed into a "*dimension* within which things can be beheld . . . as a lasting of the same." "Only sight therefore provides the sensual basis on which the mind may conceive the idea of the eternal, that which never changes and is always present."[6]

I mentioned before that language, the only medium in which the invisible can become manifest in a world of appearances, is by no means as adequate for that function as our senses are for their business of coping with the perceptible world, and I suggested that the metaphor in its own way can cure the defect. The cure has its dangers and is never wholly adequate either. The danger lies in the overwhelming evidence the metaphor provides by appealing to the unquestioned evidence of sense experience. Metaphors therefore can be used by speculative reason, which indeed cannot avoid them, but when they intrude, as is their tendency, on scientific reasoning, they are used and misused to create and provide plausible evidence for theories that are actually hypotheses that have to be proved or disproved by facts. Hans Blumenberg, in his *Paradigmen zu einer Metaphorologie*, has traced certain very common figures of speech, such as the iceberg metaphor or the various sea metaphors, through the centuries of Western thought, and thereby, almost incidentally, discovered to what an extent typically modern pseudo-sciences owe their plausibility to the seeming evidence of metaphor, which they substitute for the lacking evidence of data. His prime example is the consciousness theory of psychoanalysis, where consciousness is seen as the peak of an iceberg, a mere indication of the floating mass of unconsciousness beneath it.[7] Not only has that theory never been demonstrated but it is undemonstrable in its own terms: the moment a fragment of unconsciousness reaches the peak of the iceberg it has become conscious and lost all the properties of its alleged origin. Yet the evidence of the iceberg metaphor is so overwhelming that the theory needs neither argument nor demonstration; we would find the metaphor's use unobjectionable if we were told that we were dealing with speculations about something unknown – in the same way that former centuries used analogies for speculations about God. The only trouble is that every such speculation carries with it a mental construct in whose systematic order every datum can find its hermeneutic place with an even more stringent consistency than that provided by a successful scientific theory, since, being an exclusively mental construct without need of any real experience, it does not have to deal with exceptions to the rule.

It would be tempting to believe that metaphorical thought is only a danger when resorted to by the pseudo-sciences and that philosophic thought, if it does not claim demonstrable truth, is safe in using appropriate metaphors. Unfortunately this is not the case. The thought-systems of the great philosophers and metaphysicians of the past have an uncomfortable resemblance to the mental constructs of the pseudo-sciences, except that the great philosophers, in contrast to the cocksureness of their inferior brethren, have almost unanimously insisted on something "ineffable" behind the written words, something of which they, when they thought and did not write, were very clearly aware and which nevertheless refused to be pinned down and handed over to others; in short, they insisted that there was something that refused to lend itself to a transformation that would allow it to appear and take its place among the appearances of the world. In retrospect, we are tempted to see these ever-recurring utterances as attempts to warn the reader that he was in danger of a fatal mistake in understanding. What they enunciated were thoughts, not cognitions, not solid pieces of knowledge which, once acquired, would dispel ignorance; what, as philosophers, they were primarily concerned with were matters that escape human knowledge, although they had not escaped but even haunted human reason. And since in pursuing these questions the philosophers inevitably discovered a great number of things that are indeed knowable, namely, all the laws and axioms of correct thinking and the various theories of knowledge, they themselves very early blurred the distinction between thinking and knowing.

While Plato still held that the true *archē*, beginning and principle of philosophy, is wonder,[8] Aristotle, in the opening paragraphs of the *Metaphysics,*[9] interpreted – and was the first to do so – this same wonder as mere astonishment or puzzlement (*aporein*); through astonishment men become aware of their ignorance of things that may be known, starting with "things close at hand" and then progressing "from there to greater matters such as the sun and the moon and the stars and the genesis of all things." Men, he said, "philosophized to escape ignorance," and the Platonic wonder was no longer understood as a principle but as a mere beginning: "all men begin by wondering . . . but one must end with the opposite and with what is better [than wondering], as is the case when men learn."[10] Hence, Aristotle, though he, too, in a different context, spoke of a truth *aneu logou*, a truth that refused to be expressed in discourse,[11] would not have said with Plato: Of the subjects that concern me nothing is known, since there exists nothing in writing about them, nor will there ever exist anything in the future. People who write about such things know nothing; they do not even know themselves. For there is no way of putting these things in words like other things that one can learn. Hence, no one who possesses the true faculty of thinking (*nous*), and therefore knows the weakness of words, will ever risk framing thoughts in discourse, let alone fix them in so inflexible a form as that of written letters.[12]

We hear the same, almost in the same words, at the end of this whole development. Thus Nietzsche, certainly no Platonist, writes to his friend Overbeck: "My philosophy . . . can no longer be communicated at least not in print,"[13] and, in *Beyond Good and Evil:* "One no longer loves one's insight enough when one communicates it."[14] And Heidegger writes, not about Nietzsche but about himself, when he says: "The internal limit of all thinking . . . is that the thinker never can say what is most his own . . . because the spoken word receives its determination from the ineffable.[15] To which we may add a few remarks by Wittgenstein, whose philosophical investigations center on the ineffable in a relentless effort to *say* what "the case may *be*": "The results of philosophy are the uncovering . . . of bumps that the intellect has got by running its head up against the limits of language." These bumps are what we have called here "metaphysical fallacies"; they are what "make us see the value of the discovery." Or: "Philosophical problems arise when language goes on a holiday" (*wenn die Sprache feiert*). The German is equivocal: it can mean "to take a holiday," that is, language ceases to work, and it can mean "to celebrate," and would then signify almost the opposite. Or: "Philosophy is a battle against the bewitchment of our intelligence by language," The trouble is of course that this battle can be refought only by language.[16]

Let us return to Plato, since he is, as far as I know, the only philosopher of rank who has left us more than occasional remarks on this subject. The main thrust of the argument in the *Seventh Letter* is not against speaking but against writing. This repeats in abbreviated form the objections already raised against writing in the *Phaedrus*. There is first the fact that writing "will implant forgetfulness"; relying on the written word, men "cease to exercise memory." There is second the written word's "majestic silence"; it can neither give account of itself nor answer questions. Third, it cannot choose whom to address, falls into wrong hands, and "drifts all over the place"; ill-treated and abused, it is unable to defend itself; the best one can say for it is to call it a harmless "pastime," collecting "a store of refreshment . . . against the day 'when oblivious age comes' " or a "recreation [indulged in] as others regale themselves with drinking parties and the like."[17] But in the *Seventh Letter*, Plato goes further; he does not mention his *agrapha dogmata*, which we know about through a remark by Aristotle,[18] but implicitly denies them, too, when he explicitly asserts that "these things cannot be put into words like other things we learn."

This indeed is very different from what we read in the Platonic dialogues (though that is no reason to believe that the *Seventh Letter* is spurious). Thus we read in the *Statesman* about "likenesses" between the visible and the invisible:

> Likenesses which the senses can grasp are available in nature to those real existents . . . so that when someone asks for an account of these existents one has no trouble at all – one can simply indicate the sensible likeness and dispense with any account in words. But to the highest and most important class of existents there are no corresponding visible resemblances. . . . In these cases nothing visible can be pointed out to satisfy the inquiring mind. . . . Therefore we must train ourselves to give . . . an account in words of every

existing thing. For the existents which have no visible embodiment, the existents which are of the highest value and the chief importance, are demonstrable only in speech [*logos*] and are not to be apprehended by any other means.[19]

In the *Phaedrus*[20] Plato contrasts the written word with the spoken word as used in "the art of talking things through" (*technē dialektikē*), the "living speech, the original, of which the written discourse may fairly be called a kind of image." The art of living speech is praised because it knows how to select its listeners; it is not barren (*akarpoi*) but contains a semen whence different *logoi*, words and arguments, grow up in different listeners so that the seed may become immortal. But if in thinking we carry out this dialogue with ourselves, it is as though we were "writing words in our souls"; at such times, "our soul is like a book," but a book that no longer contains words. Following the writer, a second craftsman intervenes as we are thinking and he is a "painter," who paints in our soul those images that correspond to the written words.[21] "This happens when we have drawn these opinions and spoken assertions away from sight or any other perception, so that we now somehow *see* the images of what we first opined and spoke about."[22]

In the *Seventh Letter* Plato tells us briefly how this two-fold transformation may possibly come about, how it is that our sense perception can be *talked about* and how this talking about (*dialegesthai*) is again transformed into an image visible only to the soul. We have names for what we see, for instance, the name "circle" for something round; this name can be explained in speech (*logos*) in sentences "composed of nouns and verbs," and we say the circle is a "thing which has everywhere equal distances between its extremities and its center." These sentences can lead to the making of circles, of images (*eidōlon*) that can be "drawn and erased, turned out and destroyed," processes of course that do not affect *the* circle as such, which is different from all these circles. Knowledge and mind (*nous*) grasp the essential circle, that is, what all circles have in common, something that "lies neither in the sounds [of speech] nor in the shapes of bodies but in the soul," and this circle is clearly "different from the real circle," perceived first in nature by the eyes of the body, and different, too, from circles drawn according to verbal explanation. This circle in the soul is perceived by the mind (*nous*), which "is closest to it in affinity and likeness." And this inner intuition alone can be called truth.[23]

Truth of the evidential kind, construed on the principle of things perceived by our bodily vision, can be arrived at through the guidance (*diagōgē*) of words in the *dialegesthai*, the discursive train of thought that can be silent or spoken between teacher and disciple, "moving up and down," inquiring into "what is true and what is false." But the result, since it is supposed to be an intuition and not a conclusion, will follow suddenly after a long period of questions and answers: "when a flash of insight (*phronēsis*) about everything blazes up, and the mind . . . is flooded with light."[24] This truth itself is beyond words; names from which the thinking process starts are unreliable – "nothing prevents the things that are now called round from being called straight and the straight round"[25] – and words, the reasoned discourse of speech that seeks to explain, are "weak"; they offer no more than "a little guidance" to "kindle the light in the soul as from a leaping spark which, once generated, becomes self-sustaining."[26]

I have cited these few pages from the *Seventh Letter* at some length because they offer an otherwise unavailable insight into a possible incompatibility between intuition – the guiding metaphor for philosophical truth – and speech – the medium in which thinking manifests itself: the former always presents us with a co-temporaneous manifold, whereas the latter necessarily discloses itself in a sequence of words and sentences. That the latter was a mere instrument for the former was axiomatic even for Plato and remained axiomatic throughout the history of philosophy. Thus Kant still says: "*worauf alles Denken als Mittel abzweckt*, [*ist*] *die Anschauung*," "all thinking is a means of reaching intuition."[27] And here is Heidegger: "The *dialegesthai* has in itself a tendency towards a *noein*, a seeing. . . . It lacks the proper means of *theōrein itself. . . . This is the basic meaning* of Plato's dialectic, that it tends towards a vision, a disclosure, that it prepares the original intuition through the discourses. . . . The *logos* remains tied to vision; if speech separates itself from the evidence given in intuition, it degenerates into idle talk which prevents seeing. *Legein* is rooted in seeing, *Horan*."[28]

Heidegger's interpretation is borne out by a passage in Plato's *Philebus*[29] where the inward dialogue of me with myself is once more mentioned but now on its most elementary level: A man sees an object in the distance and, *since he happens to be alone*, he asks *himself*: What is it that appears there? He answers his own question: It is a man. If "he had someone with him he would put what he said to himself into actual speech, addressed to his companion, audibly uttering the same thoughts. . . . Whereas if he is alone he continues thinking the same thing by himself." The truth here is the seen evidence, and speaking, as well as thinking, is authentic to the extent that it follows the seen evidence, appropriates it by translating it into words; the moment this speech becomes separated from the seen evidence, for instance, when other people's opinions or thoughts are repeated, it acquires the same inauthenticity that for Plato characterizes the image as compared to the original.

Among the outstanding peculiarities of our senses is the fact that they cannot be translated into each other – no sound can be seen, no image can be heard, and so on – though they are bound together by common sense, which for this reason alone is the greatest of them all. Quoting Aquinas on the theme: "the one faculty [that] extends to all objects of the five senses." Language, corresponding to or following common sense, gives an object its common name; this commonness is not only the decisive factor for intersubjective communication – the same object being perceived by different persons and common to them – but it also serves to identify a datum that appears altogether differently to each of the five senses: hard or soft when I touch it, sweet or bitter when I taste it, bright or dark when I see it, sounding in different tones when I hear it. None of these sensations can be adequately described in words. Our cognitive senses, seeing and hearing, have little more affinity with words than the lower senses of smell, taste, and touch. Something smells *like* a rose, tastes *like* pea soup, feels *like* velvet, that is as far as we can go. "A rose is a rose is a rose."

All this, of course, is only another way of saying that truth, in the metaphysical tradition understood in terms of the sight metaphor, is ineffable by definition. We know from the Hebrew tradition what happens to truth if the guiding metaphor is not vision but hearing (in many respects more akin than sight to thinking because of its ability to follow sequences). The Hebrew God can be heard but not seen, and truth therefore becomes invisible: "Thou shalt not make unto thee any graven image or any likeness of any thing that is in heaven above or that is on the earth beneath." ***The invisibility of truth in the Hebrew religion is as axiomatic as its ineffability in Greek philosophy***, from which all later philosophy derived its axiomatic assumptions. And while truth, if understood in terms of hearing, demands obedience, truth understood in terms of vision relies on the same powerful self-evidence that forces us to admit the identity of an object the moment it is before our eyes. Metaphysics, the "awesome science" that "beholds what is insofar as it is" (*epistēmē hē theōrei to an he on*),[30] could discover a truth which "forced men by the force of necessity" (*hyp' autēs tēs alētheias anagkadzomenoi*)[31] because it relied on the same imperviousness to contradiction we know so well from sight experiences. For no discourse, whether dialectical in the Socratic–Platonic sense, or logical, using established rules to draw conclusions from accepted premisses, or rhetorical-persuasive, can ever match the simple, unquestioned and unquestionable certainty of visible evidence. "What is it that appears there? It is a man." This is the perfect *adequatio rei et intellectus*,[32] "the agreement of knowledge with its object," which even for

Kant was still the definite definition of truth. Kant, however, was aware that for this truth "no general criterion can be demanded. [It] would . . . be self-contradictory":[33] Truth as self-evidence does not need any criterion; it *is* the criterion, the final arbiter, of everything that then may follow. Thus Heidegger, discussing the traditional truth concept in *Sein und Zeit*, illustrates it as follows: "Let us suppose that someone with his back turned to the wall makes the true assumption that 'the picture on the wall is hanging askew.' The assertion is confirmed when the man who makes it turns around and perceives the picture hanging askew on the wall."[34]

The difficulties to which the "awesome science" of metaphysics has given rise since its inception could possibly all be summed up in the natural tension between *theōria* and *logos*, between seeing and reasoning with words – whether in the form of "dialectics" (*dia-legesthai*) or, on the contrary, of the "syllogism" (*syl-logizesthai*), i.e., whether it takes things, especially opinions, apart by means of words or brings them together in a discourse depending for its truth content on a primary premise perceived by intuition, by the *nous*, which is not subject to error because it is not *meta logou*, sequential to words.[35] If philosophy is the mother of the sciences, it is itself the science of the beginnings and principles of science, of the *archai*; and these *archai*, which then become the topic of Aristotelian metaphysics, can no longer be derived; they are given to the mind in self-evident intuition.

What recommended sight to be the guiding metaphor in philosophy – and, along with sight, intuition as the ideal of truth – was not just the "nobility" of this most cognitive of our senses, but the very early notion that the philosopher's quest for meaning was identical with the scientist's quest for knowledge. Here it is worth recalling the strange turn that Aristotle, in the first chapter of the *Metaphysics*, gave to Plato's proposition that *thaumazein*, wonder, is the beginning of all philosophy. But the identification of truth with meaning was made, of course, even earlier. For knowledge is indeed searching for what we are accustomed to call truth, and the highest, ultimate form of cognitive truth is indeed intuition. All knowledge starts from investigating the appearances as they are given to our senses, and if the scientist then wants to go on and find out the causes of the visible effects, his ultimate aim is to make appear whatever may be hidden behind mere surfaces. This is true even of the most complicated mechanical instruments, which are designed to catch what is hidden from the naked eye. In the last analysis, confirmation of any scientist's theory comes about through sense evidence – just as in the simplistic model I took out of Heidegger. The tension I alluded to between vision and speech does not enter here; on this level, as in the example quoted, speech quite adequately translates vision (it would be different if the content of the painting and not just its position on the wall had to be expressed in words). They very fact that mathematical symbols can be substituted for actual words and be even more expressive of the underlying phenomena that are forced by instruments to appear against their own bent, as it were, demonstrates the superior efficacy of sight metaphors to make manifest whatever does not need speech as a conveyor.

Thinking, however, in contrast to cognitive activities that may use thinking as one of their instruments, needs speech not only to sound out and become manifest; it needs it to be activated at all. And since speech is enacted in sequences of sentences, the end of thinking can never be an intuition; nor can it be confirmed by some piece of self-evidence beheld in speechless contemplation. If thinking, guided by the old sight metaphor and misunderstanding itself and its function, expects "truth" from its activity, this truth is not only ineffable by definition. "Like children trying to catch smoke by closing their hands, philosophers so often see the object they would grasp fly before them" – Bergson, the last philosopher to believe firmly in "intuition," described very accurately what really happened to thinkers of that school.[36] And the reason for the "failure" is simply that nothing expressed in words can ever attain to the immobility of an object of mere contemplation. Compared to an object of contemplation, meaning, which can be said and spoken about, is slippery; if the philosopher wants to *see* and *grasp* it, it "slips away."[37]

Since Bergson, the use of the sight metaphor in philosophy has kept dwindling, not unsurprisingly, as emphasis and interest have shifted entirely from contemplation to speech, from *nous* to *logos*. With this shift, the criterion for truth has shifted from the agreement of knowledge with its object – the *adequatio rei et intellectus*, understood as analogous to the agreement of vision with the seen object – to the mere *form* of thinking, whose basic rule is the axiom of non-contradiction, of consistency with itself, that is, to what Kant still understood as the merely "negative touchstone of truth." "Beyond the sphere of analytic knowledge . . . as a *sufficient* criterion of truth, it has no authority and no field of application."[38] In the few modern philosophers who still cling, however tenuously and doubtfully, to the traditional assumptions of metaphysics, in Heidegger and Walter Benjamin, the old sight metaphor has not altogether disappeared but has shrunk, as it were: in Benjamin truth "slips by" (*huscht vorüber*); in Heidegger the moment of illumination is understood as "lightning" (*Blitz*), and finally replaced by an altogether different metaphor, *das Geläut der Stille*, "the ringing sound of silence." In terms of the tradition, the latter metaphor is the closest approximation to the illumination arrived at in speechless contemplation. For though the metaphor for the end and culmination of the thinking process is now drawn from the sense of hearing, it does not in the least correspond to listening to an articulated sequence of sounds, as when we hear a melody, but again in an immobile mental state of sheer receptivity. And since thinking, the silent dialogue of me with myself, is sheer activity of the mind combined with complete immobility of the body – "never am I more active than when I do nothing" (Cato) – the difficulties created by metaphors drawn from the sense of hearing would be as great as the difficulties created by the metaphor of vision. (Bergson, still so firmly attached to the metaphor of intuition for the ideal of truth, speaks of the "essentially active, I might almost say violent, character of metaphysical intuition" without being aware of the contradiction between the quiet of contemplation and any activity, let alone a violent one.[39]) And Aristotle speaks of "philosophical *energeia*, activity" as the "perfect and unhindered activity which for [this very reason] harbors within itself the sweetest of all delights ("*Alla men hē ge teleia energeia kai akōlytos en heautē echei to chairein, hōste an eiē hē theōrētikē energeia pasōn hēdistē*").[40]

In other words, the chief difficulty here seems to be that for thinking itself – whose language is entirely metaphorical and whose conceptual framework depends entirely on the gift of the metaphor, which bridges the gulf between the visible and the invisible, the world of appearances and the thinking ego – there exists no metaphor that could plausibly illuminate this special activity of the mind, in which something invisible within us deals with the invisibles of the world. All metaphors drawn from the senses will lead us into difficulties for the simple reason that all our senses are essentially cognitive, hence, if understood as activities, have an end outside themselves; they are not *energeia*, an end in itself, but instruments enabling us to know and deal with the world.

Thinking is out of order because the quest for meaning produces no end result that will survive the activity, that will make sense after the activity has come to its end. In other words, the delight of which Aristotle speaks, though manifest to the thinking ego, is ineffable by definition. The only possible metaphor one may conceive of for the life of the mind is the sensation of being alive. *Without the breath of life the human body is a corpse; without thinking the human mind is dead.* This in fact is the metaphor Aristotle tried out in the famous seventh chapter of Book Lambda of the *Metaphysics*: "The activity of thinking [*energeia* that has its end in itself] is life."[41] Its inherent law, which only a god can tolerate forever, man merely now and then, during which time he is godlike, is "unceasing motion, which is motion in a circle"[42] the only movement, that is, that never reaches an end or results in an end product. This very strange notion that the authentic process of thinking, namely, the *noēsis noēseōs*, turns in circles – the most glorious justification of the circular argument in philosophy – has oddly enough never worried either the philosophers or Aristotle's interpreters – partly, perhaps, because of the frequent mistranslations of *nous* and *theōria* as "knowledge," which always reaches an end and produces an end result.[43] If thinking were a cognitive enterprise it would have to follow a rectilinear motion, starting from the quest for its object and ending with cognition of it. Aristotle's circular motion, taken together with the life metaphor, suggests a quest for meaning that for man as a thinking being accompanies life and ends only in death. The circular motion is a metaphor drawn from the life process which, though it goes from birth to death, also turns in circles as long as man is alive. This simple experience of the thinking ego has proved striking enough for the notion of the circular movement to be repeated by other thinkers, even though it stands in flagrant contradition to their traditional assumptions that truth is the result of thinking, that there is such a thing as Hegel's "speculative cognition."[44] Without any reference to Aristotle, we find Hegel saying: "Philosophy forms a circle. . . . [It] is a sequence which does not hang in the air; it is not something which begins from nothing at all; on the contrary, *it circles back into itself.*"[45] And we find the same notion at the end of Heidegger's "What is Metaphysics?" where he defines the "basic question of metaphysics" as "Why is there anything and not rather nothing?" – in a way thinking's first question but at the same time the thought to which it "always has to swing back."[46]

Yet these metaphors, although they correspond to the speculative, non-cognitive way of thinking and remain loyal to the fundamental experiences of the thinking ego, since they relate to no cognitive capacity, remain singularly empty, and Aristotle himself used them nowhere else – except when he asserts that being alive is *energein*, that is, being active for its own sake.[47] Moreover, the metaphor obviously refuses to answer the inevitable question, Why do we think?, since there is no answer to the question, Why do we live?

NOTES

[1] Hans Jonas, *The Phenomenon of Life*, New York, 1966, p. 135. His study of "The Nobility of Sight" is of unique help in the clarification of the history of Western thought.
[2] Diels and Kranz, frag. 101a.

[3] Aristotle seems to have thought along these lines in one of his scientific treatises: "Of these faculties, for the mere necessities of life and in itself, sight is the more important, but for the mind [*nous*] and indirectly [*kata symbebēkos*] hearing is the more important. . . . [It] makes the largest contribution to wisdom. For discourse, which is the cause of learning, is so because it is audible; but it is audible not in itself but indirectly, because speech is composed of words, and each word is a rational symbol. Consequently, of those who have been deprived of one sense or the other from birth, the blind are more intelligent than the deaf and the dumb." The point of the matter is that he seems never to have remembered this observation when he wrote philosophy. Aristotle, *On Sense and Sensible Objects*, 437a4–17.
[4] *Op. Cit.*, p. 152.
[5] See Hans Jonas, Chap. 3, on Philo of Alexandria, especially pp. 94–97, of *Von der Mythologie zur mystischen Philosophie*, Göttingen, 1954, which is the second part of *Gnosis und spätantiker Geist*, Göttingen, 1934.
[6] *The Phenomenon of Life*, pp. 136–147. Cf. *Von der Mythologie*, pp. 133–152.
[7] Bonn, 1960, pp. 200f.
[8] *Theaetetus*, 155d.
[9] 982b11–22.
[10] 983a14–20.
[11] See, for instance, *Nicomachean Ethics*, VI, 8, where the *nous* is the mental perception (*aisthēsis*) of the "unchangeable primary or limiting terms" for which "there exists no *logos*" (1142a25–27). Cf. 1143b5.
[12] *Seventh Letter*, 341b–343a, paraphrase.
[13] On July 2, 1885.
[14] No. 160.
[15] *Nietzsche*, Pfullingen, 1961, vol. II, p. 484.
[16] *Philosophical Investigations*, trans. G. E. M. Anscombe, New York, 1953, nos. 119, 19, 109.
[17] *Phaedrus*, 274e–277c.
[18] *Physics*, 209b15.
[19] 286a. b.
[20] 275d–277a.
[21] *Philebus*, 38e–39a.
[22] *Ibid.*, 39b.
[23] 342.
[24] *Ibid.*, 344b.
[25] *Ibid.*, 343b.
[26] *Ibid.*, 341e.
[27] *Critique of Pure Reason*, B33. For: "*Nicht dadurch, dass ich bloss denke, erkenne ich irgend ein Objekt, sondern nur dadurch, dass ich eine gegebene Anschauung . . . bestimme, kann ich irgend einen Gegenstand erkennen*" ("I do not know an object merely in that I think, but only insofar as I determine a given intuition, can I know an object") (B406).
[28] I am quoting from an early lecture-course of Heidegger's on Plato's *Sophist* (1924–25) according to a literal transcript, pp. 8, and 155, 160. See also Cornford's commentary on the *Sophist* in *Plato's Theory of Knowledge*, p. 189 and n.1, where *noien* is said to stand for the act of "intuition (*noēsis*) which *sees* directly, without . . . discursive reasoning."

[29] 38c–e.
[30] Aristotle, *Metaphysics*, 1003 a 21.
[31] *Ibid.*, 984 b 10.
[32] Thomas Aquinas, *De Veritate*, qu. I, art. 1.
[33] *Critique of Pure Reason*, B82, B83.
[34] *Sein und Zeit*, Tübingen, 1949, no. 44 (a), p. 217.
[35] See Aristotle, *Posterior Analytics*, 100b5–17.
[36] *An Introduction to Metaphysics* (1903), trans. T. E. Hulme, Indianapolis, New York, 1955, p. 45.
[37] *Ibid.*
[38] *Critique of Pure Reason*, B84 and B189–91.
[39] *An Introduction to Metaphysics*, p. 45.
[40] *Protreptikos*, During ed., B87.
[41] 1072b27.
[42] 1072a21.
[43] This mistranslation mars W. D. Ross's *Aristotle*, Meridian Books, New York, 1959, but is mercifully absent from his translation of the *Metaphysics* in Richard McKeon's *The Basic Works of Aristotle.*
[44] *Philosophy of History*, Introduction, p. 9.
[45] *Hegel's Philosophy of Right*, trans. T. M. Knox, London, Oxford, New York, 1967, addition to para. 2, p. 225.
[46] *Wegmarken*, p. 19.
[47] *Nicomachean Ethics*, 1175a12.

BIBLIOGRAPHY OF THE WORKS OF HANS JONAS

BOOKS

1. (1930) *Der Begriff der Gnosis, Inaugural-Dissertation zur Erlangung der Doktorwürde der Hohen Philosophischen Fakultät der Philipps-Universität zu Marburg*, Hubert & Co., Göttingen.

2.A (1930) *Augustin und das paulinische Freiheitsproblem. Ein philosphischer Beitrag zur Genesis der christlich-abendländischen Freiheitsidee*, Vandenhoeck & Ruprecht, Göttingen.

B (1965) *Augustin und das paulinische Freiheitsproblem. Eine philosophische Studie zum pelagianischen Streit.* Zweite neubearbeitete und erweiterte Auflage mit einer Einleitung von James M. Robinson, Vandenhoeck & Ruprecht, Göttingen.

3.A (1934) *Gnosis und spätantiker Geist.* Teil 1, *Die mythologische Gnosis.* Mit einer Einleitung *Zur Geschichte und Methodologie der Forschung*, Vandenhoeck & Ruprecht, Göttingen.

B (1954) *Idem.*, Zweite unveränderte Auflage.

C (1964) *Idem.*, Dritte, verbesserte und vermehrte Auflage.

D (1964) *Idem.*, *Ergänzungsheft zur ersten und zweiten Auflage*, S. 377–456.

4.A (1954) *Gnosis und spätantiker Geist.* Teil 2, i, *Von der Mythologie zur mystischen Philosophie*, Vandenhoeck & Ruprecht, Göttingen.

B (1966) *Idem.*, Zweite, durchgesehene Auflage.

5.A (1958) *The Gnostic Religion: The Message of the Alien God and the Beginnings of Christianity*, Beacon Press, Boston.

B (1963) *Idem.*, Second edition, enlarged and revised.

C (1969) *Het Gnosticisme*, Uitgeverij Het Spectrum N.V., Utrecht, Netherlands/Antwerpen, Belgium. Trans. from the English by A. J. M. Baljet.

D (1973) *Lo Gnosticismo*, Società Editrice Internazionale, Torino, Italy. Trans. from the English by Margherita Riccati di Ceva.

E (1977) *La Religion gnostique*, Librairie Ernest Flammarion, Paris, France. Trans. from the English by Louis Evrard and published in the series "Idées et Recherches."

6. (1963) *Zwischen Nichts und Ewigkeit. Zur Lehre vom Menschen*, Kleine Vandenhoeck-Reihe 165, Vandenhoeck & Ruprecht, Göttingen.

7.A (1966) *The Phenomenon of Life: Toward A Philosophical Biology*, Harper & Row Publishers, New York.

B (1968) *Idem.*, Delta Books, Dell Publishing Co., New York.

8.A (1970) *Wandel und Bestand. Vom Grunde der Verstehbarkeit des*

		Geschichtlichen. Wissenschaft und Gegenwart, Geisteswissenschaftliche Reihe, Heft 46, Vittorio Klostermann, Frankfurt am Main.
B	(1970)	*Idem.* This work was simultaneously printed in *Durchblicke. Martin Heidegger zum 80. Geburtstag.* Herausgegeben von Vittorio Klostermann, Vittorio Klostermann, Frankfurt am Main, pp. 1–26.
C	(1971)	*Idem.* An abridged form of this work served as the opening address to the Fifth Congress of the Fédération Internationale des Associations d'Études Classiques (FIEC) in Bonn on September 1, 1969. The author's English translation of this address, "Change and Permanence: On the Possibility of Understanding History," appears in *Die Interpretation in der Altertumswissenschaft*, herausgegeben von Wolfgang Schmid, Bouvier Verlag Herbert Grundmann, Bonn, pp. 26–54. [For the author's English version of the complete German work see "Articles and Reviews" no. 50.]
9.	(1973)	*Organismus und Freiheit. Ansätze zu einer philosophischen Biologie*, Vandenhoeck & Ruprecht, Göttingen.
10.	(1974)	*Philosophical Essays: From Ancient Creed to Technological Man*, Prentice-Hall, Englewood Cliffs, New Jersey.
11.	(1978)	*On Faith, Reason, and Responsibility: Six Essays*, Scholars Press, Missoula, Montana/Harper & Row, San Francisco.

Forthcoming Books: (titles tentative)

12	*Versuch über Ethik im Zeitalter der Technik.*
13.	*Technology and Ethics: An Essay in Moral Philosophy.*

ARTICLES AND REVIEWS

N.B. Since many of Hans Jonas' articles were eventually incorporated in the author's books, usually revised and expanded, the following abbreviations will be employed to refer to the book(s) in which the final version of an article is contained.

Book Title	Abbreviation	Book List No.
Augustin und das paulinische Freiheitsproblem	ApF	2B
Gnosis und spätantiker Geist	GsG	3–4
The Gnostic Religion	GR	5B
Zwischen Nichts und Ewigkeit	ZNE	6
The Phenomenon of Life	PL	7
Organismus und Freiheit	OF	9
Philosophical Essays	PE	10
On Faith, Reason, and Responsibility	FRR	11

NOTE on writings published in more than one language:

The word "version" in this connection (usually preceded by "author's") shall denote second-language versions by Hans Jonas himself; "translation" shall denote the work of independent translators. In the first case, the two – more freely – parallel versions are considered by the author as equally authentic regardless of temporal sequence. His two linguistic media (with a few Hebrew exceptions) are English and German, either of which may provide a first version.

1. (1929) 'Karl Mannheims Soziologie des Geistes', *Schriften der Deutschen Gesellschaft für Soziologie*, 1. Series, Vol. 6. pp. 111–114.

2. (1938) 'Edmund Husserl and the Problem of Ontology', [Hebrew] *Mosnaim*, Vol. 7, No. 5, pp. 581–589.

3. (1938) 'In Memoriam Edmund Husserl', [Hebrew] *Turim* (Tel Aviv).

4. (1946) Review: 'Karl Barth, Eine Schweizer Stimme', [German] *Yedioth* (Tel Aviv), No. 38, pp. 5f.

5. (1948) 'Origenes' *Peri archon*: ein System patristischer Gnosis', *Theologische Zeitschrift*, Vol. 4, pp. 101–119. [See No. 6.]

6. (1949) 'Die origenistische Spekulation und die Mystik', *Theologische Zeitschrift*, Vol. 5, pp. 24ff. [For the expanded definitive version of No. 5 and No. 6 see *GsG* 2, i, Kap. 5; the author's definitive English version see No. 47.]

7. (1949) 'Problems of "Knowing God" in Philo Judeaus', [Hebrew] in *Sefer Yohanan Lewy* (Jerusalem), pp. 65–84. [For the author's prior German version see *GsG* 2, i, Kap. 3.]

8. (May 25, 1950) 'Causality and Perception', *The Journal of Philosophy*, Vol. 47, No. 11, pp. 319–324. [For the expanded definitive version see *PL*, First Essay, Appendix 1; for the author's German version see *OF*, Kap. 2, I.]

9. (1950ff) "Aeon," "Augustin," "Gnosticism," "Irenaeus," [Hebrew] in *Encyclopedia Hebraica* (Jerusalem).

10. (Fall, 1950) 'Yiscor· To the Memory of Joseph Weiner', *The Chicago Jewish Forum*, Vol. 9, No. 1, pp. 1–8.

11. (1951) 'Materialism and the Theory of Organism', *University of Toronto Quarterly*, Vol. 21, pp. 39–52. [For the definitive version see *PL*, Second Essay; for the German translation see *OF*, Kap. 3.]

12. (1951) 'Is God a Mathematician?', *Measure*, Vol. 2, pp. 404–426. [For the definitive version see *PL*, Third Essay; for the author's German version see *OF*, Kap. 5.]

13. (1951) 'Comment on von Bertalanffy's General System Theory', *Human Biology*, Vol. 23, pp. 328–335.

14.A (1952) 'Gnosticism and Modern Nihilism', *Social Research*, Vol. 19, pp. 452ff. [Incorporated in *GR*, 2nd ed., 'Epilogue: Gnosticism, Existentialism, and Nihilism'; also see *PL*, Ninth Essay.]

B (1960) *Idem.* The author's German version, 'Gnosis und moderner Nihilismus', *Kerygma und Dogma*, Vol. 6, pp. 155–171. [For the author's definitive version see *ZNE* and *OF*, Kap. 11.]

15. (1952) Review 'H. A. Wolfson: Philo', *Philosophy and Phenomenological Research*, Vol. 12, pp. 442–445.

16. (1953) 'A Critique of Cybernetics', *Social Research*, Vol. 20, pp. 172–192. [For the definitive version see *PL*, Fifth Essay, 'Cybernetics and Purpose A Critique.' For the German translation see *OF*, Kap. 7.]

17. (1953) 'Motility and Emotion: An Essay in Philosophical Biology', *Proceedings of the XIth International Congress of Philosophy,* Vol. 7 (Amsterdam-Louvain), pp. 117–122. [For the definitive version see *PL*, Fourth Essay, 'To Move and to Feel: On the Animal Soul'; for the German translation see *OF*, Kap. 6.]

18. (1953–4) 'The Nobility of Sight: A Study in the Phenomenology of the Senses', *Philosophy and Phenomenological Research*, Vol. 14, pp. 507–519. [For the definitive version see *PL*, Sixth Essay. Reprinted in *The Philosophy of the Body*, Stuart Spicker (ed.), Quadrangle Books, Chicago, 1970, pp. 312–333. For the German translation see *OF*, Kap. 8.]

19. (1954) Review 'Leon Roth: Jewish Thought as a Factor in Civilization', *Review UNESCO Publications Committee* (Canada), Vol. 3, pp. 6–7.

20. (1957) 'Bemerkungen zum Systembegriff und seiner Anwendung auf Lebendiges', *Studium Generale*, Vol. 10, pp. 88–94. [Incorporated in *OF*, Kap. 4.]

21. (1958) 'Gnosticism', in *A Handbook of Christian Theology*, Meridian Books, New York, pp. 144–147.

22. (1959) 'The Practical Uses of Theory', *Social Research*, Vol. 26, pp. 127–150 A revised version appears in *Philosophy of the Social Sciences: A Reader*, Maurice Natanson (ed.), Random House, New York (1963), pp. 199–142. [For the definitive version see *PL*, Eighth Essay; reprinted in *Philosophy and Technology*, Carl Mitcham and Robert Mackey (ed.), The Free Press, New York, Collier-Macmillan, London, 1972, pp. 355–346, 377. For the author's German version see *OF*, Kap. 10.]

23. (1959) 'In Memoriam Alfred Schutz', *Social Research*, Vol 26, pp. 471–474.

24. (1959) 'Kurt Goldstein and Philosophy', *American Journal of Psychoanalysis*, Vol. 19, pp. 161–164. Reprinted in *Social Research*, Vol. 32 (1965), pp. 351–356.

25. (1960) Review 'Evangelium Veritatis', *Gnomon*, Vol. 32, pp. 327–336. [Also see No. 29.]

26.A. (1961) 'Homo pictor und die differentia des Menschen', *Zeitschrift für philosophische Forschung*, Vol. 15, No. 2, pp. 161–176. [For the definitive version see *ZNE* and *OF*, Kap 9. For the prior English version see 26B.]

B. (1962) 'Homo Pictor and the Differentia of Man', *Social Research*, Vol. 29, pp. 201–220. [For the definitive version see *PL*, Seventh Essay, 'Image-making and the Freedom of Man'.]

27. (1962) 'Immortality and the Modern Temper (The Ingersoll Lecture, 1961)', *Harvard Theological Review*, Vol. 55, pp. 1–20. [For the definitive version see *PL*, Eleventh Essay; for the author's German version see *ZNE* and *OF*, Kap. 12.]

28. (1962) Review 'J. Doresse: The Secret Books of the Egyptian Gnostics', *Journal of Religion*, Vol. 42, pp. 262–273.

29. (1962) 'Evangelium Veritatis and the Valentinian Speculation', in *Studia Patristica*, Vol. VI, F. G. Cross (ed.), (*Texte und Untersuchungen zur*

Geschichte der altchristlichen Literatur, Band 81) Akademie-Verlag, Berlin, pp. 96–111. [Also see No. 25.]

30. (1962) 'Plotin über Ewigkeit und Zeit', in *Politische Ordnung und menschliche Existenz.* Festgabe für Eric Voegelin, Verlag C. H. Beck, München, Herausgegeben von A Dempf, H. Arendt und F. Engel-Janosi, pp. 295–319.

31. (1964) 'Plotins Tugendlehre', in *Epimeleia. Die Sorge der Philosophie um den Menschen*, herausgegeben von F. Wiedmann, A. Pustet, München, pp. 143–173.

32.A. (1964) 'Philosophische Meditation über Paulus, Römerbrief, Kapitel 7', in *Zeit und Geschichte. Dankesgabe an Rudolf Bultmann zum 80. Geburtstag*, hrsg. von Erich Dinkler, J. C. B. Mohr, Tübingen, pp. 557–570. [Also see *ApF*, 2nd ed., Anhang III.]

B. (1971) *Idem*. The author's English version: 'Philosophical Meditation on the Seventh Chapter of Paul's Epistle to the Romans', in *The Future of Our Religious Past: Essays in Honour of Rudolph Bultman*, James R. Robinson (ed.), Harper & Row, New York. [Incorporated in *PE*, Eighteenth Essay, "The Abyss of the Will." For the German original see No. 32.A.]

33. (1964) 'The Anthropological Foundation of the Experience of Truth', *Memorias del XIII Congreso Internacional de Filosofia*, Vol. 5, Mexico, D. F., pp. 507–517. [For the author's expanded, definitive version see *PL*, Seventh Essay, Appendix 'On the Origins of the Experience of Truth'; for the author's German version see *OF*, Kap. 9, Anhang.]

34.A. (1964) 'Heidegger and Theology', *The Review of Metaphysics*, Vol. 18, No. 2, pp. 207–233. [For the definitive English version see *PL*, Tenth Essay.]

B. (1964) 'Heidegger und die Theologie', *Evangelische Theologie*, Vol. 24, No. 12, pp. 621–642. Reprinted in *Heidegger und die Theologie: Beginn und Fortgang der Diskussion*, herausgegeben von Gerhard Noller, Chr. Kaiser Verlag, Müchen, 1967, pp. 316–340.

35. (1965) 'Spinoza and the Theory of Organism', *Journal of the History of Philosopy*, Vol. 3, No. 1, pp. 43–57. [Incorporated in *PE*, Tenth Essay.] Reprinted in *The Philosophy of the Body*, 1970, Stuart F. Spicker (ed.), Quadrangle Books, Chicago, pp. 50–69, also in *Spinoza: A Collection of Critical Essays*, 1973, Marjorie Grene (ed.), Anchor Books, Garden City, New York, pp. 259–278.

36.A. (1965) 'Life, Death, and the Body in the Theory of Being', *The Review of Metaphysics*, Vol. 19, No. 1, pp. 1–23. [For the definitive version see *PL*, First Essay. For the author's German version see No. 36.B.]

B. (1965) 'Das Problem des Lebens und das Leibes in der Lehre vom Sein. Prolegomena zu einer Philosophie des Organischen', *Zeitschrift für Philosophische Forschung*, Vol. 19, No. 2, pp. 185–200. [For the definitive version see *OF*, Kap. 1.].

37. (1965) 'Response to G. Quispel's "Gnosticism and the New Testament" in *The Bible in Modern Scholarship*, J. Philip Hyatt (ed.), Abingdon Press, Nashville, Tennessee, pp. 279–293. [For the definitive version

see *PE*, Fourteenth Essay, 'The "Hymn of the Pearl": Case Study of a symbol, and the Claims for a Jewish Origin of Gnosticism'.]

38. (1967) 'Gnosticism', in *Encyclopedia of Philosophy*, Vol. 3, Macmillan and Free Press, New York, pp. 336–342.

39. (1967) 'Delimitation of the Gnostic Phenomenon: Typological and Historical' in *Le Origini dello Gnosticismo* (*Studies in the History of Religions XII*), Ugo Bianchi (ed.), E. J. Brill, Leiden, pp. 90–108. [Incorporated in *PE*, Thirteenth Essay, 'The Gnostic Syndrome: Typology of its Thought, Imagination, and Mood'.] For the German translation see 'Typologische und historische Abgrenzung des Phänomens der Gnosis', in *Gnosis und Gnostizismus*, herausgegeben von Kurt Rudolf, Wissenschaftliche Buchgesellschaft, Darmstadt, 1975, pp. 626–645. French translation see 'Books' 5.E (1977) in *La Religion gnostique*.

40. (Nov 1967) 'Jewish and Christian Elements in the Western Philosophical Tradition', *Commentary*, Vol. 44, pp. 61–68. For a revised version see *Creation. The Impact of an Idea*, Daniel O'Connor and Francis Oakley (eds.), Charles Scribner's Sons, New York, 1969, pp. 241–258. [For the expanded, definitive version see *PE*, Second Essay, 'Jewish and Christian Elements in Philosophy: Their Share in the Emergence of the Modern Mind'.] For the German translation see 'Judentum, Christentum und die westliche Tradition', *Evangelische Theologie*, Vol. 28 (Dezember, 1968), pp. 613–629.

41. (1968) 'Contemporary Problems in Ethics from a Jewish Perspective', *Central Conference American Rabbis Journal*, Vol. 15, No. 1, pp. 27–39. [For a revised version see *CCAR Journal Anthology on Judaism and Ethics*, 1969; for the definitive version see *PE*, Eighth Essay.]

42. (1968) 'The Concept of God after Auschwitz', in *Out of the Whirlwind*, A. H. Friedlander (ed.), Union of American Hebrew Congregations, New York, pp. 465–476. [Also in *FRR*, Second Essay.]

43. (June, 1968) 'Biological Foundations of Individuality', *International Philosophical Quarterly*, Vol. VIII, No. 2, pp. 231–251. [Incorporated in *PE*, Ninth Essay.]

44. (Spring, 1969) 'Philosophical Reflections on Experimenting with Human Subjects', *Daedalus*, Vol. 98, No. 2, pp. 219–247. Reprinted in *Philosophy in the Age of Crisis*, Eleanor Kuykendall (ed.), Harper & Row, New York, Evanston, London, 1970, pp. 81–100. For a revised version see *Experimentation with Human Subjects*, Paul A. Freund (ed.), George Braziller, New York, 1970, pp. 1–31. Reprinted in *Ethics in Medicine*, S. J. Reiser, A. J. Dyck, W. J. Curran (eds.), M. I. T. Press, Cambridge, Massachusetts, 1977, pp. 304–315. [For the definitive version see *PE*, Fifth Essay. Reprinted in *Biomedical Ethics and the Law*, James M. Humber and Robert F. Almeder (ed.), Plenum Press, New York, 1976, pp. 217–242.]

45. (1969) 'Economic Knowledge and Critique of Goals', in *Economic Means and Social Ends*, Robert L. Heibroner (ed.), Prentice-Hall, Englewood Cliffs, New Jersey, pp. 67–88. [For the definitive version see *PE*, Fourth Essay, 'Socio-Economic Knowledge and Ignorance of Goals'.]

46. (Oct., 1969) 'Myth and Mysticism: A Study of Objectification and Interiorization in Religious Thought', *The Journal of Religion*, Vol. 49, No. 4, pp. 315–329. [Incorporated in *PE*, Fifteenth Essay,]

47. (1969–70) 'Origen's Metaphysics of Free Will, Fall and Redemption: A "Divine Comedy" of the Universe', *Journal of the Universalist Historical Society*, Vol. 8, pp. 3–24. [Incorporated in *PE*, Sixteenth Essay. For the German original see No. 5 and No. 6.]

48. (Summer 1971) 'The Scientific and Technological Revolutions', *Philosophy Today*. Vol. 15, No. 2/4 pp. 79–101. [For the definitive, expanded, and revised version see *PE*, Third Essay, 'Seventeenth Century and After: The Meaning of the Scientific and Technological Revolution',]

49. (Fall, 1971) 'Rudolph Arnheim on Visual Thinking', *Journal of Aesthetics and Art Criticism*, Vol. 30, pp. 111–117. [Incorporated in *PE*, Eleventh Essay, 'Sight and Thought: A Review of "Visual Thinking" '.]

50. (Fall, 1971) 'Change and Permanence: On the Possibility of Understanding History', *Social Research*, Vol. 38, No. 3, pp. 498–528. Reprinted in *Explorations in Phenomenology: Papers of the Society for Phenomenology and Existential Philosophy*, David Carr and Edward S. Casey (ed.), Martinus Nijhoff, The Hague, Netherlands, 1973, pp. 102–132. [Incorporated in *PE*, Twelfth Essay For the original German version see 'Books' 8. A, B.]

51. (1971) 'The Soul in Gnosticism and Plotinus', in *Le Néoplatonisme*, Éditions du Centre National de la Recherche Scientifique, Paris, pp. 45–53. [Incorporated in *PE*, Seventeenth Essay.] In the same volume, see Discussion de la Conférence de M. Dorrie, pp. 29–33; Discussion des Conférences de M. Jonas et de M. Blumenthal, pp. 65–66; Discussion de la Conférence de M. Armstrong, pp. 75–76; Discussion de la Conférence de Mme. Charles, pp. 249–251; Discussion de la Conférence de M. Blumenberg, pp. 472–474.

52 (1972) Testimony before Subcommittee on Health, United States Senate: Hearings on Health, Science, and Human Rights, Nov. 9, 1971, re: S. J. Res. 75, the *National Advisory Commission on Health Science and Society Resolution*, 92nd Congress, 1st session. U. S. Government Printing Office, (73–191 -0) Washington, D.C., pp. 119–123.

53. (Spring, 1973) 'Technology and Responsibility: Reflections on the New Tasks of Ethics', *Social Research*, Vol. 40, No. 1, pp. 31–54. [Incorporated in *PE*, First Essay.] For the German translation see 'Die Natur auf der moralischen Bühne: Uberlegungen zur Ethik im technologischen Zeitalter', *Evangelische Kommentare: Monatsschrift zum Zeitgeschehen in Kirche und Gesellschaft*, Vol. 6, No. 2 (Feb 1973), pp. 73–77. For the French translation see 'Technologie et responsabilité: Pour une nouvelle éthique', in *Esprit*, N.S. 42e Année, No. 438, (Sept. 1974), pp. 163–184.

54. (Spring, 1976) 'Hannah Arendt, 1906–1975', *Social Research*, Vol. 43, No. 1, pp. 3–5. Also see *Partisan Review*, Vol. 42, No. 1, pp. 12–13 (1976) under the title 'Words Spoken at the Funeral Service for Hannah Arendt at the Riverside Memorial Chapel in New York City on Mon-

day, December 8, 1975', Also see the German translation by Dolf Sternberger in *Deutsche Akademie für Sprache und Dichtung Darmstadt* (Jahrbuch 1975), Verlag Lambert Schneider, Heidelberg, 1976, pp. 169–171.

55. (Spring, 1976) 'Responsibility Today: The Ethics of an Endangered Future', *Social Research*, Vol. 43, No. 1 pp. 77–97. [Also in *FRR*, Fourth Essay.]

56. (1976) 'On the Power or Impotence of Subjectivity', in *Philosophical Dimensions of the Neuro-Medical Sciences*, Volume 2 of the series 'Philosophy and Medicine', Stuart F. Spicker and H. Tristram Engelhardt (eds.), D. Reidel Publishing Co., Dordrecht, Holland / Boston, U.S.A., pp. 143–161. [An expanded version appears in *FRR*, Third Essay, 'Impotence or Power of Subjectivity? A Reappraisal of the Psychophysical Problem'.]

57. (August, 1976) 'Freedom of Scientific Inquiry and the Public Interest', *The Hastings Center Report*, Vol. 6, No. 4, pp. 15–17. Also published in *Biomedical Research and the Public*, prepared for the Subcommittee on Health and Scientific Research of the Committee on Human Resources, United States Senate, (May, 1977), U.S. Government Printing Office, Washington, D.C., 1977 (82–201), pp. 33–38. This publication contains the proceedings of a conference at Airlie House, Warrenton, Virginia, April 1–3, 1976.

58.A. (Oct., 1976) 'Handeln, Erkennen, Denken. Zu Hannah Arendt's philosophischem Werk', *Merkur, Deutsche Zeitschrift für europäisches Denken*, Jahrgang 30, Heft. 10, pp. 921–935. [This is the author's German version of 58.B.]

B. (Spring, 1977) 'Acting, Knowing, Thinking: Gleanings from Hannah Arendt's Philosophical Work', *Social Research*, Vol. 44, No. 1, pp. 25–43.

59. (1977) 'Im Kampf um die Möglichkeit des Glaubens. Erinnerungen an Rudolf Bultmann und Betrachtungen zum philosophischen Aspekt seines Werkes', in *Gedenken an Rudolf Bultmann*, herausgegeben von Otto Kaiser, J. C. B. Mohr (Paul Siebeck), Tübingen, pp. 41–70. [For the author's English version see *FRR*, First Essay, 'Is Faith Still Possible? Memories of Rudolf Bultmann and Reflections on Philosophical Aspects of his Work'.]

60. (1977) 'The Concept of Responsibility: An Inquiry into the Foundations of an Ethics for our Age', in *Knowledge, Value, and Belief*, H. T. Engelhardt, Jr. and Daniel Callahan (ed.), Institute of Society, Ethics and the Life Sciences, Hastings-on-Hudson, New York, pp. 169–198. [Also in *FRR*, Fifth Essay.]

61. (1977) 'A Retrospective View', in *Proceedings of the International Colloquium on Gnosticism* (Stockholm, August 20–25, 1973), Geo Widengren (ed.), Almquist & Wiksell International, Stockholm/E. J. Brill, Leiden, pp. 1–15. [Also in *FRR*, Sixth Essay: 'Beginnings and Wanderings· A Retrospective View'.]

62. (1977) Testimony in Hearing held February 4, 1977, by the Assembly Committees on Health and on Resources, Land Use and Energy of the California Legislature, at the State Capitol in Sacramento, on the subject of recombinant DNA research; transcript of Jonas' testimony (12 pp) not yet edited for publication in forthcoming committee report.

NOTES ON CONTRIBUTORS

Hannah Arendt, the late University Professor of Political Science and Philosophy, the Graduate Faculty of Political and Social Science, The New School for Social Research, New York City, New York.

Eric J. Cassell, M.D., F.A.C.P., Clinical Professor of Public Health, Department of Public Health, Cornell University Medical College, New York City, New York.

Bethia S. Currie, Ph.D., Cornwall, Connecticut.

Strachan Donnelley, Ph.D., One Hundred and Six West 78th Street, New York City, New York.

H. Tristram Engelhardt, Jr., Ph.D., M.D., Rosemary Kennedy Professor of the Philosophy of Medicine, Center for Bioethics, Georgetown University, Washington, D.C.

Murray Greene, Ph.D., Professor of Philosophy, Department of Philosophy, Baruch College of the City University of New York, New York City, New York.

Marjorie Grene, Ph.D., Professor of Philosophy, Department of Philosophy, The University of California at Davis, Davis, California.

Otto E. Guttentag, M.D., Samuel Hahnemann Professor of Medical Philosophy, Emeritus, School of Medicine, The University of California at San Francisco, San Francisco, California.

Charles Hartshorne, Ph.D., Professor of Philosophy, Department of Philosophy, The University of Texas at Austin, Austin, Texas.

Leon R. Kass, M.D., Ph.D., Henry Luce Professor in the College, The University of Chicago, Chicago, Illinois.

Richard Kennington, Ph.D., Associate Professor of Philosophy, School of Philosophy, Catholic University of America, Washington, D.C.

Paul Oskar Kristeller, Ph.D., F.J.E. Woodbridge Professor Emeritus of Philosophy, Columbia University, New York City, New York.

Michael Landmann, Ph.D., Professor Dr. of Philosophy, Department of Philosophy, Freie Universitat Berlin, Berlin, F.R.G.

Adolph Lowe, LL.D., Alvin Johnson Professor Emeritus of Economics, The Graduate Faculty of Political and Social Science, The New School for Social Research, New York City, New York.

Wilhelm Magnus, Ph.D., Professor of Mathematics, Department of Mathematics, Polytechnic Institute of New York, New York City, New York.

Jitendra Nath Mohanty, Ph.D., Professor of Philosophy, Department of Philosophy, The University of Oklahoma, Norman, Oklahoma.

Stuart F. Spicker, Ph.D., Associate Professor of Community Medicine and Health Care (Philosophy), Department of Community Medicine and Health Care, School of Medicine, The University of Connecticut Health Center, Farmington, Connecticut.

Richard M. Zaner, Ph.D., Easterwood Professor of Philosophy, Department of Philosophy, Southern Methodist University, Dallas, Texas.

INDEX

Abrams, Natalie 82–83, 85
Abramson, Arthur S. 77, 87–89
Achillini, Alessandro 36
actual entity 156
actuality, ultimate 41–42
Addis, Thomas 48
Aiken, J. 75
Alderotti, Taddeo 34
Aleichem, Sholom 60
anthropomorphism 274–275
Aquinas, St. Thomas 310
Archimedes 229
Aristotle 6, 30–32, 100, 117, 121–123, 125, 127–130, 132, 134–135, 144, 150, 180, 204–205, 233–237, 243–244, 249, 306, 311, 313–314
Arntsen, Roger 188–189
Austin, John 285
Averroës 31
Avicenna 31
Ayala, Francisco 127

Bacon, Francis 201–209, 211, 213, 221–222
Balme, D. 121
Bayne, Jacky 185–186
Benjamin, Walter 312
Berg, J. H. van den 73
Bergson, Henri 138, 140, 143, 146, 148, 312–313
Berkeley, George 123
Betts, Henry 186
Bierstedt, Albert 52
biology 121–135
Blanshard, Brand 149
Blumenberg, Hans 305
Bohr, Niels 226
Bonittis, Baverius de 36
Bradley, Francis H. 149
brain 179–183
brain death 179–180, 183, 192
Brentano, Franz 283, 288
Briggs, Kenneth A. 193
Buber, Martin 41
Burns, Chester R. 73
Butler, Samuel 105

Cantore, E. 226
Card, Claudia 80
Cardano, Girolamo 36
Cassirer, Ernst 29
Cato, Marcus Porcius 313
Celsus, Cornelius 31, 36
Charlemagne 1
Chisholm, Roderick 283
clinic 74–76, 92
coma 183–191
conscious, definition of 45
Cooley, Denton 180, 195
Copernicus, Nicolas 1
Craven, Joan 91
Cullen, William xx

Darwin, Charles 1, 97–98, 103–114, 122, 126–127, 133
death 45, 73, 81, 83
  definition of 177–178, 182, 190, 192, 196
Delbrück, Max 128
Demangeon, J. B. 75
Democritus 147
Descartes, René 46, 201–202, 206–221, 228, 234–238, 266–270, 276–280
determinism, strict 139
Dewey, John 97, 107, 228
Dinello, Daniel 82
disease 235, 251–252
  definition of 71
Dobzhansky, Theodosius 256–257
drug addiction 86

dualism 265–266, 268, 270, 273, 279, 294
Cartesian 266, 273, 283, 285, 286–287, 294
composite 46
polar 46, 49
Dusold, Carol 186–188, 190, 193, 195

EEG (electroencephalogram) 183, 191–195
Einstein, Albert 226
emancipation 3–5, 8
emancipatory revolution 4, 7
Empiricus, Sextus 30
Engels, Friedrich 183
Entralgo, Pedro Laín 64
Epictetus 214
ethics
communal 7
medical 86–87
euthanasia 73–92
active vs. passive 79, 81–82, 90
evolution 97–98, 103–104, 107, 111, 114–115

faculty, within a medieval university 33–34
Fechner, Gustav T. 145
Fichte, Johann Gottlieb 238
Ficino, Marsilio 36
Fitzgerald, P. J. 81–82
Forlivio, Jacobus de 36
Foucault, Michel 74 75
freedom 45
Freidson, E. 52–53
Freud, Sigmund 138

Galen 30–31, 36–37
Galileo, Galilei 207, 210, 220, 267
Gandhi, Mohandas 46
Gauss, C. F 228
Gaylin, Willard 179, 181–182, 191
Gebsattel, Viktor von 49
Gilson, Étienne Henri 206–207
Goldstein, Kurt 29
Goodrich, T. 80–81
Gray, Asa 97–98, 106

Hardin, Denise 190
Harvey, William 37
Hay, Gerry 190
health 43–44, 46
health care 43–44
Hegel, G. W. F. 22, 234–235, 237–260, 314
Heidegger, Martin 208, 298, 303, 306, 309, 311–312, 314
Heinemann, Henry O. 74, 76
Heisenberg, Werner 226
Heraclitus of Ephesus 303
Hippocrates 30–31, 36, 91
history
of medicine 35
of philosophy 35
Hitler, Adolph 46
Hobbes, Thomas 201, 211, 217, 221
hospital 73–76, 88–89
teaching 74
Hoyle, Fred 228
human being 45
humanitarianism 203–204
Hume, David 123–127, 139, 144, 149, 269
Husserl, Edmund 41, 143, 267, 284, 288–289, 291–293, 295
Huxley, Thomas H. 106, 257

identity
organic 159–161
physical 159–160
individuality 156–157, 167
industrial revolution 2
intentionality 283, 288
interactionism 149
intuition 312, 143
Israeli, Isaac 31

Jaspers, Karl 29, 227
Jeans, Sir James 273
John XXI, Pope 32–33
Jonas, Hans viii–x, xv–xvi, 6, 8, 29, 58, 97–98, 111, 137, 147–148, 150, 152–153, 155–168, 173–174, 177–180, 194–196, 202, 225, 227–229, 265–271, 274–280, 303, 304

Kant, Immanuel 87, 123, 125–127,

222, 228, 234, 237–241, 276, 304, 309, 311–312
Karlsson, Karoline 189
Keller, Helen 152
Kennedy, John F. 7
Klein, Marvin 187
knowledge 63–65, 70–72
Kohl, Marvin 80–81, 85
Kübler-Ross, Elisabeth 193–194
Kutner, Bernard 88

Lack, Sylvia 79, 85
Lamarck, Chevalier de 107–108, 114
Lamers, William 85
Landau, Leo 184–185, 188, 190, 193, 195
Landgrebe, Ludwig 294
Leibniz, Gottfried W. 138, 145–147, 217, 222, 238, 241
Leoniceno, Niccolò 36
life 54
  concept of 234
Locke, John 221
Luther, Martin 1

Machiavelli, Niccolò 204
Maimonides, Moses 229
Marcel, Gabriel 46
Marcuse, Herbert 5
Marx, Karl 222
Maurus 32
Maxwell, James C. 145, 150
mechanism, neo 246, 259–260
medical care 44
medical ethics 86–87
medical techniques, modern 77
medicine 30–31, 47, 57
medicine and philosophy relationship between 29–32, 34
Merleau-Ponty, Maurice 293
metabolism 270–271, 277
metaphysics 216, 311
Mill, John Stuart 126
mind-body problem 283–299
Moody, Raymond A., Jr. 193
Murphy, Patricia 189
Murray, J. F. 131

Natural Death Act of California 195
natural selection, theory of 104–112
nature 97, 103–108, 111, 114, 201–223
Newton, Isaac 228
Niebuhr, Reinhold 6
Nietzsche, Friedrich 222, 306
Nifo, Agostino 36
nihilism 223
normality 129–133

Occam's Razor 149
organism 98–102, 165
Owen, Richard 121, 122

pain 253
Pantin, C. F. A. 132–133
Paracelsus 36
Pascal, Blaise 121, 268
Peirce, Charles S. 138, 140–141, 143, 146, 149
Peter of Abano 32–33
Petrarch 36
Petrus de Hibernia 32
Philo of Alexandria 303
philosophy xv–xvi, 37, 97, 306
  of culture 29, 37
physician 32, 36, 41–42, 47, 49–55
  physician-patient relationship 50, 58, 86
Plato 6, 30, 128, 150, 306–309, 311
Plessner, Helmuth 292, 295
Pomponazzi, Pietro 36
Popper, Karl 127
Powner, David J. 191–192
positivism 29

Quinlan, Karen Ann, case of 180

Rabi, I. I. 226, 229
Rachels, James 82–83
recombinant DNA research 57, 71
Reid, Thomas 139
Rible, Clarellen 189
Riemann, Bernhard 228
robots 150–151
Royce, Josiah 149
Russell, Bertrand 7, 139, 141, 144, 149

Sadowsky, Lewis 187–188
Salutati, Coluccio 36
Sarkisov, S. A. 178, 180–183, 193
Saunders, Cicely 79, 82
Sellars, Wilfred 283–285, 298
self 45, 53
self realization 53
sense data 138
sexual reproduction 249–251
Skoda, Joseph 64
Snow, C. P. 225
Socrates 122, 128, 203
Spierer, Robert 188
Spinoza, Benedict 134, 149, 217, 237–238, 241
Strawson, Peter F. 283–284, 286–288, 291, 297–298
survival 111

technology 66–68
teleology 97–118, 125–126, 162–163
Telesio, Bernardino 36
Temkin, Owsei 47
theodicy 221
theology, faculty of 33–34
Tiene, Gaetano da 36
Tillich, Paul 41, 45
transcendental subjectivity 293
Treadway, Guy 190–191, 195
truth 308–311
Tsang, Fa 149
Twycross, R. G. 85

unconsciousness, definition of 193–194
unity 165–166
Urso 32

Vernia, Nicoletto 36
Vesalius, Andreas 37
Vetrov, Pyrotr 189
Victorius, Benedictus 36

Waddington, C. H. 126
Wald, Florence S. 91
Whewell, William 132
Whitehead, Alfred North xii, 111, 138, 141, 143–149, 152–153, 155–174
Wiener, Norbert 180–181
Wiggins, David 129
Williams, George 128
Williams, Glanville 78
Williams, Robert H. 78
Wittgenstein, Ludwig 144, 226, 307
Wordsworth, William 144
Wright, Henrik von 148
Wright, Sewall 150

Xantippe 128

Yegorov, Boris 185

Zabarella, Jacopo 36
Zono of Elea 148

GPSR Compliance

The European Union's (EU) General Product Safety Regulation (GPSR) is a set of rules that requires consumer products to be safe and our obligations to ensure this.

If you have any concerns about our products, you can contact us on ProductSafety@springernature.com

In case Publisher is established outside the EU, the EU authorized representative is:

Springer Nature Customer Service Center GmbH
Europaplatz 3
69115 Heidelberg, Germany

Batch number: 10607212

Printed by Printforce, the Netherlands